# Topics in Mining, Metallurgy and Materials Engineering

"Topics in Mining, Metallurgy and Materials Engineering" welcomes manuscripts in these three main focus areas: Extractive Metallurgy/Mineral Technology; Manufacturing Processes, and Materials Science and Technology. Manuscripts should present scientific solutions for technological problems. The three focus areas have a vertically lined multidisciplinarity, starting from mineral assets, their extraction and processing, their transformation into materials useful for the society, and their interaction with the environment.

**\*\* Indexed by Scopus (2020) \*\***

José Ignacio Verdeja González ·
Daniel Fernández-González ·
Luis Felipe Verdeja González

# Physical Metallurgy and Heat Treatment of Steel

 Springer

José Ignacio Verdeja González
University of Oviedo
Oviedo, Asturias, Spain

Luis Felipe Verdeja González ⓘ
Departamento de Ciencia de los Materiales
e Ingeniería Metalúrgica
University of Oviedo
Oviedo, Asturias, Spain

Daniel Fernández-González ⓘ
Nanomaterials and Nanotechnology
Research Center (CINN-CSIC)
El Entrego, Asturias, Spain

ISSN 2364-3293          ISSN 2364-3307  (electronic)
Topics in Mining, Metallurgy and Materials Engineering
ISBN 978-3-031-05704-5          ISBN 978-3-031-05702-1  (eBook)
https://doi.org/10.1007/978-3-031-05702-1

This Springer imprint is published by the registered company Springer Nature Switzerland AG
The registered company address is: Gewerbestrasse 11, 6330 Cham, Switzerland

*This book is dedicated to the memory of José Antonio Pero-Sanz Elorz (1934–2012), who founded the research group in Iron and Steelmaking, Metallurgy and Materials at the University of Oviedo (Asturias, Spain) in the middle of the 70s of the last century. Several books have been published as a result of this long period of research and teaching, including Physical Metallurgy of Cast Irons (Springer, 2018), Structural Materials: Properties and Selection (Springer, 2019), Operations and Basic Processes in Ironmaking (Springer, 2020) and Operations and Basic Processes in Steelmaking (Springer, 2021).*

# Acknowledgements

Daniel Fernández-González acknowledges the grant (Juan de la Cierva-Formación program) FJC2019-041139-I funded by MCIN/AEI/ 10.13039/501100011033 (Ministerio de Ciencia e Innovación, Agencia Estatal de Investigación).

# Prologue

The book *Solidification and Solid-State Transformations of Metals and Alloys* (Elsevier, 1st ed., Boston, USA, 2017) was the first of a series of books dealing with the Fe-C diagram. This first book was about microstructures, transformations, and properties of metals and alloys, with emphasis in the fundamentals of Materials Science and Engineering. The following book *Physical Metallurgy of Cast Irons* (Springer, 1st ed., Cham, Switzerland, 2018) was completely dedicated to the cast irons as an important material in different applications, being the most produced metallic material after steels. The series continued with the book *Structural Materials: Properties and Selection* (Springer, 1st ed., Cham, Switzerland, 2019), which focused on the different families of materials (metals, ceramics, polymers, and composites) and how to choose the suitable one for a specific application. The next book was *Operations and Basic Processes in Ironmaking* (Springer, 1st ed., Cham, Switzerland, 2020). This was the real beginning of the study of the field of steels. It was aimed at studying the ironmaking, that is to say, from the mineral to pig iron passing by the agglomeration techniques (pelletizing and sintering) or the production of ironmaking coke. The following book was *Operations and Basic Processes in Steelmaking* (Springer, 1st ed., Cham, Switzerland, 2021). It was aimed at studying the steelmaking from conversion of the pig iron into steel in the basic oxygen furnace to the finishing processes (secondary metallurgy and solidification of steel), including casting, hot-working and cold-working operations (deep drawing, stretching, and wiring). The production of steel in the electric steelworks is also described as an important option in the production of steel. The present book *Physical Metallurgy and Heat Treatment of Steel* is a continuation of the above-indicated books dealing with the solid-state transformations in the Fe-C system as well as with the different heat treatments used in production of different types of steels.

The main objective of the series of books is to offer a collection to students, workers, operators, and all interested people about the field of structural materials, and particularly those derived from the Fe-C diagram. We have included in all the cases different exercises, problems, and case studies, all of them solved, that provide to the reader a support to understand the concepts described in the theoretical part. This book is located in the field of science oriented to Materials Engineering. Even when this book could be also used as handbook, these pages want to be also a formulation of the knowledge acquired in the topic during more than 50 years of research and teaching, setting the fundamentals of the physical metallurgy (crystallography, microstructure, and phase transformations) that condition the relations between composition, microstructure, and properties of steels.

As it was already pointed out some time ago by our colleague Prof. F. B. Pickering, the development of the industrial production cannot be—in fact it is not—the result of the simple empiric activity. Even when there is certain empiric character "new steels are being developed nowadays considering the fundamentals of a good rational understanding of the metallurgical phenomena". This scientific understanding is key not only for a rational election within the existing materials but also for their design according to the requirements by means of the suitable thermal, thermochemical, and thermomechanical treatments. *Physical Metallurgy and Heat Treatment of Steel* wants to facilitate this understanding to the current and future professionals.

This book, as it was already indicated, is to a large extent the result of the teaching activity. On the one hand, of academic courses—graduate or postgraduate—given in the Schools of Mines of Oviedo (Oviedo/Uviéu, Asturias, Spain) and Madrid, in the Schools of Industrial Engineering of San Sebastián (Gipuzkoa, Spain), of Piura (Perú), of the Universidad Panamericana (México), or of the Universidad of Montevideo (Uruguay). However, this book is also the result of courses given to professionals of metallurgical engineering in: Asociación Técnica de Fundidores (INASMET), Centro de Estudios e Investigaciones Técnicas de Gipuzkoa (CEIT), Petroperú, Dirección General de Asesoramiento Técnico de Rosario (Argentina), Fundación Universidad-Empresa (Madrid), Instituto Tecnológico de Costa Rica, Instituto de Estudios Empresariales de Montevideo (Uruguay), Arcelor Mittal S. A. (Spain), Organización de las Naciones Unidas para el Desarrollo Industrial (Argentina), etc.

With this regard, it is convenient to mention here "particular attention was paid to the impact caused between the attendants regarding the quality of the exposition, said, to the adopted systematics and understandability of the concepts", so "it is necessary to consider, when reading the book, that this was generated from oral courses and this explains the references, the language and the thematic boundaries". And, as this book is the result of courses promoted by public or private organizations, we want to express here our gratitude to the Dirección General de Cooperación Internacional del Ministerio de Educación y Ciencia of Spain, to the Agencia Española de Cooperación Internacional (AECI), a la Organización de las Naciones Unidas para el Desarrollo Industrial (ONUDI), and Arcelor Mittal S. A.

Regarding the personal chapter, we want to acknowledge the professors Paul Lacombe, Ramón Durán Balcells, José Apraiz Barreiro, and Rafael Calvo Rodés. We also want to thank professors F. B. Pickering (University of Sheffield), T. Gladman (University of Leeds), J. A. García Poggio and P. Tarín (Universidad Politécnica de Madrid), G. Béranger (University of Compiègne), ASM International, United States Steel Corporation, Maney Publishing, EDP Science, and all those that have made possible the publication of this book in one or other manner.

Finally, we want to use this book, one more time, as tribute to the memory of the Professor José Antonio Pero-Sanz Elorz, who was the pioneer of Metallotechnics in the School of Mines of Oviedo (Oviedo, Asturias, Spain), and Founder of the Research Group in Ironmaking, Steelmaking, Metallurgy, and Materials of the University of Oviedo in the middle of the last century.

July 2022

# Contents

# Chapter 1
# Solid-State Transformations in the Iron Carbon System

## 1.1 Introduction

Steels are iron and carbon alloys in different proportions. The carbon content in these alloys can reach 2.11 wt.%, if the alloy does not contain, additionally, other alloying elements that could reduce this maximum carbon content.

Iron is the fourth most abundant element in the Earth crust, preceded only by the oxygen, silicon, and aluminum. It is estimated that the iron weight proportion in the Earth crust is 50000 parts per million, and that the global iron content in our planet (if the nucleus is included) is around 40%.

Only five metals are produced in quantities of several millions of tons every year in the XXIst century: iron, aluminum, copper, zinc, and lead. From the total production of metals, that of steel is greater than 1800 million tons (the production approached 1900 Mt in 2021, according to World Steel Association). This is equivalent to >90% of the annual production of metallic materials and exceeds almost 9 times the annual production, in weight, of organic polymeric materials. After concrete, steel is the most used structural material. The ranking of production and consumption of metals for those after steel is aluminum (>90 Mt in 2021, including primary and secondary aluminum), copper (21.3 Mt in 2021), zinc (12.8 Mt in 2021), and lead (4.7 Mt in 2021). This produced quantities are still affected by the COVID-19 pandemic situation.

These five metals and other three—eight in total: iron, aluminum, copper, zinc, lead, nickel, magnesium, and tin—represent 99% of the annual production (in weight) of metallic materials worldwide. With respect to the economic importance of this production (tons multiplied by the price/ton), the ranking of importance would be iron, aluminum, copper, gold, nickel, tin, zinc, lead, and others.

Besides, it is possible to indicate that the tendency in the production and consumption of the main structural metals has changed significantly in the last 20 years given the important industrial and economic growth of China (Verdeja et al. 2020). It is

J. I. Verdeja González et al., *Physical Metallurgy and Heat Treatment of Steel*,
Topics in Mining, Metallurgy and Materials Engineering,
https://doi.org/10.1007/978-3-031-05702-1_1

possible to indicate that the production and consumption of certain metals has grown faster than in the case of other metals.

The economic prevalence of the steel can be also justified if we consider that it is a cheap material. For that reason, in a possible election within different materials, when we use as criterion of selection compounded properties—integrating mechanical properties, density, and unitary cost—steels are almost always advantageous as metallic structural materials. That is to say, materials that are interesting due to the bulk mechanical properties (stiffness, elasticity, mechanical resistance, toughness, fatigue behavior, resistance to creep, etc.) and surface characteristics (behavior in front of oxidation, corrosion, friction, wear, etc.) (Calvo-Rodés 1956; Pero-Sanz et al. 2018).

Most of the metallic materials of advanced characteristics are the result of the direct answer to the challenge of new properties required by part of different industries (transportation, energy, construction, etc.). With a general character, from the materials engineering point of view—"new features searching for materials" (not new materials searching for applications) (Flinn and Trojan 1979)—it is possible to say that the challenge of the XXIst has four important points: better features; greater quality; lower environmental impact; and economy. Regarding new or better features, the challenge is translated into new processes (or improved processes) and new alloys (or the redesign of the conventional alloys). On the other hand, quality is related with the possibilities of processes monitoring and control of warehouses, the new laboratory techniques—analyses and tests—, as well as the mathematical modeling of processes and shapes. In the case of lower environmental impact, this refers to the production of materials with generation of less pollutants and wastes generated in the process, among other issues. Referring to economy, the challenges are aimed at savings in the obtaining of materials, simplification of processes, and obtaining of near net shape products (Verdeja et al. 2021).

We can still consider the iron as one of the most promising materials in the competition between materials that started in the second half of the twentieth century. There is every reason to believe that, in the case of great productions, the answer of the Fe–C alloys to the challenge of the advanced materials will follow a growing tendency. It is sufficient to consider (Attwood 1994) that more than 50% of the steels currently used were not industrially known in 1985.

On another note, from a merely physical point of view, iron is a very promising material regarding the possibilities of invention and development of new features of structural type. In fact, this metal due to its allotropy, the dual character of the Fe–C equilibrium (stable/metastable), the non-equilibrium states, and the influence of the substitutional and interstitial elements allow to sum up and advance in research about almost all the properties of utilization that the Physical Metallurgy offers.

The vigor of such investigation—and in broad sense, the knowledge of steels—lies on the crystalline structures, terminal and intermediate phases, solidification, recovery/recrystallization, textures, and solid-state transformations by allotropy and solubility changes. We estimate that this should be the start point for the science and engineering of steels.

## 1.2 Crystalline Structures of the Iron

Iron is the chemical element with atomic number 26. The atomic weight is 55.847 g/mol and the density at room temperature is 7.9 g/cm$^3$. Figure 1.1 illustrates the microstructure of the pure iron at room temperature. It is possible to appreciate by means of X-ray diffraction that each one of the cells—called grains—that can be seen in the micrograph of Fig. 1.1 is formed by an ensemble of elemental cells of crystallography body centered cubic (see Fig. 1.2a), with two Fe atoms in each

**Fig. 1.1** Pure iron at room temperature

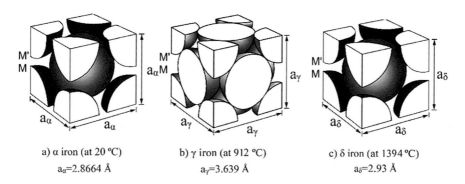

a) α iron (at 20 °C)
$a_\alpha$=2.8664 Å

b) γ iron (at 912 °C)
$a_\gamma$=3.639 Å

c) δ iron (at 1394 °C)
$a_\delta$=2.93 Å

**Fig. 1.2** Crystalline structures of the iron

elemental cell, one in the center of the cube and 1/8 of atom in each one of the corners of the cube, being the atom of the corners shared with the contiguous cells. The atomic radius of this variety of Fe (called alpha iron) is $a_\alpha \cdot 3^{1/2}/4$. The lattice parameter ($a_\alpha$) of this cell at 20 °C is 2.8664 Å (1 Å $= 10^{-8}$ cm $= 0.1$ nm).

If the temperature slowly rises from 20 °C to 911 °C, this alpha iron gradually expands (the linear thermal expansion coefficient—mean in the range of temperatures 0–800 °C—is $14.8 \cdot 10^{-6}$ °C$^{-1}$) and the lattice parameter of the alpha iron is $2.898 \cdot 10^{-8}$ cm at 911 °C. However, in this slow heating, when the temperature of 912 °C is reached, the iron contracts due to an allotropic transformation of its crystalline structure that passes from body centered cubic to face centered cubic (see Fig. 1.2b, gamma iron). Each elemental cell of alpha iron contains four iron atoms ($6 \cdot 1/2 + 8 \cdot 1/8$). The lattice parameter ($a_\gamma$) of this cell at 913 °C is 3.639 Å and the atomic radius is $a_\gamma/8^{1/2}$.

During the allotropic transformation of the alpha iron into gamma iron, the temperature remains constant at 912 °C, and the transformation takes place as a result of the classical mechanism of nucleation and growth: by thermal fluctuations several atoms get together to give face centered cubic nucleus (and they start to grow when the size of these gamma nuclei exceeds a certain critical value). This gives as a result a new microstructure, different from that of Fig. 1.1 (with other grain boundaries).

The contraction that the iron experiences when it passes from alpha iron to gamma iron is due to two cells of alpha—in each one of these cells there are two atoms of iron—give a gamma cell (with four atoms in each cell). The contraction in volume at 912 °C is 1%—the linear contraction would be 0.33%—as it is possible to deduce from Eq. 1.1.

$$\frac{a_\gamma^3 - 2 \cdot a_\alpha^3}{a_\gamma^3} \tag{1.1}$$

If once the iron was transformed into the variety gamma, the slow heating continues above 912 °C, the iron continues the expansion—with a thermal linear expansion coefficient of $23 \cdot 10^{-6}$ °C$^{-1}$ (average value between 916 and 1394 °C)—until reaching the temperature of 1394 °C. At this temperature, a new allotropic transformation of the iron takes place: it transforms from gamma into other crystallographic structure that is called delta iron, which is body centered cubic (as the alpha iron, but with greater lattice parameter, $a_\delta$ equal to 2.98 Å). If the heating continues above 1394 °C, delta iron continues to expand, with a thermal linear expansion coefficient of $16.7 \cdot 10^{-6}$ °C$^{-1}$ (average value between 1394 and 1502 °C). The melting of delta iron takes place at 1538 °C.

The gamma-delta transformation takes also place as by nucleation and growth. The resulting microstructure, when the gamma state structure is erased, is called primary because it is the first structure that appears during the solidification of a pure iron melt. While the gamma structure, which is obtained in solid state during the cooling from the primary structure, is called secondary. The structure of Fig. 1.1 is called tertiary.

The alpha iron is nonmagnetic above 770 °C and magnetic below this temperature. The change of magnetism, associated to the spin of the atoms, does not alter the crystalline structure, which is still body centered cubic: there is no phase transformation and, for that reason, neither the microstructure of the alpha iron is modified. Consequently, the term beta iron that was previously used to designate the nonmagnetic alpha iron seems to be not correct.

## 1.3   Solid Solutions in the Iron

Solid solutions in the iron—habitually called solid solutions—are obtained when other atoms randomly replace the iron atoms in its crystalline lattice (substitutional solid solutions). If the atoms have small atomic radius, smaller than 0.1 nm (as in the case of the H, O, N, C, and B), they do not replace the iron atoms, but they are inserted in the interatomic spaces of the iron lattice (interstitial solid solutions). In both cases, the bond, which keeps in cohesion all the atoms, is of metallic type (otherwise, we would talk about intermetallic compounds and not about solid solutions).

### 1.3.1   Substitutional Solid Solutions

Several conditions are necessary to have a metal completely soluble in the Fe, limitlessly, that is to say, it does not matter the proportion between this metal and the iron. First, the crystalline systems in one and the other metal must be the same. Second, both metals must have the same valence. Additionally, they must have a similar electrochemical character and their atomic diameters must not differ by more or less 15% (Hume-Rothery rules).

It is possible to see in Table 1.1 that Mn, Ni, and Co are completely soluble in the $Fe_\gamma$: they form fully solid solution with the iron at 1200 °C, 1300 °C, and 1300 °C, respectively. On the contrary, these same elements give always limited solid solutions, or partial solutions, in the alpha iron. We should also advise that elements as the chromium and vanadium, which are completely soluble in the delta iron, are only partially soluble in the gamma iron. Therefore, we should notice that when the iron changes its allotropic state, the solubility is also modified. On the other hand, in a certain allotropic state of the iron, the solubility of the elements in the iron can decrease with the temperature due to the subsequent contraction of the lattice parameter (thermodynamic conditions) and, for that reason, due to the reduction of the atomic diameter of the solvent iron (either if we talk about gamma iron or alpha iron).

**Table 1.1**  Solubility of several elements in the iron

| In gamma iron | | In delta or alpha iron | |
|---|---|---|---|
| Solute | Solubility limits | Solute | Solubility limits |
| H | 0.0009% at 1394 °C (0.0004% at 912 °C) | H | 0.0003% at 910 °C (0.0001% at 100 °C) |
| C | 2.11% at 1148 °C (0.77% at 727 °C) | C | 0.02% at 727 °C (0.005% at 575 °C) |
| N | 2.8% at 650 °C (2.35% at 590 °C) | N | 0.095% at 585 °C (0.05% at 500 °C) |
| S | 0.065% at 1370 °C (0.010% at 914 °C) | S | 0.0.020% at 914 °C |
| Mn | 100% at 1200 °C | Mn | 3.0% at 770 °C |
| Ni | 100% at 1300 °C | Ni | 7.0% at 500 °C |
| Co | 100% at 1300 °C | Co | 75% at 770 °C |
| Cr | 12% at 1130 °C (7% at 830 °C) | Cr | 100% at 1400 °C |
| Mo | 2.5% at 1150 °C (decreases until 910 °C) | Mo | 30% at 1400 °C (10% at 910 °C) |
| W | 3.5% at 1100 °C (decreases until 910 °C) | W | 35.5% at 1554 °C (6% at 888 °C) |
| V | 1.4% at 1154 °C (decreases until 910 °C) | V | 100% at 1400 °C |
| Ti | 0.7% at 1175 °C (decreases until 910 °C) | Ti | 9% at 1291 °C (2.2% at 880 °C) |
| Al | 0.625% at 1150 °C (decreases until 910 °C) | Al | 36% at 1094 °C (33.33% at 100 °C) |
| Si | 1.9% at 1150 °C (decreases until 910 °C) | Si | 10.9% at 1275 °C (3.5% at 500 °C) |

## 1.3.2  Interstitial Solid Solutions

When the difference of atomic diameters between solvent iron and the solute element is noticeably big, the atoms of small diameter can, as it was mentioned in previous paragraphs, randomly insert in the interatomic spaces of the iron (Pero-Sanz et al. 2017).

The size of the interatomic space that makes possible the insertion of the solute atom (MM' in Fig. 1.2) depends on the crystalline state of the iron. A solute element can insert in the center of the edges of the gamma iron (or in the center of the cube, which is an equivalent position), without deforming the lattice, the relation between atomic radius of the solute and the iron, $r_i/r_\gamma$, is 0.414. Relation that is obtained when we eliminate $a_\gamma$ in the following equations:

$$r_{Fe_\gamma} = \frac{a_\gamma}{8^{1/2}} \tag{1.2}$$

and

$$2 \cdot r_i + 2 \cdot r_{Fe_\gamma} = a_\gamma \qquad (1.3)$$

Analogously, it is possible to calculate the relation of atomic radius or diameters $r_i/r_{Fe_\alpha}$ to make possible the insertion of one atom in the alpha iron without distorting its lattice. In this case, the relation is 0.154 as

$$r_{Fe_\alpha} = a_\alpha \cdot \frac{3^{1/2}}{4} \qquad (1.4)$$

and:

$$2 \cdot r_i + 2 \cdot r_{Fe_\alpha} = a_\alpha \qquad (1.5)$$

We warn that interstitial solid solutions are always limited solutions—because, generally, the diameter of the inserted atom (i.e., for carbon $r_C = 0.772$ Å and for nitrogen $r_N = 0.53$ Å) is usually greater than the available space—and that the solubility of any element of small diameter is bigger in the gamma iron than in the alpha iron.

### 1.3.2.1 Austenite

We call austenite to the octahedral interstitial solid solution of the carbon in the $Fe_\gamma$. This one could admit carbon in solution until the moment when all the possible positions in the edges and also in the center of the unit cell would be occupied. The saturation of the $Fe_\gamma$ by the carbon would be reached with four atoms of carbon in each cell (because there are also four atoms of iron per cell, this would mean 50 at. % carbon); that is to say, about 18 wt. % C. However, the real maximum carbon content in the austenite, 2.11 wt. %, is much lower than the presumable 18%.

This apparent anomaly is since the real relation between atomic diameters of the carbon and iron broadly exceeds the value of 0.414. The relation of diameters of the carbon and iron is 0.63 according to the Goldsmidt relation (this relation assumes that each atom of one element is tangent to twelve equals to it to compare packing metals atoms between them). Therefore, the insertion of each carbon atom in the $Fe_\gamma$ cell involves an important distortion of the lattice (the greater the greater the carbon quantity). Concretely, the parameter of the austenite varies as a function of the carbon weight percentage, $C_1$, according to the following equation:

$$a_\gamma = 3.548 + 0.044 \cdot C_1 \qquad (1.6)$$

The austenite is saturated by the carbon with 2.11 wt. % C at 1148 °C. However, if the gamma iron would have, apart from the carbon, atoms of other elements in solid solution distorting the lattice—for instance, Cr, V, W, or Si—the maximum solubility of the carbon would be smaller than 2.11% C. Ferrous alloys with carbon content smaller or equal to the maximum limit of carbon saturation in the austenite are called steels.

### 1.3.2.2  Ferrite

We say ferrite to the octahedral interstitial solid solution of the carbon in the body centered cubic iron. We can deduct from the previously indicated that its carbon content is noticeably smaller than that admitted by the face centered cubic iron (austenite).

The maximum carbon content in the delta ferrite is 0.09 wt. % at 1495 °C. The alpha ferrite—habitually called only ferrite—admits a maximum of 0.02 wt. % C (200 ppm) at 727 °C. Below this temperature, the carbon content in the alpha ferrite decreases until reaching almost zero: 20 ppm C (parts per million) at room temperature.

**Exercise 1.1** Calculate the maximum theoretical solubility in molar and weight fraction of the carbon in the $Fe_\gamma$ if all the interstitial spaces were occupied. Make the same calculations in the case of the $Fe_\alpha$. Compare the values with the real. Consequences.

*Solution starts*:

Austenite crystallizes in the fcc (face centered cubic system) and contains four iron atoms: $8 \cdot 1/8$ in the corners, and $6 \cdot 1/2$ in the center of the faces. The maximum carbon solubility in the austenite is 2.11 wt. %. We calculate the atomic weight of carbon:

$$\text{at.\% C} = \frac{\frac{2.11}{12}}{\frac{2.11}{12} + \frac{97.89}{55.85}} \cdot 100 = 9.12\% (\simeq 10\%) \tag{1.7}$$

We calculate now the maximum theoretical solubility of the carbon in the austenite if all the sites were occupied by carbon atoms. There are 12 sites corresponding to the edges shared by four cells and 1 site in the cell, which means that $12 \cdot 1/4 + 1$ are the carbon atoms in the cell if all the interstitial sites were occupied by them:

$$\text{at.\% C} = \frac{4}{4 + 4} \cdot 100 = 50\% \tag{1.8}$$

Ferrite crystallizes in the bcc (body centered cubic system) and contains three iron atoms: $8 \cdot 1/8$ in the corners, and 1 in the center of the cube. The maximum carbon solubility in the ferrite is 0.02 wt. %. We calculate the atomic weight of carbon:

$$\text{at.\% C} = \frac{\frac{0.02}{12}}{\frac{0.02}{12} + \frac{99.98}{55.85}} \cdot 100 = 9.30 \cdot 10^{-2}\% \tag{1.9}$$

Ferrite could have the composition of $FeC_3$ if all the octahedral sites were occupied by carbon atoms, 75% atomic. Thus, the maximum ideal degree of occupation of carbon in the ferrite would be equal to

$$\frac{9.30 \cdot 10^{-4}}{0.75} \simeq 1.24 \cdot 10^{-3} \tag{1.10}$$

The inverse value is

$$\frac{1}{1.24 \cdot 10^{-3}} \simeq 806 \tag{1.11}$$

As average value, there is, as maximum, 1 carbon atom for each approximately 800 octahedral interstices (sites in the ferrite lattice).

This is the reason of having $D_C^\alpha$ greater than $D_C^\gamma$ (where $D_C^\alpha$ and $D_C^\gamma$ are the coefficients of diffusion of the carbon in the ferrite and austenite, respectively). The source of distortion that originates a carbon atom octahedrally inserted in the ferrite together with the possibility of finding around it many empty sites that make possible its greater capacity of migration (diffusion).

*End of the question.*

Ferrite is the softest constituent of steels. If iron does not contain in solution other elements, the mechanical characteristics of the ferrite at room temperature are 90 HB, 300 MPa of ultimate tensile strength and 40% elongation. It is magnetic below 770 °C and nonmagnetic above this temperature.

Elements in solid solution hardens the ferrite. This hardening depends on the content in these solubilized elements and is noticeably greater in the interstitial solid solutions (C and N) than in the substitutional solid solutions. The hardener characteristic of the common elements—from the greatest to the smallest—follows the next order: C, N, P, Sn, Si, Cu, Mn, and Mo (Bain 1924; Bain and Paxton 1966). Nickel and Aluminum hardly harden the ferrite. Chromium, possibly due to the removal of interstitial solutes, apparently softens the ferrite for contents smaller than 2%. See hardening of alpha ferrite by solid solution for different elements in Fig. 1.3.

The increase of the hardness is accompanied, in general, by a loss of toughness. Nitrogen and phosphorus noticeably increase the ductile–brittle transition temperature (DBTT) and, consequently, increase the brittleness of the ferrite. For a theoretical nitrogen content of 1% solubilized in the ferrite, the DBTT would be increased in 700 °C. 1% of phosphorus would increase the DBTT in 400 °C. 1% Sn would increase the DBTT in 150 °C and 1% Si in 44 °C. Cu, Mn, Mo, and Cr dissolved in the ferrite have little influence in the toughness.

We are going to define in the following sections the main mechanical parameters that characterize the steels: hardness, tensile test, and toughness, with solved exercises. The aim is to establish, in a progressive manner, the important relationship that exists between structure and mechanical properties in steels.

**Fig. 1.3** Hardening of the
ferrite by solid solution of
the several elements

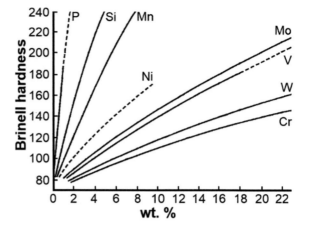

### 1.3.3  Hardness

#### 1.3.3.1  Brinell Hardness

Hardness of metals is a measure of their resistance to permanent or plastic deformation. Hardness is, to a person concerned with the mechanics of material testing, most likely to mean the resistance to indentation, and it often means for the design engineer an easily measured and specified quantity which indicates something about the strength and heat treatment of the metal.

The first widely accepted and standardized hardness test was proposed by J. A. Brinell in 1900. The Brinell test consists in indenting the metal surface with a 10 mm diameter steel ball or WC ball at a load of 3000 kg (for steels, $P/D^2 = 30$, where P is the applied load and D is the diameter of the ball). The load is applied for a standard time, usually 30 s, and the diameter of the indentation is measured with a low power microscope after removal of the load. The average of two readings of the diameter of the impression at right angles should be made. The surface on which the indentation is made should be relatively smooth and free from dirt or scale (previously prepared by grinding, for example). The Brinell hardness number (BHN) is obtained as the load P divided by the surface area of the indentation. This is expressed by the formula:

$$\text{BHN} = \frac{2 \cdot P}{\pi \cdot D^2 \cdot \left[1 - \sqrt{1 - \left(\frac{d}{D}\right)^2}\right]} = \frac{P}{\pi \cdot D \cdot t} \tag{1.12}$$

where P is the applied load (kg), D is the diameter of the ball (mm), d is the diameter of the indentation (mm) and t is the depth of the impression.

It will be noticed that the units of the BHN are kilograms per square millimeter. However, the BHN is not a satisfactory physical concept since the equation does not

**Fig. 1.4**  Basic parameters in
Brinell test

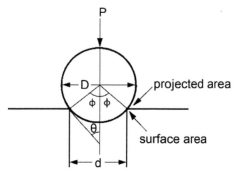

give the mean pressure:

$$\bar{p} = \frac{4 \cdot p}{\pi \cdot d^2} \qquad (1.13)$$

over the surface of the indentation, that is the pression over the projected area of the impression rather than the surface area, which is called the Meyer hardness.

From Fig. 1.4, it can be seen that

$$d = D \cdot \sin \phi \qquad (1.14)$$

that gives an alternative expression for Brinell hardness number:

$$BHN = \frac{2 \cdot P}{\pi \cdot D^2 \cdot (1 - \cos \phi)} \qquad (1.15)$$

In order to obtain the same BHN with non-standard load or ball diameter, it is necessary to produce geometrically similar indentations. Geometric similitude is achieved so long as the included angle $\phi$ remains constant. Equation (1.15) shows that for $\phi$ and BHN to remain constant, the load and ball diameter must be varied in the ratio:

$$\frac{P_1}{D_1^2} = \frac{P_2}{D_2^2} = \frac{P_3}{D_3^2} = constant \qquad (1.16)$$

For example, considering that the BHN of ferrite is 90 and that of the perlite is 240, in equilibrium state, it is possible to calculate, applying the lever rule, the hardness of equilibrium of a hypoeutectoid carbon steel:

$$BHN \left( kg/mm^2 \right) = 90 \cdot f_\alpha + 240 \cdot f_p \qquad (1.17)$$

for a $C_1$ carbon content, since:

$$f_\alpha \simeq \frac{0.77 - C_1}{0.77 - 0.02} \tag{1.18}$$

and:

$$f_p = \frac{C_1 - 0.02}{0.77 - 0.02} \tag{1.19}$$

Thus:

$$\begin{aligned}
\text{BHN } (\text{kg/mm}^2) &= 90 \cdot \left( \frac{0.77 - C_1}{0.77 - 0.02} \right) + 240 \cdot \left( \frac{C_1 - 0.02}{0.77 - 0.02} \right) \\
&\simeq 120 \cdot (0.77 - C_1) + 320 \cdot (C_1 - 0.02) \\
&= 92.4 - 6.4 + 200 \cdot C_1 = 86 + 200 \cdot C_1
\end{aligned} \tag{1.20}$$

The ultimate tensile strength can be calculated as follows considering that the $\sigma_u$ (MPa) of ferrite is 300 and that of the perlite is 800, in equilibrium state. Applying the lever rule, it is possible to calculate the ultimate tensile strength of equilibrium hypoeutectoid carbon steel:

$$\sigma_u(\text{MPa}) = 300 \cdot f_\alpha + 800 \cdot f_p \tag{1.21}$$

for a $C_1$ carbon content, applying the lever rule, since:

$$f_\alpha \simeq \frac{0.77 - C_1}{0.77 - 0.02} \tag{1.22}$$

and:

$$f_p = \frac{C_1 - 0.02}{0.77 - 0.02} \tag{1.23}$$

Thus:

$$\begin{aligned}
\sigma_u(\text{MPa}) &= 300 \cdot \left( \frac{0.77 - C_1}{0.77 - 0.02} \right) + 800 \cdot \left( \frac{C_1 - 0.02}{0.77 - 0.02} \right) \\
&\simeq 400 \cdot (0.77 - C_1) + 1067 \cdot (C_1 - 0.02) \\
&= 308 - 21.34 + 667 \cdot C_1 = 286.7 + 667 \cdot C_1
\end{aligned} \tag{1.24}$$

*Solution ends.*

**Exercise 1.2** Calculate the Brinell hardness number ($P = 187.5$ kg, $D = 2.5$ mm) of three steels whose hardness prints are 1, 1.26, and 1.54 mm, respectively. Calculate also the ultimate tensile strength and the approximate composition of the steels.

*Solution starts*:

First, the constant test parameter $P/D^2$ is calculated:

$$\frac{P}{D^2} = \frac{187.5 \text{ kg}}{(2.5 \text{ mm})^2} = 30 \qquad (1.25)$$

The BHN is now calculated as follows:

$$\text{BHN} = \frac{2 \cdot P}{\pi \cdot D^2 \cdot \left[1 - \sqrt{1 - \left(\frac{d}{D}\right)^2}\right]} = \frac{2 \cdot 187.5 \text{ kg}}{\pi \cdot (2.5 \text{ mm})^2 \cdot \left[1 - \sqrt{1 - \left(\frac{1 \text{ mm}}{2.5 \text{ mm}}\right)^2}\right]} = 229 \qquad (1.26)$$

From the formula:

$$\text{BHN} \left(\text{kg/mm}^2\right) = 86 + 200 \cdot C_1 \qquad (1.27)$$

$$229 = 86 + 200 \cdot C_1 \rightarrow C_1 = 0.715\%C \qquad (1.28)$$

And:

$$\sigma_u(\text{MPa}) = 2867 + 667 \cdot C_1 \qquad (1.29)$$

$$\sigma_u(\text{MPa}) = 286.7 + 667 \cdot 0.715 = 763.6 \text{ MPa} \qquad (1.30)$$

The same procedure is repeated for the other values of hardness indentations:
The BHN is calculated as follows:

$$\text{BHN} = \frac{2 \cdot P}{\pi \cdot D^2 \cdot \left[1 - \sqrt{1 - \left(\frac{d}{D}\right)^2}\right]} = \frac{2 \cdot 187.5 \text{ kg}}{\pi \cdot (2.5 \text{ mm})^2 \cdot \left[1 - \sqrt{1 - \left(\frac{1.26 \text{ mm}}{2.5 \text{ mm}}\right)^2}\right]} = 140 \qquad (1.31)$$

From the formula:

$$\text{BHN} \left(\text{kg/mm}^2\right) = 86 + 200 \cdot C_1 \qquad (1.32)$$

$$140 = 86 + 200 \cdot C_1 \rightarrow C_1 = 0.27\%C \qquad (1.33)$$

And:

$$\sigma_u(\text{MPa}) = 2867 + 667 \cdot C_1 \qquad (1.34)$$

$$\sigma_u(\text{MPa}) = 286.7 + 667 \cdot 0.27 = 466.80 \text{ MPa} \qquad (1.35)$$

And now for the d = 1.54 mm:
The BHN is calculated as follows:

$$\text{BHN} = \frac{2 \cdot P}{\pi \cdot D^2 \cdot \left[1 - \sqrt{1 - \left(\frac{d}{D}\right)^2}\right]} = \frac{2 \cdot 187.5 \, \text{kg}}{\pi \cdot (2.5 \, \text{mm})^2 \cdot \left[1 - \sqrt{1 - \left(\frac{1.54 \, \text{mm}}{2.5 \, \text{mm}}\right)^2}\right]} = 90 \quad (1.36)$$

From the formula:

$$\text{BHN} \left(\text{kg/mm}^2\right) = 86 + 200 \cdot C_1 \tag{1.37}$$

$$90 = 86 + 200 \cdot C_1 \rightarrow C_1 = 0.02\%C \tag{1.38}$$

And:

$$\sigma_u(\text{MPa}) = 286.7 + 667 \cdot C_1 \tag{1.39}$$

$$\sigma_u(\text{MPa}) = 286.7 + 667 \cdot 0.02 = 300 \, \text{MPa} \tag{1.40}$$

*Solution ends.*

### 1.3.3.2  Vickers Hardness

The Vickers hardness test uses a square-base diamond pyramid as an indenter. The included angle between opposite faces of the pyramid is 136 °. This angle was chosen because it approximates the most desirable ratio of indentation diameter to the ball diameter in the Brinell hardness test. It is because the most Brinell test lies between 0.25 and 0.50. For the diamond-pyramid indenter, a value of d/D = 0.375 (average value of 0.25 and 0.50) was used, which results in cone angle of 136°. In Fig. 1.4 of the Brinell test:

$$\sin \phi = \frac{d}{D} = 0.375 \rightarrow \phi \simeq 22° \rightarrow \theta = 180 - 90 - 22 = 68° \rightarrow 2\theta = 136°$$
$$(1.41)$$

Because of the nature and shape of the indenter, this is frequently called the diamond-pyramid hardness test. The diamond-pyramid hardness number (DPH), or Vickers hardness number (VHN, or VPH), is defined as the load divided by the surface area of indentation. In practice, this area is calculated from microscopic measurements of the lengths of the diagonals of the impressions. The DPH (VHN) may be determined from the following equations:

**Table 1.2** Hardness of several materials

| Constituent | Vickers hardness |
| --- | --- |
| High-carbon content pearlite | 240–425 |
| Austenite highly alloyed in chromium | 350–400 |
| Orthoclase (6 in Mohs scale) | 500–600 |
| Martensite of high carbon content | 770–800 |
| Quartz (7 in Mohs scale) | 800–1000 |
| Cementite carbides $K_C$ | 1060–1240 |
| Topaz (8 in Mohs scale) | 1300 |
| Carbides $K_2$ | 1500–1800 |
| Corundum (9 in Mohs scale) | 2000 |
| Tungsten carbide | 2400 |
| Silicon carbide | 2500 |
| Diamond (10 in Mohs scale) | 8000 |

$$\text{DPH (VHN)} = \frac{2 \cdot P \cdot \sin(\theta/2)}{d^2} = 1.8544 \cdot \frac{P}{d^2} = 0.9272 \cdot \bar{p} \qquad (1.42)$$

where $\bar{p}$ is the Meyer hardness, P is the applied load in kilograms, d is the average length of the diagonal in mm and $\theta$ is the angle between opposite face of diamond, 136 °.

The Vickers hardness test has received wide acceptance for research work because it provides a continuous scale of hardness for a given load (in steel is frequent to use $P = 30$ kg), from very soft metals with a DPH of 5 to extremely hard materials with a DPH of 2500 (see Table 1.2). As the indentations (prints) made by the pyramid indenter are geometrically similar no matter what their size, the VHN should be independent of load and proportional to the Meyer hardness which measured, as said, the average pressure exerted by the load over the material. However, this test requires a more careful surface preparation than Brinell test and allows great chance for personal error in the determination of the diagonal length due to "sinking" or "barreled" defects of indentations with respect to a perfect's one (Fig. 1.5).

It can be demonstrated in the case of binary and non-hardened low alloyed steels that:

**Fig. 1.5** Types of diamond-pyramid indentations. **a** Perfect indentation; **b** pin-cushion indentation due to sinking in; **c** barreled indentation due to ridging

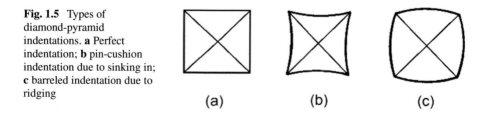

(a)          (b)          (c)

$$BHN \simeq VHN \tag{1.43}$$

for hardness values lower than around 400, so it is possible to estimate the average carbon content and strength Fe–C binary steels cooled in equilibrium conditions. Furthermore:

$$3 \cdot \sigma_u \simeq BHN \simeq VHN \ \left(kg/mm^2\right) \tag{1.44}$$

where $\sigma_u$ is the ultimate tensile strength measured in engineering tensile test $(kg/mm^2)$.

**Exercise 1.3** Calculate the values of the Vickers hardness number (P = 30 kg) of three steels whose hardness prints have an average length of diagonal of 0.45, 0.55, and 0.75 mm, respectively. Estimate the ultimate tensile strength and the approximate carbon content of these steels.

*Solution starts*:
   In the case of $\overline{d}$(average length of diagonals) = 0.45 mm, the value of the VHN is:

$$DPH \ (VHN) = 1.8544 \cdot \frac{P}{d^2} = \frac{1.854 \cdot 30 \ kg}{(0.45 \ mm)^2} \simeq 275 \frac{kg}{mm^2} \tag{1.45}$$

Since:

$$VHN \simeq BHN = 275 \rightarrow \sigma_u = \frac{275 \ kg/mm^2}{3} = 92 \frac{kg}{mm^2} \simeq 900 \ MPa \tag{1.46}$$

It is not possible to find the %C because BHN > 240 and cooling condition from austenitic state has not been specified (heat-treated steel).

– $P = 30$ kg; $\overline{d} = 0.55$ mm:

$$DPH \ (VHN) = 1.8544 \cdot \frac{P}{d^2} = \frac{1.854 \cdot 30 \ kg}{(0.55 \ mm)^2} \simeq 184 \frac{kg}{mm^2} \tag{1.47}$$

Since:

$$VHN \simeq BHN = 184 \rightarrow \sigma_u = \frac{184 \ kg/mm^2}{3} = 61 \frac{kg}{mm^2} \simeq 598 \ MPa \tag{1.48}$$

From the formula:

$$BHN \ \left(kg/mm^2\right) = 86 + 200 \cdot C_1 \tag{1.49}$$

$$184 = 86 + 200 \cdot C_1 \rightarrow C_1 = 0.49\%C \tag{1.50}$$

And:

$$\sigma_u(MPa) = 286.7 + 667 \cdot C_1 \tag{1.51}$$

$$\sigma_u(MPa) = 286.7 + 667 \cdot 0.49 = 613.5 \text{ MPa} \tag{1.52}$$

– $P = 30$ kg; $\bar{d} = 0.75$ mm:

$$DPH \text{ (VHN)} = 1.8544 \cdot \frac{P}{d^2} = \frac{1.854 \cdot 30 \text{ kg}}{(0.75 \text{ mm})^2} \simeq 99 \frac{\text{kg}}{\text{mm}^2} \tag{1.53}$$

Since:

$$VHN \simeq BHN = 99 \rightarrow \sigma_u = \frac{99 \text{ kg/mm}^2}{3} = 33 \frac{\text{kg}}{\text{mm}^2} \simeq 323 MPa \tag{1.54}$$

From the formula:

$$BHN \left(\text{kg/mm}^2\right) = 86 + 200 \cdot C_1 \tag{1.55}$$

$$99 = 86 + 200 \cdot C_1 \rightarrow C_1 = 0.065\%C \tag{1.56}$$

And:

$$\sigma_u(MPa) = 286.7 + 667 \cdot C_1 \tag{1.57}$$

$$\sigma_u(MPa) = 286.7 + 667 \cdot 0.065 = 330 \text{ MPa} \tag{1.58}$$

*Solution ends.*

### 1.3.3.3  Rockwell Hardness

The most widely used hardness test for hardened steel is the Rockwell hardness test. Its general acceptance is due to its speed, freedom from personal errors, ability to distinguish small hardness differences in hardened steel, and the small size of the indentation, so that finished heat-treated parts can be tested without damage. This test utilizes the depth of indentation, under constant load, as a measure of hardness. A

**Table 1.3** Different types of Rockwell C tests typically used in industry

|                            | Ball |       |      |      |       | Cone |     |     |
| -------------------------- | ---- | ----- | ---- | ---- | ----- | ---- | --- | --- |
| Name of the test           | B    | E     | F    | G    | K     | A    | C   | D   |
| Diameter of the ball (mm)  | 1.59 | 3.175 | 1.59 | 1.59 | 3.175 | –    | –   | –   |
| Load (kg)                  | 100  | 100   | 60   | 60   | 150   | 60   | 150 | 100 |

minor load of 10 kg is first applied to seat the specimen. This minimizes the amount surface preparation needed and reduces the tendency for ridging or sinking in by the indenter. The major load, 140 kg in HRC scale, is then applied, and the depth of indentation is automatically recorded on a dial gage in terms of arbitrary numbers. The dial contains 100 divisions, each division representing a penetration of 0.00008 in ($\simeq$0.002 mm). The dial is reversed so that a high hardness, which corresponds to a small penetration, results in a high hardness number. This agrees with the other hardness number described previously, but unlike the Brinell and Vickers hardness designation, which have units of kilograms per square millimeter, the Rockwell hardness number is purely arbitrary.

Hardened steels were tested, ordinarily on the C scale (HRC), $P = 140 + 10 = 150$ kg major load combined with a 120° diamond cone with slight rounded point, called Brale indenter. Next Table 1.3 reproduces the load and the geometry of the indenters currently used. The useful range of this scale is from about HRC = 20 to HRC = 70. Softer materials are usually tested on the B scale with 1/16 inches diameter steel bar and 100 kg major load. The range of this scale is from HRB = 0 to HRB = 100. The A scale diamond penetrator, 60 kg major load, provides the most extended Rockwell hardness scale, which is usable for materials from annealed brass to cemented carbides. The Rockwell hardness test is very useful and reproducible. However, several simple precautions are necessary, which are also applicable to the other hardness tests (seating, cleaning, perpendicular application of load, specimen thickness, spacing between indentations, speed of application, etc., such as are collected in standards).

### 1.3.4   Tensile Test

The engineering tension test is widely used to provide basic design information on the strength and ductility of materials (steels) and as an acceptance test for the specification of materials. In the tension test, a standardized specimen of $L_0$ length and $A_0$ cross section is subjected to a continuously slowly increasing uniaxial tensile force while simultaneous observations are made of the elongation of the specimen. An engineering stress–strain curve is constructed from the load-elongation measurements (Fig. 1.6). The significant points on the engineering stress–strain curve, comprising their zones, appearance of yield point, necking diffuse point, localized necking, etc. are indicated in Fig. 1.7. The stress used in this stress–strain curve is the average

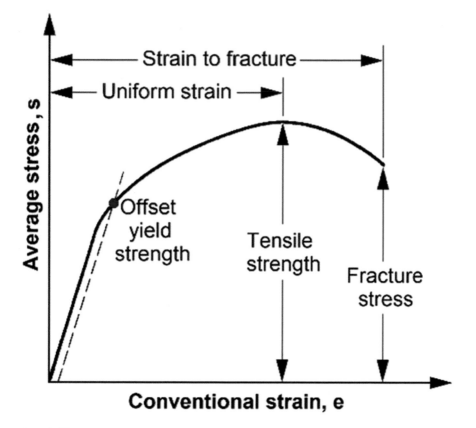

**Fig. 1.6**  The engineering stress–strain curve

longitudinal stress in the tensile specimen. It is obtained by dividing the load by the original area of the cross section of the specimen:

$$s = \frac{P}{A_0} \qquad (1.59)$$

The strain used for the engineering stress–strain curve is the average linear strain, which is obtained by dividing the elongation of the gage length $\Delta L$ (the relationship $L_0 = 5 \cdot d$, where d is the bar diameter for round products, is frequently used in steels), by its original length $e = \Delta L/L_0 = (L - L_0)/L_0$.

Since both the stress and the strain are obtained by dividing the load and elongation by constant factors, the load-elongation (displacement) curve has the same shape as the engineering stress–strain curve. The two curves are frequently used interchangeably.

The shape and magnitude of the stress–strain curve of a metal (steel) depend on its composition, heat treatment, prior history of plastic deformation and the strain

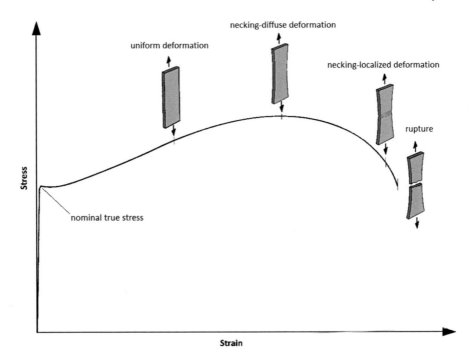

**Fig. 1.7** Stress–strain curve in the case of a flat specimen

rate, temperature, and state of stress imposed during the testing. The parameters which are used to describe the stress–strain curve of a metal (steel) are the tensile strength (UTS, ultimate tensile strength, $s_u = P_{max}/A_0$), yield stress or yield point (at 0.2% or 0.1% of deformation offset, when the transition of elastic to plastic zone is continuous), the percent elongation (either uniform or at fracture), and reduction of area (either uniform or at fracture). The first two are strength parameter, the last two indicate the ductility.

The general shape of the engineering curve (see Fig. 1.6) requires further explanations. In the so-called elastic region, stress is linearly proportional to strain, but its slope is lower than Young's modulus, E, because some grains in the gage length have already entered in plasticity. When the load exceeds a value corresponding to the yield strength, the specimen undergoes uniform (on the whole) plastic deformation. It is permanently deformed if the load is removed. The stress to produce continuous plastic deformation increases normally with increasing plastic strain, i.e., the steel strains-hardens. The volume of the specimen remains constant during plastic deformation, $\Delta L = A_0 \cdot L_0$ and as the specimen elongates, it decreases uniformly along the gage length in cross-sectional area. Initially the strain hardening more than compensating for this decrease in area (geometrical softening) and the engineering stress (proportional to the load P) continues to rise with deformation (increasing strain). Eventually, a point is reached when the decrease in specimen cross-sectional area is greater than the increase in deformation load arising from strain hardening. This condition will be reached first at some point in the specimen that is slightly weaker than the rest (diffuse necking). All further plastic deformation is concentrated in this region, and the specimen begins to neck or thin down locally (localized necking). Because the cross-sectional area now is decreasing for more rapidly than the deformation load is increased by strain hardening, the actual load required to deform the specimen falls off and the engineering stress described by Eq. 1.59 likewise continues to decrease until fracture occurs.

**Exercise 1.4** Data collected in Table 1.4 was obtained in the tensile testing of a steel rounded bar of 15 mm in diameter and calibrated length $L_0 = 50$ mm.
   The diameter of the specimen in the fracture zone was 12.45 mm. The following parameters are required:

- Represent the F-$\Delta$L curve and the engineering curve s-e.
- Calculate the proportionality modulus of the steel.
- Calculate $\sigma_{y_{0.2\%}}$, $\sigma_u$, uniform elongation, elongation at breaking, uniform ductility, and ductility at breaking of the steel.

   If it was a pearlitic steel, what consequences can be deduced?

*Solution starts*:
   We plot in Figs. 1.8 and 1.9 the data in Table 1.4.
   The area of the cross section of the specimen is

**Table 1.4** Stress–strain data for the steel rounded bar

| Load (kN) | 70 | 120 | 150 | 160 | 170 | 200 | 220 | 233 | 233 | 220 |
|-----------|------|------|------|------|------|------|------|------|------|------|
| $\Delta$L (mm) | 0.25 | 0.40 | 0.50 | 0.60 | 0.75 | 1.75 | 3.00 | 5.00 | 6.50 | 8.00 |

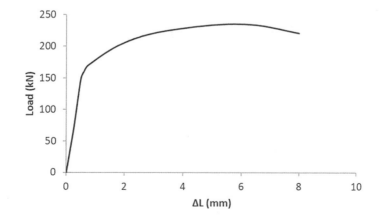

**Fig. 1.8** Load–displacement curve

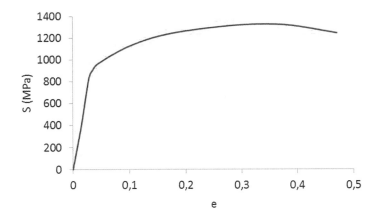

**Fig. 1.9** Stress–strain curve S (F/So)-e($\Delta$L/L)

$$A = \frac{\pi}{4} \cdot (15 \text{ mm})^2 = 176.7 \text{ mm}^2 \tag{1.60}$$

According to the curve, the maximum force is 235 kN:

$$\sigma_u = \frac{235000 \text{ N}}{176.7 \text{ mm}^2} = 1330 \text{ MPa} \tag{1.61}$$

The proportionality modulus is

$$E = \frac{300 \cdot 10^3 \frac{N}{mm} \cdot 50 \text{ mm}}{176.7 \text{ mm}^2} = 84.9 \text{ GPa} \tag{1.62}$$

that is smaller than $\simeq$210 GPa, which is the Young's modulus of the steel.

0.2% of 50 mm is equal to 0.1 mm. We draw a line parallel to the linear zone of the curve from the abscissa of 0.1 mm. This line intersects the curve for a load of 164 kN. Therefore, the yield strength is

$$\sigma_{0.2\%} = \frac{164000 \text{ N}}{176.7 \text{ mm}^2} = 928 \text{ MPa} \tag{1.63}$$

Uniform deformation: we draw a line parallel to the linear zone of the curve from the maximum of the stress–strain curve, and we read in abscissas 4.1 mm:

$$e_u = \frac{4.1}{50} \cdot 100 = 8.2\% \tag{1.64}$$

Total elongation: we draw a line parallel to the elastic zone of the stress–strain curve from the point of breaking, this line intersects the abscissa in 7.3 mm. Therefore, the elongation is

$$e_T = \frac{7.3}{50} \cdot 100 = 14.6\% \tag{1.65}$$

The uniform ductility and the ductility at breaking are

$$S_u = \frac{\left[ \frac{\pi \cdot 15^2}{4} - \frac{\pi \cdot 14.3^2}{4} \right]}{\frac{\pi \cdot 15^2}{4}} = 0.0912 \tag{1.66}$$

$$S_r = \frac{\left[ \frac{\pi \cdot 15^2}{4} - \frac{\pi \cdot 12.45^2}{4} \right]}{\frac{\pi \cdot 15^2}{4}} = 0.6889 \tag{1.67}$$

The consequences that are deduced if the steel is pearlitic are: it might be a wire rod used in the manufacture of pre-stressed concrete, which is in as-hot rolled condition, air cooled (normalizing).

*Solution ends.*

## 1.3.5  Toughness

Toughness is the premium property of metals and alloys. It expresses the quantity of work that a metal is able to absorb before breaking in the impact test. $K_{IC}$ versus $\sigma_y$ (critical stress intensity factor versus yield strength) is represented in Fig. 1.10. It is possible to check the hegemony of steels and non-ferrous alloys in the toughness/strength ratio, only exceeded by some composites of polymeric or ceramic matrix (which are more expensive than steels).

The great toughness of metallic materials with respect to other possible competitors results from their great ductility, consequence of the difference that exists between the metallic bond on the one side, and the ceramic and ionic bonds on the other side. In ductile materials, there are linear defects, called dislocations, whose movement under the action of applied stresses, plastically deforms (permanently) the parts, avoiding (in greater or lower extent) the nucleation-growth of cracks, by blunting or arresting them.

However, brittle fractures can appear (without apparent deformations) in some widely used metallic materials for the manufacture of tanks, pressure vessels, pipelines, bridges, etc., as in the case of ferritic steels (bcc structure) of great resistance ($\sigma_y > 1000$ MPa). Three basic factors contribute to brittle-cleavage type of fracture. They are (1) a triaxial state of stress, (2) a low temperature, and (3) a high strain or rapid rate of loading.

So, specimens used to show the steel ductile–brittle fracture are currently Charpy's one. Charpy's specimen has a square-cross section (10 mm × 10 mm) and contains a 45° V-notch, 2 mm depth with a 0.25 mm root radius. The specimen undergoes a three-point high velocity bending test: is loaded in the opposite side to the notch

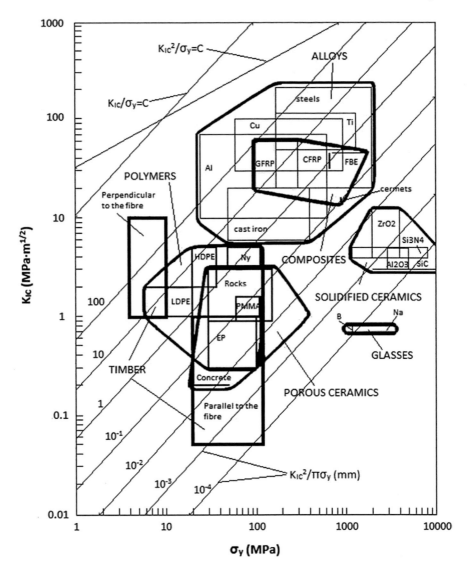

**Fig. 1.10** Chart for the selection of materials ($K_{IC}$, $\sigma_y$). Guidelines corresponding to the common geometries of loading are drawn on the diagram (Pero-Sanz et al. 2019)

by the impact of the heavy swinging pendulum (P = 30 kg, h = 1 m), so the impact speed is approximately 4.8 m/s. The specimen is forced to bend and sometimes to fracture at a high strain rate, on the order of $10^3$ s$^{-1}$. The principal measurement from the impact test is the energy absorbed in bending (fracturing) the specimen. The energy absorbed, usually expressed in foot-pounds, kg·m or J is read directly from a calibrated dial, CVN (impact energy absorbed in a Charpy test with V-notch). The notched-bar impact test is most meaningful when conducted over a range of temperature so that temperature at which the ductile-to-brittle transition takes place can be determined as the energy absorbed decreases with decreasing temperature. Due to a standardized nature of test, the DBTT (Ductile–Brittle Transition Temperature) or ITT (Impact Transition Temperature) (°C) corresponds to a CVN of 27–30 J absorbed. This test, in despite of non-desirable result scatter, the difficulties of preparing perfectly reproducible notches and the proper placement of the specimen in the impact machine, is frequently used for quality control and material acceptance purposes. Furthermore, relationships between $K_{IC}$ and CVN can be found today.

Structure–mechanical properties relationships have been emphasized by Pickering (Pickering and Gladman 1963) who proposed in respect to DBTT (°C) for steels the following formulae:

– for weldable steels (%C < 0.25%):

$$DBTT\ (°C) = -19 - 11.5 \cdot d^{-0.5} + 44 \cdot (\%Si) + 700 \cdot \sqrt{N_f} + 2.2 \cdot f_p(\%) \quad (1.68)$$

– for ferrite-pearlite steels:

$$DBTT\ (°C) = f_\alpha \cdot \left(-46 - 11.5 \cdot d^{-0.5}\right) + (1 - f_\alpha)$$
$$\cdot \left(-335 + 5.6 \cdot S_0^{-0.5} - 13.3 \cdot p^{-0.5} + 3480 \cdot 10^3 \cdot t\right) + 48.7$$
$$\cdot (\%Si) + 762 \cdot \sqrt{N_f} \quad (1.69)$$

where d is the mean linear intercept of ferrite (mm), $S_0$ is the mean interlamellar spacing of pearlite (mm), p is the mean size of pearlite colonies (mm), t is the thickness of cementite (mm), $f_\alpha$ is the fraction of ferrite, $f_p$ is the fraction of pearlite, % Si is the silicon content (wt. %), and $N_f$ is the free nitrogen content (wt. %).

Figure 1.11 shows the effect of carbon content both on the impact transition temperature for steel and on the upper-shelf energy as well.

**Exercise 1.5** Impact transition temperature in steels.

*Solution starts*:

As the ITT (°C) of ferritic steels is greatly dependent on the grain size, with fine-grained steels having better (lower) transition temperatures, this has led to specifications for pipelines (for transportation of natural gas, oil, and petroleum) welded plate began to require aluminum-killed fine-grain steels. Finer grain sizes were obtained by low-temperature finishing (controlled rolling) of the plate. The initial controlled-rolling practices consisted of stopping the rolling sequence (after roughing) when the plate thickness was about twice (or more) the desired thickness, allowing the plate to cool about 70 °C, and then continuing (finishing) the rolling sequence to the final gage.

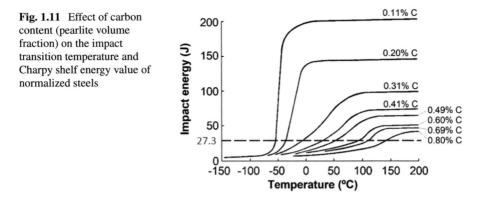

**Fig. 1.11** Effect of carbon content (pearlite volume fraction) on the impact transition temperature and Charpy shelf energy value of normalized steels

This practice produced a fine non-recrystallized austenite grain size (due to the presence of microalloying elements on the steels, such as niobium $\simeq 0.04\%$) and therefore, a finer ferrite-pearlite structure on subsequent allotropic $\gamma \rightarrow \alpha$ transformation (see Chap. 6 in Verdeja et al. 2021, controlled rolling of the steel API-X-70).

By the late 1960s, the discovery of large oil and gas fields in Alaska and other Artic regions created a demand for API (American Petroleum Institute) grades up to X-70 (70 ksi minimum yield strength) pipelines with lower transition temperature and better weldability (lower carbon contents) than those available in that moment. The first pipelines manufactured to the afore-mentioned toughness requirements were put into operation. In spite of all precautions, several failures occurred with relatively long propagating ductile fractures (one or two lengths of the pipe). Consideration of these failures suggested that there was a minimum Charpy V-notch (CVN) shear-fracture energy necessary to decelerate and stop even a ductile fracture in a high-pressure operating gas-transmission pipeline. Subsequent full-scale shear-fracture propagation test, sponsored by the AGA (American Gas Association) and other tests sponsored by the American Iron and Steel Institute (AISI) showed that this was indeed the case. Fracture Mechanics analyses coupled with empirical correlations were used to analyze these test results. The AISI works indicated that the minimum toughness required to arrest a running shear fracture in a buried gas-transmission pipeline was related to the circumferential stress, $\sigma_c$, and the pipe diameter, D, as follows:

$$\text{CVN} \geq 0.016 \cdot \sigma_c^{1.5} \cdot D^{0.5} \tag{1.70}$$

where CVN is 2/3 of the size Charpy V-notch energy (ft-lb), $\sigma_c$ is the circumferential stress (ksi) equal to $p \cdot D/2 \cdot t$ (t is the thickness of the pipe) and D is the diameter of the pipe (in).

**Calculate, in a pipe of 40 inches in diameter and thickness of 1 inch, through which natural gas circulates at 1.4 ksi of pressure, the minimum value of CVN required to delay the most possible the propagation of the ductile fracture.**

$$\sigma_c = \frac{p \cdot D}{2 \cdot t} = \frac{1.4 \text{ ksi} \cdot 40 \text{ in}}{2 \cdot 1 \text{ in}} \simeq 28 \text{ ksi} \tag{1.71}$$

If we apply the Eq. 1.70, we obtain

$$\frac{2}{3} \cdot CVN \geq 0.016 \cdot 28^{1.5} \cdot 40^{0.5} \simeq 15 \text{ ft} \cdot \text{lb} \tag{1.72}$$

since 1 ft·lb $\simeq$ 0.14 kg·m, we have

$$CVN \geq 22.4 \text{ ft} \cdot \text{lb} \simeq 3.1 \text{ kg} \cdot \text{m} \simeq 30 \text{ J} \tag{1.73}$$

which approximately coincides with the ductile–brittle transition temperature expressed by the Pickering Eq. 1.68.

**Calculate the decrease in the DBTT when the steel grain size decreases from 8 ASTM ($\simeq$20 $\mu$m) to 12 ASTM ($\simeq$5 $\mu$m).**

The value of DBTT according to the Eq. (1.68) is

$$DBTT \; (°C) = -11.5 \cdot d^{-0.5} \tag{1.74}$$

For d = 20 $\mu$m:

$$DBTT \; (°C) = -11.5 \cdot 0.0020^{-0.5} \simeq -81.3 \, °C \tag{1.75}$$

For d = 5 $\mu$m:

$$DBTT \; (°C) = -11.5 \cdot 0.005^{-0.5} \simeq -162.6 \, °C \tag{1.76}$$

The indicated decrease in the grain size produces a descent of the DBTT greater than $-80$ °C. HSLA steels (as the API grade X-70), with low pearlite contents, low $N_f$, low silicon content and micro-alloyed with titanium, niobium, and vanadium are of Artic quality (DBTT temperature lower than $-80$ °C) (Paxton 1979).

*Solution ends.*

## 1.4  Alphagenous and Gammagenous Character of the Elements Solubilized in the Iron

Some elements—such as Si, W, Mo, Ti, and others—are called alphagenous because they allotropically stabilize the body centered cubic iron when they form solid solution with it.

This stability is translated into the fact that the transformation of the delta iron into gamma iron, by equilibrium cooling, starts at a temperature—habitually designated with $A_5$—lower than that of the pure iron (1394 °C). The $A_5$ temperature is lower the greater the quantity of alphagenous elements solubilized in the alloyed iron. Moreover, the allotropic $\delta \rightarrow \gamma$ transformation does not finish at the temperature $A_5$, but takes place progressively in an interval of temperatures, during the equilibrium cooling below $A_5$ (phase rule applied to a binary system). The end of the $\delta \rightarrow \gamma$ transformation is reached at a temperature, called $A_4$, which equally depends on the quantity of alphagenous element solubilized in the iron (Porter and Easterling 1981).

The stability is also observed, later, when the equilibrium cooling continues, in the $\gamma \rightarrow \alpha$ transformation. This allotropic change starts at a temperature, called $A_3$, greater than that of the pure iron (912 °C). And the $\gamma \rightarrow \alpha$ transformation takes place in an interval of temperatures; from $A_3$ to the final temperature of transformation into alpha iron, called $A_1$. Both the temperature $A_3$ and $A_1$ are greater the greater is the content of alphagenous elements in the iron.

Curiously, some elements—such as the chromium—can be alphagenous if the $\delta \rightarrow \gamma$ transformation is concerned and gammagenous (stabilize the gamma phase in the $\gamma \rightarrow \alpha$ transformation) for certain contents of this element. It is illustrative of this matter the Fe–Cr equilibrium diagram shown in Fig. 1.12 (Pero-Sanz et al. 2018).

It is possible to appreciate in this diagram that, for Cr > 13 wt. %, ferrochromium solidifies in delta phase (solution in the body centered cubic iron) and there is not any allotropic transformation until the room temperature (transformation lines

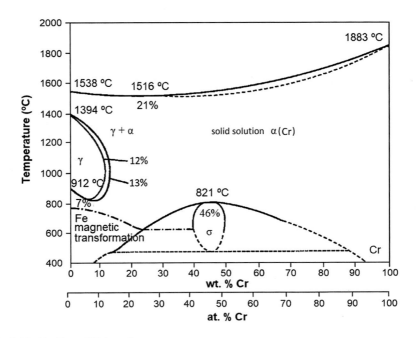

**Fig. 1.12** Fe–Cr equilibrium diagram

drawn in the diagram for percentages greater than 13%—where chromium is always alphagenous—are lines of sigma phase precipitation).

The Fe–Cr binary diagram also illustrates—see the gamma loop—that chromium is always alphagenous for contents greater than 7 wt. %, both regarding the $\delta \to \gamma$ transformation and the $\gamma \to \alpha$ transformation. On the contrary, for %Cr < 7 wt. %, chromium is always deltagenous—reduces the $A_5$ and $A_4$ critical temperatures—but it is also gammagenous: when the chromium content is increased in the alloys from 0% Cr to 7% Cr, the corresponding $A_3$ and $A_1$ critical points decrease.

We call gammagenous to the elements—such as the C, N, Mn, Ni, and Co—that solubilized in the iron stabilize the gamma allotropic variety. In the case of the iron alloyed in this manner, the $\delta \to \gamma$ transformation takes place above 1394 °C, in an interval of temperatures, with critical temperatures $A_5$ and $A_4$ increasing the greater the percentage of these gammagenous elements solubilized in the iron. Regarding the $\gamma \to \alpha$ transformation that always takes place below 912 °C, the critical temperatures $A_3$ and $A_1$ decrease with the percentage of gammagenous element.

It is particularly illustrative of the Fe–Mn equilibrium diagram of Fig. 1.13. Manganese is a strongly gammagenous element and makes disappearing the $A_5$ and $A_4$ critical points, and also $A_3$, for manganese contents greater than 30 wt. %. For these percentages, the solidification of the ferromanganese takes place in gamma phase, without appearance of delta phase, and the solid does not experience any transformation until the room temperature.

**Exercise 1.6** The Hadfield steel (1.3% C, 12% Mn, balance Fe) has an austenitic structure at room temperature. Its lattice parameter is 3.62 Å (0.362 nm). Calculate the density of the steel. Carbon atomic weight is 12 g/mol. Manganese atomic weight is 54.9 g/mol. Iron atomic weight is 55.85 g/mol.

*Solution starts*:

In 100 g of steel, there are 1.3 g of carbon, 12.0 g of manganese, and 86.7 g of iron.

The number of moles of carbon, manganese, and iron in 100 g of steel are

$$n_C = \frac{1.3 \text{ g}}{12 \text{ g/mol}} = 0.11 \text{ mol} \tag{1.77}$$

$$n_{Mn} = \frac{12 \text{ g}}{54.9 \text{ g/mol}} = 0.22 \text{ mol} \tag{1.78}$$

$$n_{Fe} = \frac{86.7 \text{ g}}{55.85 \text{ g/mol}} = 1.55 \text{ mol} \tag{1.79}$$

The atomic fractions (moles) of carbon, manganese, and iron are

$$x_C = \frac{0.11}{0.11 + 0.22 + 1.55} = 0.059 \tag{1.80}$$

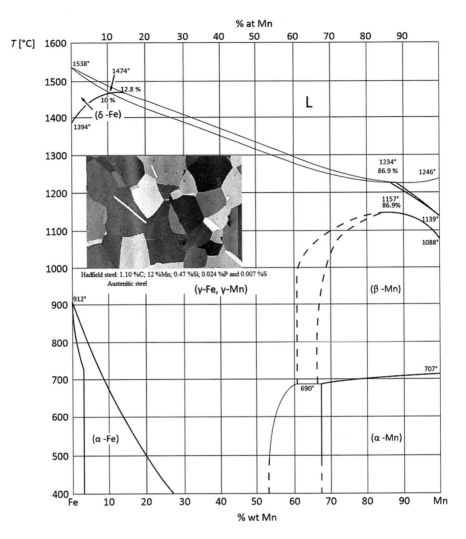

**Fig. 1.13** Fe–Mn diagram

$$x_{Mn} = \frac{0.22}{0.11 + 0.22 + 1.55} = 0.117 \tag{1.81}$$

$$x_{Fe} = \frac{1.55}{0.11 + 0.22 + 1.55} = 0.824 \tag{1.82}$$

The density is

$$\rho = 4 \text{ at}_{(Fe+Mn)} \cdot \left[ \frac{11.7 \text{ atoms Mn}}{(11.7 + 82.4)_{(Fe+Mn)}} \cdot 54.9 \text{ grams} + \frac{82.4 \text{ atoms Fe}}{(11.7 + 82.4)_{(Fe+Mn)}} \cdot 55.85 \text{ grams} \right.$$

$$+ \left. \frac{5.9 \text{ atoms C}}{(11.7 + 82.4)_{(Fe+Mn)}} \cdot 12 \text{ grams} \right] \cdot \frac{1}{6.022 \cdot 10^{23} \cdot (3.62 \cdot 10^{-8} \text{ m})^3}$$

$$= 7.93 \text{ g/cm}^3 \tag{1.83}$$

Note: If carbon was in substitutional solid solution, which is not the case (it is interstitial solid solution), the density of the steel would be

$$\rho = 4 \cdot [0.059 \cdot 12 + 0.117 \cdot 54.9 + 0.824 \cdot 55.85]$$

$$\cdot \frac{1}{6.022 \cdot 10^{23} \cdot (3.62 \cdot 10^{-8} \text{ m})^3} = 7.4 \text{ g/cm}^3 \tag{1.84}$$

Carbon does not compete with the iron and manganese for the substitutional sites (4 places/unit cell), but the elemental cell acts as carrier for those possible atoms in interstitial solid solution (H, C, N, O, and B).

Hadfield steel is used in cross-hatched in railways and mining (crushing operations, jab crusher lining, balls mills, hammer mills). It resists adequately the wear with abrasives of hardness lower than that of the quartz and it has a noticeable toughness (wear resistance consequence of the austenitic structure).

*End of the solution*

The equilibrium cooling involves a decrease of the temperature so slow that, in the interval of temperatures where the transformation takes place, for each intermediate temperature between the beginning and end of the transformation, the equilibrium phases, each one of the phases—both $\delta$ and $\gamma$, or in the corresponding case $\gamma$ and $\alpha$—must have uniform chemical composition in all points, both inside and in the periphery of the considered phase, before the decrease of temperature could continue (phase rule and lever rule).

When, on the contrary, in a binary alloy, the cooling does not take place with the required slowness for the equilibrium, the critical temperatures—$A_5$, $A_4$, $A_3$, $A_1$—are smaller than those of equilibrium. In this case, they are usually identified—as historical heritage of the first investigations carried out by French researchers—with a subscript "r" (refroidissement, equivalent to cooling) or "c" (chauffage, equivalent to heating) whether the transformation took place during the cooling or during the heating, respectively. Therefore, as example, always $A_{3r} < A_3 < A_{3c}$ and the difference between $A_{3r}$ and $A_{3c}$ for the same alloy is called thermal hysteresis.

## 1.5  The Metastable Fe–C Equilibrium Diagram

We call metastable to the Fe–C binary diagram whose simple constituents are austenite, ferrite, and cementite (for further information about the stable Fe–C diagram we suggest the book *Physical Metallurgy of Cast Irons* published by Springer, Pero-Sanz et al. 2018).

Cementite is a carbide with formula $Fe_3C$ (6.67% wt. % C), and bonds (Figs. 1.14 and 1.15) mainly non-metallic and orthorhombic crystallography with twelve iron atoms and four carbon atoms per unit cell.

As it is an intermetallic compound—it is not a solid solution of carbon in the iron—its brittle behavior as well as its great hardness (68 HRC) are logical, see Exercise 1.7. It is the hardest constituent of steels. Its melting point (1227 °C) is hardly reached because at high temperatures cementite sublimates into iron and carbon—from 450 °C if the temperature is kept for thousands of hours or for shorter if the temperature is greater. The cementite is ferromagnetic below 210 °C.

**Exercise 1.7** Calculate the density of the cementite and compare it with the theoretical value 7694 kg/m$^3$.

Data: a = 5.08 Å; b = 4.51 Å and c = 6.73 Å.

*Solution starts*:

Cementite ($Fe_3C$) crystallizes in the orthorhombic system and each cell has 4 atoms of C and 12 atoms of Fe, thus:

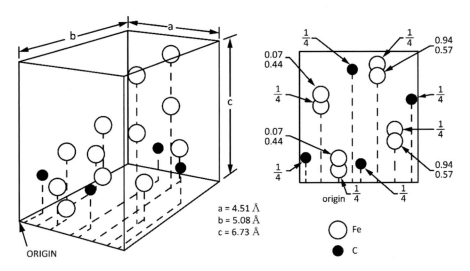

**Fig. 1.14** Orthorhombic structure of the cementite

**Fig. 1.15** Detail of the relative position of the Fe and C atoms in cementite

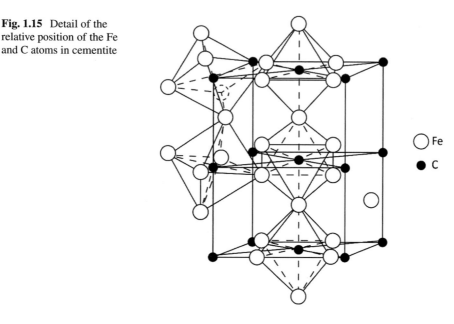

○ Fe

● C

$$\rho_{Fe_3C} = \frac{12 \text{ atoms Fe} \cdot \left(55.85 \frac{g}{mol\,Fe}\right) + 4 \text{ atoms C} \cdot \left(12 \frac{g}{mol\,C}\right)}{6.022 \cdot 10^{23} \frac{atoms}{mol} \cdot 5.08 \cdot 10^{-8}\,cm \cdot 4.51 \cdot 10^{-8}\,cm \cdot 6.73 \cdot 10^{-8}\,cm} = 7.733\ g/cm^3$$

(1.85)

which differs a 0.5% with respect to the theoretical value.

*Solution ends.*

Figure 1.16 illustrates the Fe–C metastable diagram. It shows that any iron and carbon melt—$0 < \%C < 4.3$—starts its solidification at lower temperature than that of the pure iron (1538 °C) the greater the carbon content.

We should also pay attention in Fig. 1.16 to the fact that the primary solidification structure of binary ferrous alloys whose carbon content is greater than 0.53% C is gamma at the end of the solidification (whose solidus temperature, $T_S$, decreases when the carbon content in the steel is increased). If we had started from a melt with 2.11% C, the solidus temperature would have been 1148 °C, and the primary structure of solidification of this alloy comprises austenite of maximum carbon content. We remind that binary alloys with carbon content equal or smaller than 2.11% are called steels.

The alloys of the Fe–C metastable diagram with more than 2.11% C are called white cast irons. Their solidification ends at 1148 °C with the formation of a eutectic matrix constituent, of 4.3% C, called ledeburite, formed by austenite with 2.11% C and cementite (52% austenite, 48% cementite, normal eutectic).

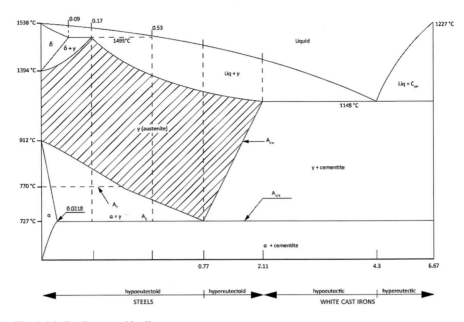

**Fig. 1.16** Fe–C metastable diagram

On the other hand, all steels whose carbon content is in the range 0.09–0.53% C experience a peritectic reaction at 1495 °C (delta ferrite of 0.09%C+ liquid of 0.53% C, to give austenite of 0.17% C). From them, once finished the peritectic reaction, those with a carbon content in the range 0.17–0.53% continue their solidification below 1495 °C, and their structure at the end of the solidification is fully gamma.

Regarding steels with 0.09 < %C < 0.17 after undergoing peritectic reaction, their primary structure of solidification is biphasic: delta + gamma, with greater proportion of gamma the greater the carbon percentage. Later, during the cooling, the fraction of delta phase is completely transformed into gamma. The starting and finishing temperatures for this δ → γ transformation also appear in the diagram.

We should also appreciate, if we observe the γ field in Fig. 1.16, that all the binary steels, once solidified, it does not matter the carbon percentage, have gamma structure from certain temperatures (for instance, below 1148 °C).

We will pay preferential attention in the following paragraphs to the transformations, by equilibrium cooling from the gamma state, of binary steels—with only iron and carbon—with different carbon content.

## 1.5.1   Steels with 2.11% C

The austenitic cell of 2.11% C, when the temperature decreases below 1148 °C, contracts and destabilizes due to the distortion produced by the carbon inserted in the cell. The lower the temperature, the austenite saturates in carbon whether carbon contents are lower than 2.11% C (Porter and Easterling 1981).

The excess of carbon atoms that supersaturates the cell at each temperature $T_1$ < 1148 °C migrate toward the grain boundaries where it precipitates as cementite. Therefore, in the equilibrium, at each temperature $T_1$, the inner of the austenitic grain is depleted in carbon and, however, the austenite is stable at this temperature. Therefore, it is necessary to continue descending the temperature to continue the precipitation of cementite. This way, carbon diffuses toward the grain boundaries, and the remaining austenite impoverishes in carbon (this avoids the growing distortion caused by the carbon when, with the temperature, diminishes the lattice parameter of the austenite).

If the slow descent of the temperature corresponding to the equilibrium cooling continues, carbon atoms excesses corresponding to the austenite react with the iron atoms and precipitate as cementite in growing quantities as the temperature decreases from 1148 °C to 727 °C, line $A_{cm}$ in the diagram of Fig. 1.16. When 727 °C are reached, that austenite with 2.11% C has been uniformly impoverished in carbon—it has now 0.77% C—and also cementite has formed in the grain boundaries. The weight fraction of the precipitated cementite with respect to the total weight—addition of the cementite and the remaining austenite, of 0.77%—is at 727 °C:

$$Fe_3C \ (wt.\%) = \frac{2.11 - 0.77}{6.67 - 0.77} = 0.227 \ (22.7\%) \tag{1.86}$$

Until this temperature, the $\gamma \rightarrow \alpha$ allotropic transformation does not take place due to the strongly gammagenous character of the carbon. The stability of the $Fe_\gamma$ due to this dissolved 0.77% C is such that the $\gamma \rightarrow \alpha$ allotropic transformation, by equilibrium cooling of this remaining austenite of 0.77% C, starts 185 °C below 912 °C (that correspond to the pure iron). Moreover, the temperature remains constant during this transformation at 727 °C of the 0.77% C austenite ($A_3$ and $A_1$ coincide).

As it was indicated in the previous paragraph, the austenite of 0.77% C does not experience $\gamma \rightarrow \alpha$ allotropic transformation due to slow cooling before reaching the 727 °C. However, keeping the steel at 727 °C for a sufficiently long time, the carbon starts to migrate from the austenite and the formation of cementite clusters starts (habitually in the austenite grain boundaries). The zones close to the cementite clusters impoverish in carbon the surrounded austenite and allotropically transform into ferrite clusters. Cementite clusters continue to grow at the expense of carbon absorbed from the contiguous austenite. The result is an ensemble of cementite lamellae surrounded by others of ferrite that forms a complex constituent, of pearly reflections, that Sorby called pearlite.

This pearlitic reaction—austenite of 0.77% C = cementite + ferrite of 0.02% C—, which takes place at constant temperature (it starts and finishes at 727 °C), is called eutectoid reaction by analogy with the eutectic reactions, although eutectics are reactions that are produced at the end of the solidification and appear, for that reason, as matrix constituent. On the contrary, a eutectoid constituent is a dispersed constituent and adopts convex shapes that are suitable to be surrounded by matrix phases.

The weight percentages of cementite and ferrite in the pearlite are, respectively, 11.25% and 88.75%, as it is possible to deduce from the lever rule:

$$\% \text{ cementite} = 100 \cdot \frac{0.77 - 0.02}{6.67 - 0.02} \tag{1.87}$$

$$\% \text{ferrite} = 100 \cdot \frac{6.67 - 0.77}{6.67 - 0.02} \tag{1.88}$$

Figure 1.17 shows the microstructure obtained in a 2.11% steel after being slowly cooled down to 726 °C. It is possible to appreciate that the cementite precipitated from the austenite appears as matrix constituent surrounded by the pearlite. This model of structure—similar to a cementite sponge with holes filled in by pearlite— allows forecasting that this steel, if tested in tension, will break with almost zero elongation. It is hard but few tough.

This free cementite—it is not part of the pearlite—is called proeutectoid cementite (it is formed before the eutectoid pearlite following the $A_{cm}$ precipitation line). The percentage of proeutectoid cementite in this steel of 2.11% C is

**Fig. 1.17** Coarse lamellar pearlite

$$\% \text{ proeutectoid cementite} = 100 \cdot \frac{2.11 - 0.77}{6.67 - 0.77} = 22.7 (\text{matrix constituent})$$

$$(1.89)$$

and the percentage of pearlite is 77.3% (dispersed constituent).

The same as in the case of the 2.11% C steel—the same would happen in any hypereutectoid steel $C_1$ (0.77% < $C_1$ < 2.11% C). In this new assumption, and by equilibrium cooling from the gamma state, the steel of $C_1$ would begin to precipitate cementite when the $A_{cm}$ temperature of saturation of the austenite by carbon is reached ($A_{cm}$ is lower the lower the $C_1$). Below $A_{cm}$, the austenite of the steel continuously impoverishes during the cooling and at 727 °C, the steel contains 0.77% C austenite and cementite. However, in this case, as opposed to that indicated in the case of the 2.11% steel, the fraction of proeutectoid cementite with respect to the total weight is

$$\% \text{ proeutectoid cementite} = 100 \cdot \frac{C_1 - 0.77}{6.67 - 0.77} \qquad (1.90)$$

The microstructure would be similar to that of Fig. 1.17, although with less proeutectoid cementite than in the case of the 2.11% C steel.

In hypereutectoid steels, the critical points for the allotropic transformation of the austenite are called in technical literature $A_{123}$ to point out that the $A_3$, $A_1$, and magnetic transformation $A_2$ temperatures coincide (Porter and Easterling 1981). In all these cases, $A_{123}$ coincides with the eutectoid transformation, 727 °C, and, for that reason, $A_{123}$ is usually called $A_e$ (eutectoid critical temperature). We can also consider as critical points—but for the precipitation—the $A_{cm}$ temperatures, which are different for each hypereutectoid steel as a function of the carbon content.

## 1.5.2   Transformations by Equilibrium Cooling of Hypoeutectoid Steels (<0.77% C)

We take as example a steel with 0.20% C. This steel is completely austenitic at temperatures higher than 912 °C. If the temperature slowly descends below 912 °C, the austenite is still not transformed as a consequence of the gammagenous character of the carbon until reaching the $A_3$ temperature of the steel, where the $\gamma \rightarrow \alpha$ transformation starts. This beginning transformation takes place with volume increase, and this allows determining this $A_3$ temperature by dilatometric test.

The $\gamma \rightarrow \alpha$ transformation does not take place completely at $A_3$. At this temperature, the formation of ferrite starts, habitually in the austenitic grain boundaries, and the carbon ceded by the formed ferrite migrates—by diffusion—toward inside of the grain, and the still not formed austenite is enriched in carbon. The austenite like this enriched in carbon will require a further descent of the temperature to continue the $\gamma \rightarrow \alpha$ transformation.

If the slow decent of the temperature continues, the amount of ferrite increases, and the quantity of non-transformed austenite decreases. At each temperature and considering that the initial carbon content of the steel remains constant in all the grains, it is possible to calculate the proportions of austenite and ferrite.

When the temperature of 727 °C is reached, there is still non-transformed part of that austenite that originally had 0.2% C, and whose current carbon content, by progressive enrichment in this element, is 0.77% C. Particularly at this temperature, as it is possible to deduce when we apply the lever rule in the diagram of Fig. 1.16, there is still a weight percentage of eutectoid austenite—100·(0.2–0.02)/(0.77–0.02)—equal to 24%. That is to say, a percentage of austenite that, by sufficiently holding at this temperature, finally transforms into pearlite.

We notice that all hypoeutectoid steels, with carbon content in the range 0.02–0.77, end the $\gamma \rightarrow \alpha$ transformation at the eutectoid temperature ($A_e = 727$ °C). This explains that the critical point $A_1$ of all these steels is the eutectoid temperature ($A_e = 727$ °C).

Any of these steels of carbon percentage in the range 0.02–0.77 will have, after equilibrium cooling from the austenitic state to 726 °C, a ferritic-pearlitic structure similar to that observed in Fig. 1.18. This microstructure is formed by pearlite colonies surrounded by grains of free ferrite called proeutectoid ferrite (because it was formed before the eutectoid temperature). The grains of proeutectoid ferrite have smaller size—and more numerous—the lower the $A_3$ temperature of the steel is.

In hypoeutectoid binary steels, the proeutectoid ferrite clearly appears as matrix constituent—in the contours of the pearlite—when steels have carbon contents in

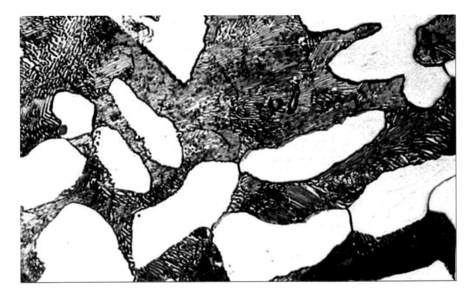

**Fig. 1.18** Binary steel with 0.38% C. Hypoeutectoid steel slowly cooled down to room temperature: structure of ferrite and pearlite. As-cast steel. 600x

the range 0.55–0.77. In steels with <0.55% C, ferrite, which is always the matrix constituent, appears as crystals mixed with pearlite colonies.

Considering as basis this ferritic-pearlitic morphology of the microstructure, it is possible to estimate that the ultimate tensile strength in tension of a ferritic-pearlitic steel is approximately given by

$$\sigma_u(\text{MPa}) = \frac{f_p(\%) \cdot 800 + f_f(\%) \cdot 300}{100} \tag{1.91}$$

where $f_p$ is the weight percentage of pearlite in the steel, and $f_f$ is the weight percentage of ferrite, considering that the pearlite ultimate tensile strength is 800 MPa and that of the ferrite 300 MPa. On the other hand, for a hypoeutectoid steel with $C_1\%$ of carbon, the pearlite, and proeutectoid ferrite percentages are, respectively, as it is possible to deduce if we apply the lever rule:

$$\%\text{pearlite} = 100 \cdot \frac{C_1 - 0.02}{0.77 - 0.02} \tag{1.92}$$

$$\% \text{ proeutectoid cementite} = 100 \cdot \frac{0.77 - C_1}{0.77 - 0.02} \tag{1.93}$$

Consequently, in a first approximation, the ultimate tensile strength of a ferritic-pearlitic steel as a function of the carbon percentage $C_1$ is

$$\sigma_u(\text{MPa}) = 300 + 650 \cdot C_1 \tag{1.94}$$

Thus, for instance, the ultimate tensile strength of a steel with 0.2% C is 430 MPa, approximately.

In fact, if we increase the cooling rate with respect to that of equilibrium, the value of $\sigma_u$ increases. The reason is that the size of the proeutectoid ferrite grain decreases, the percentage of pearlite grows (the percentage of proeutectoid ferrite decreases), and the interlamellar spacing between cementite layers in the pearlite is reduced. This everything increases the resistance.

**Exercise 1.8** Iron has an atomic weight of 55.85 g/mol and a lattice parameter of 2.86 Å at room temperature ($\alpha$-bcc = 2.898 Å at 912 °C; $\gamma$-fcc = 3.639 Å at 912 °C; $\gamma$-fcc = 3.656 Å at 950 °C; $\gamma$-fcc = 3.68 Å at 1390 °C; $\delta$-bcc = 2.94 Å at 1425 °C; $\delta$-bcc = 2.930 Å at 1432 °C; $\delta$-bcc = 2.935 Å at 1534 °C). Calculate:

(a)   Density.
(b)   Volume change associate with the allotropic transformation of $\alpha$-Fe into $\gamma$-Fe at 912 °C.
(c)   Linear expansion coefficient of $\alpha$-Fe.

Data: bcc (T < 912 °C), fcc (912 °C < T < 1392 °C) and bcc (T > 1392 °C).

*Solution starts*:

(a) Calculation of the density

Density of α-Fe (at room temperature).

α-Fe has a bcc crystalline system; thus, the density can be calculated using the following equation, making the appropriate unit transformations as well:

$$\rho_{bcc} = \frac{2 \cdot \text{atomic weight}}{N_A \cdot a^3} = \frac{2 \cdot 55.85 \text{ g/mol}}{6.022 \cdot 10^{23} \frac{1}{\text{mol}} \cdot \left(2.86 \text{ Å} \cdot \frac{10^{-8} \text{ cm}}{1 \text{ Å}}\right)^3} = 7.927 \text{g/cm}^3$$

(1.95)

Density of γ-Fe (at 950 °C).

γ-Fe has a fcc crystalline system; thus, the density can be calculated using the following equation, making the appropriate unit transformations as well:

$$\rho_{fcc} = \frac{4 \cdot \text{atomic weight}}{N_A \cdot a^3} = \frac{4 \cdot 55.85 \text{ g/mol}}{6.022 \cdot 10^{23} \frac{1}{\text{mol}} \cdot \left(3.656 \text{ Å} \cdot \frac{10^{-8} \text{ cm}}{1 \text{ Å}}\right)^3} = 7.589 \text{g/cm}^3$$

(1.96)

Density of δ-Fe (at 1425 °C).

δ-Fe has a bcc crystalline system; thus, the density can be calculated using the following equation, making the appropriate unit transformations as well:

$$\rho_{bcc} = \frac{2 \cdot \text{atomic weight}}{N_A \cdot a^3} = \frac{2 \cdot 55.85 \text{ g/mol}}{6.022 \cdot 10^{23} \frac{1}{\text{mol}} \cdot \left(2.94 \text{ Å} \cdot \frac{10^{-8} \text{ cm}}{1 \text{ Å}}\right)^3} = 7.298 \text{g/cm}^3$$

(1.97)

(b) Volume change

A bcc unit cell has two atoms while an fcc unit cell has four, meaning that two bcc crystals will form one fcc crystal. Considering the number of crystals of each type, the volume change associated with the allotropic transformation of ferrite (α-Fe) into austenite (γ-Fe) can be calculated as follows:

$$\frac{\Delta V}{V} = \frac{a_{\gamma-Fe}^3 - 2a_{\alpha-Fe}^3}{2a_{\alpha-Fe^3}} = \frac{(3.639 \text{ Å})^3 - 2 \cdot (2.898 \text{ Å})^3}{2 \cdot (2.898 \text{ Å})^3} \simeq -0.01 = -1\% \quad (1.98)$$

It is important to consider isothermal density changes caused by allotropic transformations during heat treatments.

(c) Linear expansion coefficient

For the α-Fe:

Fitting the linear expansion coefficient to this exercise:

$$a_{912\,°C} = a_{23\,°C} \cdot (1 + \alpha \cdot \Delta T) \tag{1.99}$$

and solving for α:

$$\alpha = \frac{1}{\Delta T} \cdot \left( \frac{a_{912\,°C}}{a_{23\,°C}} - 1 \right) = \frac{1}{(912 - 23)} \cdot \left( \frac{2.898}{2.866} - 1 \right) = 12.56 \cdot 10^{-6}{}^{°}C^{-1} \tag{1.100}$$

For the γ-Fe:

Fitting the linear expansion coefficient to this exercise:

$$a_{1390\,°C} = a_{912\,°C} \cdot (1 + \alpha \cdot \Delta T) \tag{1.101}$$

and solving for α:

$$\alpha = \frac{1}{\Delta T} \cdot \left( \frac{a_{1390\,°C}}{a_{912\,°C}} - 1 \right) = \frac{1}{(1390 - 912)} \cdot \left( \frac{3.68}{3.64} - 1 \right) = 22.99 \cdot 10^{-6}\ {}^{°}C^{-1} \tag{1.102}$$

Nearly twice that of the α-Fe.
For the δ-Fe:
Fitting the linear expansion coefficient to this exercise:

$$a_{1534°C} = a_{1432°C} \cdot (1 + \alpha \cdot \Delta T) \tag{1.103}$$

and solving for α:

$$\alpha = \frac{1}{\Delta T} \cdot \left( \frac{a_{1534\,°C}}{a_{1432\,°C}} - 1 \right) = \frac{1}{(1534 - 1432)} \cdot \left( \frac{2.935}{2.930} - 1 \right) = 16.73 \cdot 10^{-6}{}^{°}C^{-1} \tag{1.104}$$

Similar to that of the α-Fe.

*Solution ends.*

**Exercise 1.9** Some binary steels have such carbon weight contents that the obtained ferrite after the beginning of the transformation by cooling from the gamma state would be always magnetic (we remind that it is nonmagnetic above 770 °C). Calculate the minimum carbon content $C_1$ that the steel should have to achieve the above-mentioned.

*Solution starts*:

Austenite becomes nonmagnetic below 770 °C. If we consider similarity of triangles, we calculate the minimum carbon content, $C_1$:

$$\frac{912 - 727}{0.77} = \frac{770 - 727}{0.77 - C_1} \rightarrow C_1 = 0.6\% \tag{1.105}$$

**Answer also to the following questions:**

**1. Calculate the temperatures of beginning and end of the solidification of the steel of the enunciation.**

We consider the steel with 0.6% C and the Fe–C binary diagram (Fig. 1.19). We consider similarity of triangles and $T_L$ and $T_S$ as straight lines. First, we determine the liquidus temperature ($T_L$):

$$\frac{1495 - 1148}{4.3 - 0.53} = \frac{1495 - T_L}{0.60 - 0.53} \rightarrow T_L = 1488\,°C \tag{1.106}$$

We calculate now the solidus temperature ($T_S$):

$$\frac{1495 - 1148}{2.11 - 0.17} = \frac{1495 - T_S}{0.60 - 0.17} \rightarrow T_S = 1418\,°C \tag{1.107}$$

**Calculate the weight percentage of proeutectoid ferrite and eutectoid ferrite of the steel of the enunciation with equilibrium cooling** (Fig. 1.19).

We consider the steel with 0.6% C and the Fe–C binary diagram. We use the lever rule. First, we determine the proeutectoid ferrite ($\alpha_{pro}$):

$$\alpha_{pro} = \frac{0.77 - 0.6}{0.77 - 0.02} = 0.2267(22.67\%) \tag{1.108}$$

**Fig. 1.19** Scheme of the cooling process

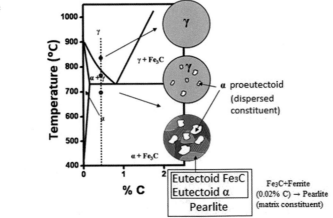

We calculate now the eutectoid ferrite ($\alpha_{eut}$):

$$\alpha_{eut} = \frac{6.67 - 0.6}{6.67 - 0.02} \cdot (1 - 0.2267) = 0.7059(70.59\%) \qquad (1.109)$$

*End of the exercise.*

### 1.5.3 Hypoeutectoid Steels with <0.02% C

The $\gamma \rightarrow \alpha$ transformation starts at the corresponding $A_3$ temperature. However, in the case of these steels, during the equilibrium cooling, the greater enrichment of the austenite by the carbon is always smaller than 0.77% C. For that reason, the allotropic transformation ends at $A_1$ temperatures greater than 727 °C. The microstructure of these steels, at the end of the allotropic transformation, is completely ferritic (it is not ferritic-pearlitic).

We should remember that at 20 °C, the maximum carbon content that the ferrite admits in solution is 20 ppm. Therefore, in all steels, both hypereutectoid and hypoeutectoid, if the temperature is slowly and continuously reduced from 726 °C to the room temperature, the carbon dissolved in the ferrite—both of the pearlitic ferrite and, in the corresponding case, of the proeutectoid ferrite—when the solubility line is exceeded reacts with iron atoms and precipitates in the form of the so-called tertiary cementite.

This cementite is easily identifiable when using an optical microscope in the case of very low carbon steels (%C < 0.02). It usually adopts the shape of small worms (this cementite is also called "vermicular" cementite). This cementite precipitates in the ferrite grain boundaries or in the triple points where three grains are in contact (Fig. 1.20). However, if from the soluble state—completely ferritic—a steel with %C < 0.02 is rapidly cooled, the precipitation can appear later, very fine and dispersedly, inside of the grains.

## 1.6 Influence of Alloying Elements in the Fe–C Metastable Diagram

When a steel of $C_1$ carbon content is not simply binary—that is to say, it does not only bear iron and carbon—steel alloying elements also modify the critical points of the diagram in Fig. 1.16 for this carbon content.

It is not sufficient with the fact that the chemical analysis could indicate the existence of alloying elements to ensure that these elements modify the critical points of the diagram. These elements must be in solid solution inside of the iron to have influence in the modification of the diagram. If, on the other hand, these elements

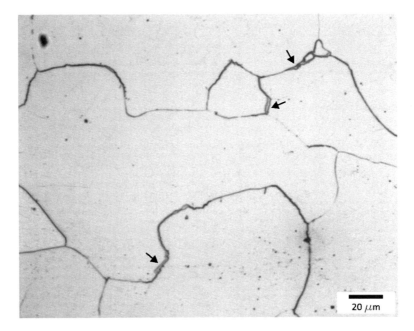

**Fig. 1.20** Tertiary cementite in grain boundaries and triple points of ferrite (arrows)

form second phases that precipitated—for example, carbides, nitrides, carbonitrides, etc.—, the critical points in the diagram would be the same that in the case of a binary steel.

In Sect. 1.4, we have referred to the alphagenous and gammagenous character of the elements solubilized in the pure iron. However, regarding the influence of the alloying elements in the diagram of Fig. 1.16, it seems also convenient to consider their aptitude to enter in solid solution in the gamma iron and in the alpha iron (and, complementarily, to consider the carbide-forming or the non-carbide-forming character of these elements).

Regarding the most commonly used alloying elements, and the impurities as Sn, Sb, and As, Table 1.5 collects the gammagenous and alphagenous behavior of these elements as well as their solubility in gamma or alpha pure iron. It is possible to use the Fe–X and C-X diagrams for any element X. We suggest Massalski et al. (1986) for further detail.

Regarding manganese, nickel, and cobalt—commonly used gammagenous elements—, they do not alter the maximum limit of austenite saturation by the carbon it does not matter their content: 2.11% C. The reason is that the austenite cell is not destabilized, or distorted, when iron atoms are replaced by manganese, nickel, or cobalt atoms, because there is total solubility of these elements in the face centered cubic iron, as said above (Sect. 1.3.1, Table 1.1).

On another note, nickel and cobalt have not affinity for the carbon to form carbides. Manganese is hardly carbide-forming element, in any case, the atoms of manganese

**Table 1.5** Alphagenous and gammagenous behavior of some elements. Solubility in iron

| Gammagenous | Solubility in pure $Fe_\gamma$ | Solubility in $Fe_\delta$ or pure $Fe_\alpha$ |
|---|---|---|
| Carbide-forming: C | Partial | Partial |
| Few carbide-forming: Mn | Total | Partial |
| Non-carbide-forming: Ni, Co | Total | Partial |
| Non-carbide-forming: N, Zn, Cu, Au | Partial | Partial |
| Alphagenous | Solubility in pure $Fe_\gamma$ | Solubility in $Fe_\delta$ or pure $Fe_\alpha$ |
| Carbide-forming: | | |
| Partially alphagenous: Cr, V | Partial | Total in $Fe_\delta$ |
| Partially alphagenous: W, Mo, Ti, Nb | Partial | Partial |
| Non-carbide-forming: Si, P, S, Al, Sn, Sb, As | Partial | Partial |

can replace some iron atoms in the cementite to form carbides, type $K_c$, which have a stability very similar to that of the cementite (they are decomposed during the heating at a temperature nearly that of the cementite). Therefore, it is possible to say that using these alloying elements—if they were previously solubilized in the gamma iron—the $A_{cm}$ temperatures of the metastable Fe–C binary diagram of hypereutectoid austenite are nearly not modified.

Non-carbide-forming elements, with no activity for the carbon, have graphitizing behavior: silicon, phosphorus, aluminum, nickel, and copper (in decreasing order of graphitizing behavior). As a result of a dilution and affinity effect, they favor during the solidification the individual formation of either graphite and $Fe_\gamma$ or $Fe_\alpha$, which impedes the reaction between the carbon and iron atoms to form cementite. Thus, cast irons with suitable silicon content (greater than 1.5% Si) solidify without formation of ledeburite. They give a eutectic aggregate of austenite and graphite (gray cast irons) for 2% silicon. The proportion of graphitizing elements always produce a compromise between the stable solidification (gray cast irons) and the metastable solidification (white cast irons), which is sometimes solved with the simultaneous presence of graphite and ledeburite (mottled cast irons) (Pero-Sanz et al. 2018). Once solidified the ferrous alloy, graphitizing elements also favor—if they are in sufficient quantity—the stable transformations of the austenite to give ferrite and graphite as constituents instead of cementite. Therefore, for instance, the austenite of a gray cast iron can transform into ferrite and graphite, or—alternatively—into ferrite and cementite (pearlitic matrix gray cast irons).

The $A_3$ temperature of an alloyed steel with manganese, nickel, and cobalt—whose carbon content was smaller than that corresponding to the eutectoid—is lower than that of a non-alloyed steel with the same carbon content. As example, it is possible to see in Fig. 1.21 the modification of the gamma loop of the Fe–C diagram for different manganese contents.

It is also possible to see in Fig. 1.21 that the temperature and carbon percentage in the eutectoid austenite—intersection of the $A_3$ and $A_{cm}$ transformation lines—vary with the manganese percentage in the steel. Therefore, when the manganese

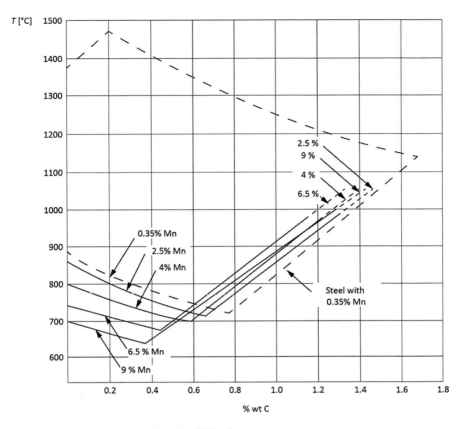

**Fig. 1.21**  Simplified sections of the Fe–C-Mn diagram

percentage grows, the eutectoid temperature decreases and the pearlite has a carbon content, $C_x$, smaller than 0.77%.

As it is possible to appreciate, if we apply the lever rule, this pearlite has greater proportion of ferrite than in the case of the non-alloyed pearlite. For that reason, this pearlite is usually called diluted pearlite. Additionally, cementite and ferrite lamellae of the diluted pearlite are thinner than in the case of the non-alloyed pearlite—and with smaller interlamellar spacing—result of being transformed at temperatures below 727 °C. This way, alloyed pearlite—even when its carbon percentage, $C_x$, was <0.77% C—is harder and more resistant than the pearlite of 0.77% C.

The following metallographic (microstructure) consequences can be deduced from the considerations indicated in previous paragraphs: a hypoeutectoid binary steel—for example, 0.45% C—can be hypereutectoid if the manganese content is as high as 6.5%, as it is possible to deduce from Fig. 1.21. Besides, if the content in gammagenous elements is sufficiently high, it is possible that the eutectoid transformation could take place below the room temperature, and, for that reason, steels are austenitic above this temperature.

However, if the content in gammagenous elements allows the eutectoid transformation at temperatures greater than the room temperature (for example, 0.35–9% Mn, see Fig. 1.21), the hypoeutectoid steel microstructure cooled in equilibrium conditions—for example, the steel with 1.5% Mn and 0.2% C—would continue being ferritic-pearlitic: ferrite as matrix constituent and pearlite as disperse constituent. However, the proportion of proeutectoid ferrite would be smaller than that obtained in the 0.2% C plain steel—and greater the pearlite proportion—in this case, as it is possible to deduce if we apply the lever rule. Moreover, there is a decrease of the $A_3$ temperature due to the presence of manganese and, consequently, a reduction of the critical size of the proeutectoid ferrite nucleus, which gives origin to a grain refinement of this proeutectoid ferrite.

As opposed to the gammagenous elements (Ni, Mn, and Co), alphagenous elements, as Cr, W, Mo, Ta, V, Nb, Ti, Si, Al, P, Sn, Sb, and As, distort and destabilize the cell of the austenite when they enter in solid solution because they do not crystallize in the face centered cubic system, and they are only partially soluble in the $Fe_\gamma$. Thus, the saturation limit of the carbon in the austenite alloyed with these alphagenous elements is lower than 2.11% C. We should also appreciate that the saturation of the austenite by the carbon, to form cementite or other carbides during the cooling from the austenitic state, starts at higher temperatures than the corresponding $A_{cm}$ of the binary steels.

As always happens, the gamma loop of an alloyed steel will be limited by the locus of the $A_{cm}$ and $A_3$ temperatures that correspond to different carbon contents (we remind that alphagenous elements rise the $A_3$ temperatures). As example that illustrates the influence of alphagenous elements, see the gamma loop of steels alloyed with different silicon percentages in Fig. 1.22. For the same carbon content, the $A_3$ critical point appears at greater temperatures the greater the silicon content. The eutectoid point also appears at higher temperatures and lower carbon contents when the silicon percentage increases in the alloy. We also point out that hypoeutectoid binary steels (for instance, 0.6% C and 0% Si) would be, on the contrary, eutectoid if the silicon content was 2%, and hypereutectoid for greater silicon contents.

We also indicate that for a certain temperature (for instance, 1100 °C), the maximum carbon content in the austenite decreases when the silicon content is increased (1.6% C when 0.25% Si; 1.4% C when 2% Si; 1.2% C when 4% Si). In other words, regarding the maximum carbon content that can saturate the austenite (2.11% C at 1148 °C for a non-alloyed steel), its percentage decreases when the silicon content is increased, while the temperature required to achieve this maximum solubility grows with the silicon.

Figures 1.23 and 1.24 represent, respectively, sections of the Fe–C-Mo and Fe–C-Ti ternary diagrams. The considerations that could be made with respect to these elements are analogous to that made for the silicon. For example, when the content in alphagenous elements is increased: line $A_3$ shift toward above; the maximum carbon content admitted by the austenite decreases; $A_{cm}$ displaces above; $A_e$ grows; the eutectic carbon content decreases; the proportion of pearlite increases; and diluted pearlite is obtained (but as opposed to the case of the gammagenous elements, it

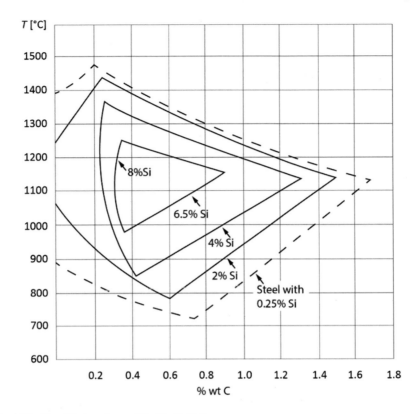

**Fig. 1.22** Simplified sections of the Fe–C-Si diagram

is not possible to say that a refinement of the pearlite lamellae is obtained as $A_e$ is greater than 727 °C):

Regarding chromium, we should remind that it is gammagenous (reduces $A_3$) or alphagenous (increases $A_3$) depending on its content, <7% or >7%, respectively. It is always limitedly soluble in the austenite and is carbide-forming (Fig. 1.25).

As conclusion of this section dedicated to the influence of the alloying elements in the Fe–C metastable diagram, the $A_3$ temperature of an alloyed steel can be approximately calculated as a function of the chemical composition of the steel by means of the equation proposed by Andrews (1965), which weights the gammagenous (−) or alphagenous (+) behavior of the alloying elements. This equation collects the gammagenous character of the chromium for low contents of this element, and it is not useful in the case of high chromium steels:

$$A_3(°C) = 912 - 203 \cdot \%C^{1/2} - 30 \cdot \%Mn - 15.2 \cdot \%Ni - 11 \cdot \%Cr - 20 \cdot \%Cu$$
$$+ 44.7 \cdot \%Si + 31.5 \cdot \%Mo + 13.1 \cdot \%W + 104 \cdot \%V + 120 \cdot \%As$$
$$+ 400 \cdot \%Ti + 400 \cdot \%Al + 700 \cdot \%P \qquad (1.110)$$

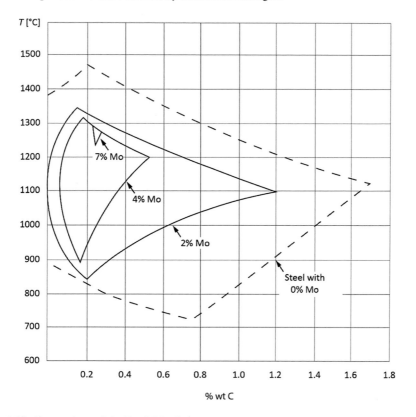

**Fig. 1.23**  Gamma loop of the Fe–C-Mo diagram

The temperature at which the eutectoid transformation takes place by equilibrium cooling, as well as the carbon contents in the austenite, vary as indicated by the Bain curves (Fig. 1.26). As a function of the steel composition, it is also possible to approximately calculate the eutectoid temperature, $A_e$, as proposed by Andrews (1965):

$$A_e(°C) = 727 - 20.7 \cdot \%Mn - 16.9 \cdot \%Ni + 29.1 \cdot \%Si$$
$$+ 16.9 \cdot \%Cr + 290 \cdot \%As + 6.38 \cdot \%W \qquad (1.111)$$

## 1.7  Non-equilibrium Transformations by Isothermal Cooling of the Austenite

We are going to describe in this chapter the possible transformations of the austenite by isothermal cooling, at different temperatures, below the eutectoid temperature.

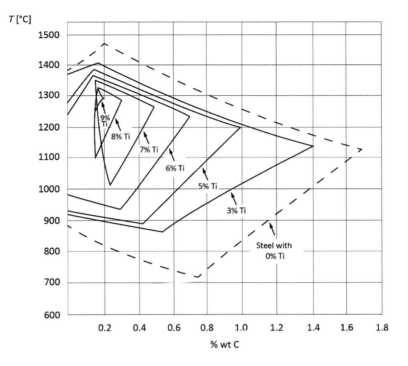

**Fig. 1.24** Gamma loop of the Fe–C–Ti diagram

**Fig. 1.25** Gamma loop of the Fe–C–Cr diagram projected in the plane for different chromium percentages-constant-. ▶: carbon saturation limit in the austenite (%C, T(°C)); ●: pearlitic eutectoids (%C, T(°C))

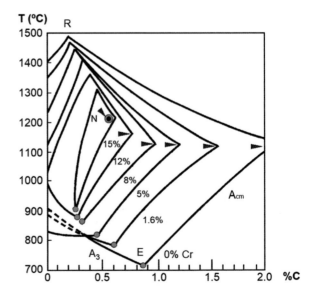

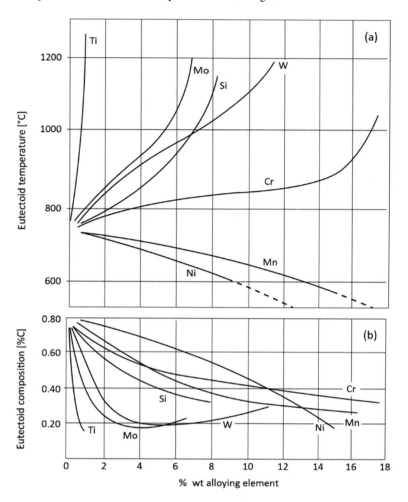

**Fig. 1.26** Influence of elements on eutectoid temperature (a) and composition (b)

It is a non-equilibrium cooling (equilibrium cooling are those that follow the Fe-Fe$_3$C diagram). Isothermal transformations, by analogy, will help in the better understanding of the continuous transformations and the constituents that can be formed, from the austenite, when different cooling rates are used, in the continuous cooling from the austenitic state.

### 1.7.1  Metallography and Kinetics of the Pearlitic Transformations

When the eutectoid binary austenite, of 0.77% C, from temperatures greater than $A_e$ is immersed in a medium—see molten salts—maintained at constant temperature, close but lower than $A_e$, the austenite starts to transform after a certain time, and finally the structure is completely pearlitic. This process takes place by nucleation and growth as the time progresses at the chosen temperature.

During the permanence at this constant temperature, the appearance of multiple nucleus of cementite takes place first, generally, in the former austenite grain boundaries. Some carbon atoms coming from the austenite leave the interstitial places of the lattice and diffuse out, enrich some points, and react there with iron atoms to give $Fe_3C$.

In the contours of this cementite nucleus, austenite—already without carbon—can allotropically transform into ferrite at this temperature and, therefore, the cementite cluster is enclosed between two nuclei of ferrite. The carbon diffusion drives the pearlitic transformation, whose constituents are cooperatively formed. Even though the carbon—by diffusion at this temperature—governs the transformation and, for that reason, the cementite nuclei are the drivers of the pearlitic transformation in reality, it is possible to claim that the pearlite nuclei—of ferrite + cementite—are cooperatively formed.

The growth of the pearlite nuclei at this temperature is also cooperative. The cementite grows toward inside of the austenite grain at the expenses of the carbon atoms in the cementite/austenite and in the ferrite/cementite interfaces. The ferrite/cementite interfaces, impoverished in this manner in carbon, will allotropically transform into alpha and also displaces to the interior of the austenitic grain. Definitely, "pearlite nuclei" grow as lamellar aggregates of cementite and ferrite (Fig. 1.27).

The epitaxy orientations between the cementite and austenite nucleus are: $(100)_c//(1\bar{1}1)_\gamma$, $(010)_c//(110)_\gamma$, $(001)_c//(\bar{1}12)_\gamma$; and the cementite and ferrite interfaces are also crystallographically related inside of each pearlite nucleus. Interfaces between ferrite/cementite are crystallographically oriented too.

The lower the chosen temperature for this isothermal transformation, the smaller the size of the cementite nucleus and, consequently, also smaller the ferrite size. As these nuclei cooperatively grow, smaller the separation, $S_0$, between lamellae of cementite enclosed by ferrite, thinner the cementite and ferrite lamellae are, and, definitely, the pearlite is thinner, resistant and harder.

Therefore, pearlite adopts different morphologies at each temperature of isothermal transformation of the austenite, which are usually known with the names of coarse pearlite, fine pearlite (sometimes called sorbite), and troostite.

Pearlite is coarse within 726 and 650 °C, formed by alternated lamellae of ferrite and cementite separated between them ($S_0$, interlamellar spacing) 0.25–0.50 μm, and the ultimate tensile strength is 800 MPa. Pearlite is fine within 600 and 650 °C, where

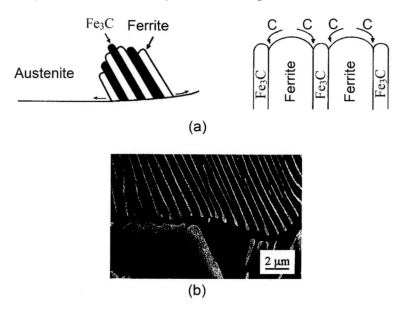

**Fig. 1.27** Transformation of the eutectoid austenite into pearlite: **a** scheme, **b** micrograph

the layers are separated between them approximately 0.10–0.20 $\mu$m, and the ultimate tensile strength takes values in the range 900–1400 MPa. Troostite, formed at lower temperatures (500–550 °C) is nearly indistinguishable using optical microscope (with interlamellar distances of approximately 0.10 $\mu$m), and adopts nodular, radial, and dark shapes (see it in Fig. 1.28 at the previous austenitic grain boundaries), with ultimate tensile strength in the range 1400–1700 MPa, Brinell hardness in the range 400–500 and elongation between 5 and 10%.

Figure 1.29 summarizes the kinetics of this transformation for different temperatures lower than $A_e$. One of the curves gives information about temperatures and times required for the austenite to transform into only 1% of pearlite. The other curve corresponds to the almost complete transformation (99%) of the austenite into pearlite. Transformation is slow at a temperature close to $A_e$. Transformation is faster at lower temperatures. Finally, the transformation slows down at temperatures even lower.

**Exercise 1.10** Lamellar growth of pearlite.

*Solution starts*:

Two phases, ferrite, and cementite can grow cooperatively behind an essential planar grain boundary of the austenite. As the ferrite phase grows, carbon diffuses a short distance laterally where it is incorporated into the carbon-rich phase, cementite. Similarly, the $Fe_\alpha$ atoms rejected ahead of cementite diffuse to the tips of the adjacent $\alpha$ lamellae. The rate at which the eutectic grows will depend on how fast this diffusion can occur and this in turn will depend on the interlamellar spacing, $S^*$. Thus, small

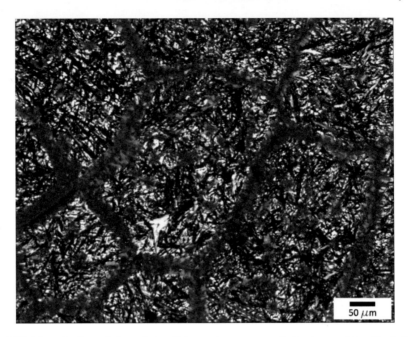

**Fig. 1.28**  Troostite (in grain boundaries, light gray) over martensite background (needles in black color)

**Fig. 1.29**  Kinetics of the solid-state transformation: C curve

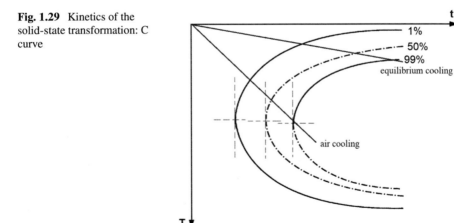

interlamellar spacings should lead to rapid growth. However, there is a lower limit λ determined by the need to supply the $\alpha/Fe_3C$ interfacial energy $\gamma_{\alpha/Fe_3C}$.

For an interlamellar spacing $S^*$, there is a total $2/S^*$ m$^2$ of $\alpha/Fe_3C$ interface per m$^3$ of eutectoid. Thus, the free energy change associated with a mol of austenite is given by

$$\Delta G(S^*) = -\Delta G(\infty) + \frac{2 \cdot \gamma_{\alpha/Fe_3C} \cdot V_m}{S^*} \qquad (1.112)$$

where $V_m$ is the molar volume of eutectoid and $\Delta G(\infty)$ is the free energy decrease for very large values of $S^*$. Since eutectoid transformation will not take place if $\Delta G(S)$ is positive, $\Delta G(\infty)$ must be large enough to compensate for the interfacial energy term, i.e., the austenite/eutectoid interface must be undercooled below the equilibrium eutectoid temperature ($T_e = 727\ °C = 1000\ K$). If the total undercooling is $\Delta T$, it can be shown that $\Delta G(\infty)$ is given approximately by

$$\Delta G(\infty) \simeq \frac{L \cdot \Delta T}{T_E} \qquad (1.113)$$

where L is the latent heat or enthalpy term of the pearlite transformation. The minimum possible spacing (S) is obtained by using the relation:

$$\Delta G(S^*) = 0 \qquad (1.114)$$

where

$$S^* = \frac{2 \cdot \gamma_{\alpha/Fe_3C} \cdot V_m \cdot T_E}{L \cdot \Delta T} \qquad (1.115)$$

and the growth rate of pearlite would be approximate to

$$V \simeq \frac{D}{S} \cdot \left(1 - \frac{S^*}{S}\right) k \cdot D \cdot \Delta T \cdot \frac{1}{S} \cdot \left(1 - \frac{S^*}{S}\right) \simeq k \cdot \frac{D}{S} \cdot \Delta G(\infty)$$

$$\cdot \left(1 - \frac{\Delta G}{\Delta G^*}\right) = k \cdot \frac{D}{S} \cdot \Delta G(\infty) \cdot \left(1 - \frac{S^*}{S}\right) \qquad (1.116)$$

The maximum growth rate corresponds to the most probable interlamellar spacing. If we derive with respect to S and we equalize to 0, we obtain

$$S_0 = 2 \cdot S^* = \frac{4 \cdot \gamma_{\alpha\beta} \cdot V_m \cdot T_E}{L \cdot \Delta T} \qquad (1.117)$$

So, the optimum lamellar spacing is $\propto 1/\Delta T$, and it should inversely vary with undercooling below the eutectoid temperature. Taking for pearlite:

$$\frac{\Delta H}{V_m}(\text{where } \Delta H = L) = 20 \cdot 10^6 \cdot J/m^3 \qquad (1.118)$$

$$T_E = 1000\ K \qquad (1.119)$$

$$\gamma_{\alpha/Fe_3C} = 100 \cdot \frac{\text{erg}}{\text{cm}^2} \cdot \left( 100 \cdot 10^{-3} \frac{\text{J}}{\text{m}^2} \right) \rightarrow S_0$$

$$= \frac{0.4 \text{J/m}^2 \cdot 1000 \text{K}}{20 \cdot 10^6 \text{J/m}^3 \cdot \Delta T} \cdot \frac{4 \cdot 10^2}{20 \cdot 10^6} \cdot \frac{10^6 \mu\text{m}}{1\text{m}} = \frac{20}{\Delta T} \mu\text{m} \qquad (1.120)$$

If:

$$S_0 = 0.1 \mu\text{m} \rightarrow \Delta T = \frac{2U}{0.1} \simeq 200 \text{ K (very fine pearlite, fast cooling rates)}$$

$$(1.121)$$

If:

$$S_0 = 1 \mu\text{m} \rightarrow \Delta T = \frac{20}{1} \simeq 20 \text{ K (coarse pearlite, near equilibrium cooling)}$$

$$(1.122)$$

*Solution ends.*

This behavior is coherent. When the chosen temperature for the transformation is reduced, the critical size of the pearlite nuclei also decreases. For that reason, it seems logical that their formation begins after shorter times if the carbon diffusivity is fast enough. However, we should also take into account that this diffusivity also decreases with the temperature. Therefore, when the carbon diffusion is slow—as happens at temperatures far from the $A_e$—longer times are required to form the pearlite nuclei and, although they were small, they take time to appear and, they slowly grow, which is translated into the fact that 1% pearlite takes more time to form. And that the pearlite is finer.

Thus, the curve of 1% austenite transformed into pearlite adopts the form of "C curve" typical of the phase transformations by nucleation and growth of any alloy. The same happens with the curve of 99% of transformed austenite. In general, the vertical maximums for 1% transformation and 99% transformation do not coincide at the same temperature. If we consider that both the nucleation rate and the growing rate are the product of two mathematical functions: one decreases with the undercooling (diffusivity), and the other grows because it corresponds to smaller nucleus sizes.

If the composition of the initial austenite is not the eutectoid (0.77% C), first, the appearance of proeutectoid products would take place (also by nucleation and growth): ferrite or cementite depending on the type of steel (hypoeutectoid or hypereutectoid). In these isothermal transformations, proeutectoid constituents also originate their own "C curve" of transformation. If the steel is hypoeutectoid, the proeutectoid constituent ferrite will be first formed for temperatures between the corresponding $A_{3r}$ and the isothermal transformation temperature that was chosen. The remaining austenite is transformed into pearlite.

If the temperature of isothermal transformation is lower than that corresponding to the pearlitic nose (vertical maximum), the austenite is usually directly transformed, without previous formation of proeutectoid ferrite, and gives diluted pearlite: formed by lamellae of cementite and ferrite but with greater proportion of ferrite in the pearlite.

Similar considerations could be made in the case of the hypereutectoid steels. However, as opposed to the hypoeutectoid steels, the C curve that precedes the pearlitic transformation corresponds to the formation of proeutectoid cementite.

**Exercise 1.11**  Effect of pearlite dilution in $S_0$ (optimum interlamellar spacing).

*Solution starts*:

The morphology of the pearlite has complex effects, however in that a decrease in the interlamellar spacing, $S_0$, is detrimental to the impact properties, but the simultaneous observed decrease in the pearlite-cementite plate thickness t, Fig. 1.30, is beneficial. Thus, there is a balance between a decrease in the interlamellar spacing adversely affecting impact toughness, probably through its effect in increasing strength, and the simultaneous decrease in carbide plate thickness which improves the impact resistance, because the thinner carbide plates can bend rather than crack and initiate cleavage failure. Frequently these two opposing effects cancel out one another, so there is no overall effect of the pearlite morphology, but either effect can predominate and cause either an improvement or deterioration in the impact properties with decreasing interlamellar spacing. The effects are further complicated by any changes which occur in the pearlite colony size.

Because $S_0$ and t have opposite effect (see Eqs. 1.68 and 1.69) there will be an optimum interlamellar spacing for the best impact toughness, Fig. 1.31. It is also clear that the dilution of pearlite will be important, because increasing carbon content of

**Fig. 1.30**  Relationship between interlamellar spacing and cementite lamella thickness in pearlite

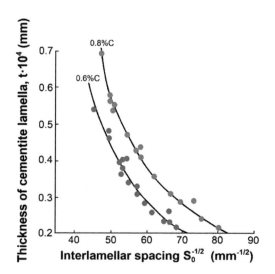

**Fig. 1.31** Effects of pearlite interlamellar spacing and cementite lamella thickness on contribution to impact transition temperature, showing occurrence of the optimum interlamellar spacing

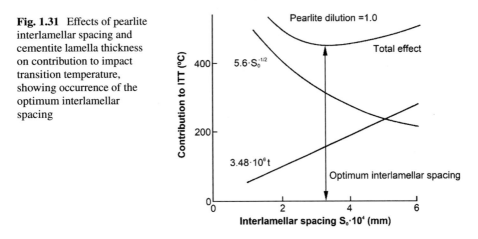

the pearlite will either change the value of $S_0$ or increase t, both of which will impair the impact properties. As said, the pearlite dilution factor D is defined as

$$D = \frac{0.77}{(\% \text{ C})_e} \qquad (1.123)$$

D will be unity for a fully pearlitic Fe–C steel of eutectoid composition but will be greater than unity (in some cases it can reach a value 2 corresponding to approximately 0.38% C low alloyed steel) for a fully pearlitic hypoeutectoid steel such as can be produced by low transformation temperatures approaching the nose of the pearlite "C" curve. It can be shown that the optimum interlamellar spacing for the best impact properties is related to the dilution factor by

$$\frac{t}{S_0} \text{(fraction of cementite)} = \frac{(\%C)_e}{6.67} \qquad (1.124)$$

$$D = \frac{0.77}{(\%C)_e} \rightarrow t = \frac{0.77}{6.67} = \frac{S_0}{D} = \frac{0.115 \cdot S_0}{D} \text{(t and } S_0 \text{ expressed in mm)} \quad (1.125)$$

$$\Delta\text{ITT} = 5.6 \cdot S_0^{-0.5} + 3.48 \cdot 10^6 \cdot t = 5.6 \cdot S_0^{-0.5} + \frac{3.48 \cdot 10^6 \cdot 0.115 \cdot S_0}{D} \qquad (1.126)$$

Deriving with respect to $S_0$ and equalizing to 0 (finding a minimum)

$$-\frac{5.6}{2} \cdot S_0^{-1.5} + \frac{4 \cdot 10^5}{D} = 0 \rightarrow S_0^{-1.5} = \frac{8}{5.6} \cdot \frac{10^5}{D} \qquad (1.127)$$

$$S_0 = \frac{D^{2/3}}{\left(1.42 \cdot 10^5\right)^{2/3}} = 3.53 \cdot 10^{-4} \cdot D^{-2/3} \qquad (1.128)$$

for D $= 1 \rightarrow S_0 = 0.353\,\mu$m(maximum toughness).
for D $= 2 \rightarrow S_0 = 0.22\,\mu$m(maximum toughness).
*Solution ends.*

## 1.7.2   Bainitic Transformations

When a previously austenitized steel is transformed by isothermal cooling, if the temperature of transformation is lower than that of the C curve of the pearlitic zone, the low diffusivity of the carbon in the austenite at this temperature will impede the carbon atoms to migrate by diffusion to concentrate in certain regions and give cementite clusters. Perlite is not formed.

However, austenite, which is very far from its equilibrium conditions (that is to say, the A$_3$ of the steel) at this temperature, experiences a strong driving force for its transformation into alpha phase, due to the great undercooling. The thermal difference, between temperatures greater than A$_3$ to the temperature of isothermal transformation, can be sufficient to activate the formation of ferrite nuclei, by simple allotropic transformation of part of the austenite. This austenite is transformed to give a structure that is called bainite.

### 1.7.2.1   Upper Bainite

If the temperature chosen for the isothermal transformation of the austenite is in the range 550–400 °C, ferrite habitually nucleates in the austenitic grain boundaries and these nuclei grow as needles of ferrite toward inside of this grain (and they also laterally develop). Carbon ceded by the ferrite (until having only 0.02% C) migrates toward the ferrite/austenite interface and locally enriches the austenite.

The precipitation of cementite in like this formed ferrite does not take place until after a certain time. Meanwhile, the initial ferrite needles can grow until they have relatively important dimensions.

When sufficient carbon accumulates in the ferrite/austenite interface, the precipitation of cementite starts, and also the formation of a complex constituent called upper bainite also begins. This constituent comprises also cementite and ferrite, but with a characteristic morphology derived from the fact that the driving constituent of the bainitic transformations is the ferrite (as opposed to the pearlitic transformation where the driving constituent is the cementite).

Using optical microscope, at 1000 times magnification, the morphology of the upper bainite usually has a similar aspect to that observed in Fig. 1.32. It is possible to identify precipitated carbides when the electronic microscope is used: they appear in the flanks of the ferrite needles or lamellae, nearly parallel to the axis and with elongated shapes (Fig. 1.32). The separation between needles or lamellae depends on both the size of the starting austenite grains and the transformation temperature.

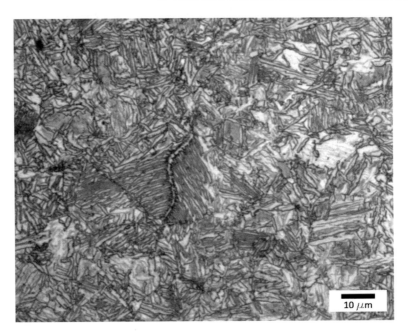

**Fig. 1.32** Upper bainite

The distance between ferrite lamellae is smaller than in the pearlite (below 0.05 μm) in the bainite, due to its low transformation temperature. Smaller is also the size of the precipitated carbides and greater the dispersion and density of crystalline dislocations inside of the ferrite. Therefore, for the same chemical composition, the hardness of the bainite is greater than that of the very fine pearlite, as the isothermal undercooling increases.

### 1.7.2.2   Lower Bainite

The possibility of diffusion of the carbon atoms, ceded when the ferrite is formed, is reduced for transformation temperatures comprised in the range 400–250 °C. Ferrite adopts the form of thin lamellae supersaturated in carbon, and carbides precipitate inside of these lamellae in the planes (100). This complex constituent, which was formed as described, is called lower bainite (Fig. 1.33).

There are real epitaxy relationships between the austenitic phase and the ferrite formed in the lower bainite: $\{111\}_\gamma//\{110\}_\alpha$ and $\langle 110\rangle_\gamma//\langle 111\rangle_\alpha$. The process of lower bainite formation is very similar to that of the martensite formation (see Sect. 1.8). The difference is that, while the carbon in the martensite is retained in an "alpha distorted" cell (tetragonal), formed by shearing in gamma phase, there is also shearing in the lower bainite, but carbon is simultaneously expelled from the cell. Evidence of the fact that shearing is produced in the lower bainite is the relief

**Fig. 1.33** Lower bainite over martensite background (light color)

that is observed in the polished surface, while in both the proeutectoid ferrite and the pearlitic ferrite this relief is not observed.

The lower bainite is harder than the upper bainite for the same chemical composition. The bainite hardness depends on the carbon content of the steel and the temperature at which it was formed. As orientation, it is possible to say that the hardness values vary in the range 40–60 HRC. In some cases, especially in high carbon steels (0.6–0.7% C), the lower bainite is tougher—for the same hardness—than the structure of tempered martensite, and this explains the objective of obtaining this type of structure by the austempering heat treatment (see Chap. 2, Fig. 2.56).

The kinetics of bainite formation, upper and lower, are also summarized in C-shaped curves due to—as in the pearlitic curves—a concurrence between the critical size of the ferrite nucleus (origin of that of the bainite) and the carbon diffusivity. C curves—TTT curves, Temperature Time Transformation curves—of the pearlitic and bainitic zones of a certain steel are observed well separated like is observed in the steel of Fig. 1.34.

In some steels, pearlitic and bainitic transformations are clearly distinguishable. In other cases, the situation is not so because, additionally, pearlitic and bainitic transformations are overlapped. Steven and Haynes (1956) give a temperature $B_S$ below which transformation is bainitic:

$$B_S(°C) = 830 - 270 \cdot \%C - 90 \cdot \%Mn - 37 \cdot \%Ni - 70 \cdot \%Cr - 83 \cdot \%Mo$$
$$(1.129)$$

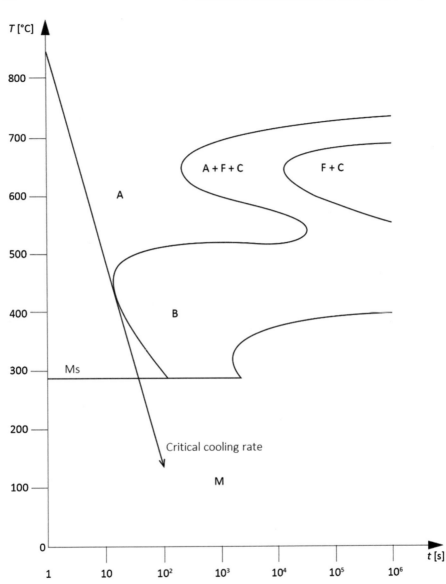

**Fig. 1.34** TTT curve of a steel with 0.57% C, 0.70% Mn, 0.20% Si, 0.70% Cr, 1.70% Ni, 0.30% Mo, 0.10% V for an austenitization temperature of 880 °C (CCR≃44 °C/s)

The same authors point out that $B_S$ equation is adequate with approximation of $\pm\,25$ °C and 90% confidence interval, for steels whose composition is in the range: 0.10–0.55% C, 0.20–1.70% Mn, <5% Ni, <3.5% Cr, <1% Mo.

### 1.7.3 TTT Curves (Transformation-Temperature-Time Curves)

Figure 1.34 represents the TTT curve of the steel 0.57% C, 0.70% Mn, 0.20% Si, 0.70% Cr, 1.70% Ni, 0.30% Mo, 0.10% V and ASTM austenitic grain size equal to 7.

The morphology of the TTT curves and their separation from the y-axis (distance from the origin of times) is conditioned by intrinsic factors—as the content in carbon and other alloy elements—and by extrinsic factors—as the austenitic grain size, as well.

#### 1.7.3.1   Intrinsic Factors

Any element that forms solid solution with the austenite—either substitutional solid solution (Mn, Ni, Cr, etc.) or interstitial solid solution (C, B, N, etc.)—generally delays the isothermal transformations, either pearlitic or bainitic. This seems logical because these elements represent a barrier for the carbon diffusion and, for that reason, the cementite clusters (in the pearlitic zone) or ferrite clusters (in the bainitic zone) will require more time to appear.

Gammagenous elements—specially manganese and nickel—reduce the temperatures of the austenitic transformation $A_3$ and $A_e$ and, therefore, they decrease the temperatures of the pearlitic transformations. Manganese and nickel also delay the same the pearlitic and bainitic noses.

Carbide-forming elements—chromium, molybdenum, vanadium, and others—delay more the pearlitic transformations than the bainitic transformations. They are alphagenous elements. They rise the $A_3$ and $A_e$ transformation temperatures and, for that reason, they should facilitate the carbon diffusion in the reactions and accelerate them, but this effect is counterbalanced by the opposite action that determines their affinity for the carbon to produce carbides different from the cementite. Therefore, the nucleation of the pearlitic cementite is delayed. Generally, the pearlitic curve is displaced up and to the right, and its domain is reduced.

Boron, for contents in the range 0.0005–0.003% forming interstitial solid solution in the austenite, noticeably delays both the proeutectoid and pearlitic transformations.

The influence of carbon is, in theory, similar to that of any element. In hypoeutectoid steels, pearlitic and bainitic transformations are shifted to the right when the carbon content is increased. Eutectoid steels have a pearlitic curve farer from the origin of times than in the case of both hypoeutectoid and hypereutectoid steels. In these ones, the formed proeutectoid cementite accelerates, by heterogeneous nucleation, the appearance of the cementite clusters that are the drivers for the pearlitic transformation.

### 1.7.3.2  Extrinsic Factors

The TTT curve mainly depends on the chemical composition of the steel, but there are also some other extrinsic factors that—although to a lesser extent—have influence on it. For instance, the austenitic grain size and the austenitization temperature from which the cooling begins. For the same chemical composition, austenite transformations by nucleation and growth start earlier the smaller the austenitic grain size. This situation seems reasonable because the driving clusters of the pearlite—cementite—or in its case of the bainite—ferrite—are preferentially formed at the grain boundaries. Consequently, the smaller the austenitic grain size, the greater the number of existing grain boundaries (greater number of suitable sites for the nucleation), and earlier the transformations in the pearlitic or/and bainitic zones will start.

The austenitization temperature has a twofold influence in the kinetical delay of the transformations. On the one hand, if the steel has not inhibition to the grain size, the increase of the temperature produces an increase of the austenitic grain size, with the consequences that were previously mentioned: delay the kinetics of the transformation and displacement of the TTT curves to the right (in the direction of growing times). On the other hand, if the austenitization temperature is very high, austenite when homogenizes in chemical composition becomes more stable and, consequently, pearlitic and bainitic transformations are also delayed: a non-homogenous austenite has a greater probability of that, in certain points of its mass, pearlitic and bainitic reactions could start earlier.

We can say as conclusion of this section that knowing the isothermal TTT curve of each steel always represents a support for the heat treatments. This curve can be only properly applied in isothermal treatments, although it is currently used for continuous cooling: overlapping to the TTT curve, the curve of the cooling rate at which the austenite should be cooled, it is possible to forecast with significant approximation the structures resulted from this cooling. If the cooling rate is very slow, pearlitic structures are obtained. If the cooling rate was greater, bainitic structures would be obtained. If the cooling rate was sufficient to not intersect the TTT curve in any point, martensite would be obtained.

## 1.8  Transformation of the Austenite into Martensite

It was already mentioned in previous sections that austenite, depending on the temperature at which the isothermal transformation takes place—always by nucleation and growth (by diffusion)—produces structures of ferritic-pearlitic, cementitic-pearlitic, pearlitic, or bainitic type.

Austenite almost instantaneously transforms (without diffusion) by means of a fast cooling—which could avoid the formation of pearlite or bainite—into martensite, which has identical carbon content (and the same alloying contents if there were available in the steel). The martensitic transformation involves a cooling below certain temperature Ms (martensite start temperature), which is characteristic of each steel. Martensite is the constituent of the quenched steels.

## *1.8.1  Crystalline Structure and Hardness of the Martensite*

Crystallographically, the martensite exhibits body centered tetragonal structure, with carbon atoms in interstitial position in certain interatomic spaces that exist in the vertical edges and in the center of the base of the cell (Fig. 1.35a). It is like a ferrite cell deformed preferentially according to the c axis. The parameters $a_M$ and $c_M$ of the martensite, as a function of its carbon $C_1$ percentage in the original austenite are

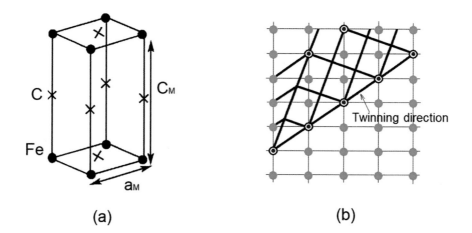

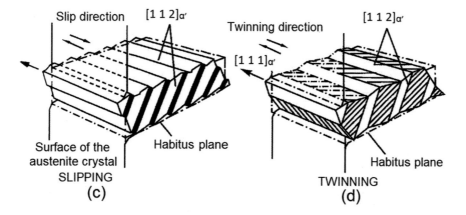

**Fig. 1.35**  Martensite: **a** crystallography of the martensite: ● iron atoms, × sites for the insertion of the carbon; **b** deformation by twining; **c** Martensitic transformation by slipping; **d** Martensite transformation by twining

$$c_M \left( \overset{\circ}{A} \right) = 2.861 + 0.116 \cdot C_1 \qquad\qquad (1.130)$$

$$a_M \left( \overset{\circ}{A} \right) = 2.861 - 0.013 \cdot C_1 \qquad\qquad (1.131)$$

We can check that for 0% C, the parameters $c_M$ and $a_M$ are 2.861 Å, and coincide with the value of the parameter of the pure $Fe_\alpha$ and this justifies the metastable character of the martensite. In fact, if the temperature is risen (by tempering of the martensite), it is possible to remove the interstitial carbon atoms, which would react with the iron atoms to give as final products of the martensite decomposition: ferrite and cementite.

Figure 1.36 summarizes, as a function of the carbon percentage, the variation of the crystalline parameters of both the austenite ($a_\gamma$) and martensite formed from this austenite ($a_M$ and $c_M$). We should notice that the parameter $c_M$ of the martensite grows with the carbon percentage, while the parameter $a_M$ slightly decreases. Consequently, the tetragonality of the martensite cell grows with the carbon percentage.

The great hardness of the martensite—see Fig. 1.36—, and the subsequent low toughness for more than 0.25% C, is due to mainly the carbon, not due to the other possible alloying elements solubilized in the austenite (which would pass to the martensite, also in substitutional solid solution).

**Fig. 1.36** Parameters and hardness of the martensite as a function of the carbon percentage (austenite parameters are collected in the upper zone of the figure)

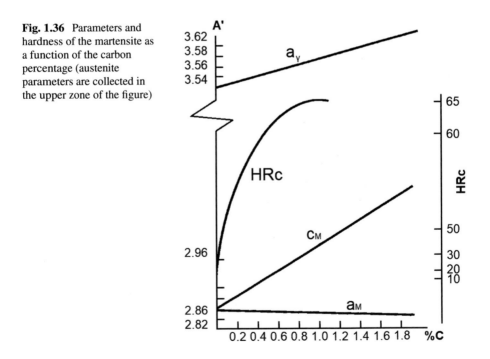

Alloying elements hardly increase in 3 unities the value of the HRC. The explanation of the great hardness of the martensite, growing with the carbon percentage—of around 60 HRC for 0.4% C—, as being only consequence of the increase of the tetragonality conferred by the carbon in the martensite cell has lasted for nearly 60 years. Nowadays, it is pointed out as explanation of this hardness the significant number of dislocations that are available in the martensite: around $10^{11}$ to $10^{12}$ per $cm^2$ (similar values to those of the heavily cold deformed steel). Deformations by slipping are produced in the martensite (see Fig. 1.35b) if carbon percentage is greater than 0.5%. Apart from the slipping, deformations by twining are also produced (see Fig. 1.35b) when the austenite yield strength is high (0.5 < %C < 1.0).

Martensite appears as needles (see Fig. 1.37) for the greatest carbon contents when observed with the optical microscope, because the limit surface between the martensite and the non-transformed austenite is curved. On the contrary, when %C < 0.4, martensite does not appear as needles but massively as thick lamellae (0.5 μm).

The hardness of the steel can be estimated by its easiness for the filing when a durometer is not available. If its resistance to the file is sufficient to avoid the penetration of the file, it is possible to say that the steel has almost full martensitic structure, that is to say, it is quenched. We collect in Table 1.6, approximately, as a function of the easiness to file a steel, its hardness—Rockwell (HRC), Brinell (HB), and Vickers (HV)—and its ultimate tensile strength ($\sigma_u$).

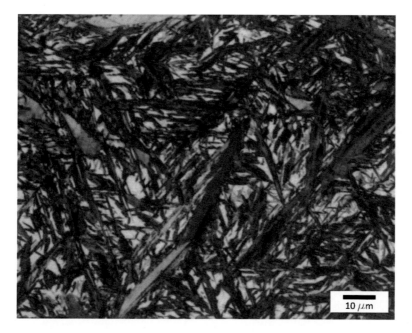

**Fig. 1.37** Acicular morphology of the martensite for 1.5% C steel over a background of retained austenite (white)

**Table 1.6** Approximate correlations between the easiness for the abrasive penetration of the steel by filing and its hardness (Brinell, HB; Vickers, HV; and, Rockwell, HRc) and its ultimate tensile strength ($\sigma_u$)

| Easiness of penetration | HB | HV | HRc | $\sigma_u$ (MPa) |
|---|---|---|---|---|
| The file easily scratches the surface very easily: the material is very soft | 100 | 100 | 56 HRb | |
| The file produces shavings if some additional pressure is applied | 200–230 | 200–230 | 20 | 750 |
| Material exhibits the first resistance for penetration to the file | 300 | 300 | 30 | 980 |
| Material is enough hard: shavings are difficultly produced by the file | 400 | 400 | 45 | 1280 |
| Shavings are hardly produced by the file: the metal is only slightly softer than the file | 500 | 550 | 55 | 1800 |
| The file slides over the surface without producing shavings | 600–650 | 70 | 60 | 2100 |

As it was already mentioned, the transformation of austenite into martensite takes place without diffusion (without process of nucleation and growth) by simple shearing. Carbon atoms hardly displace. The distances that they go through from the position that they occupied in the original austenite are smaller than $a_\gamma$. This way, the formation of the martensite is almost instantaneous.

Figure 1.38 illustrates the mechanism proposed by Bain (1924) and Bain and Paxton (1966) to explain the martensitic transformation: it is possible to appreciate between two cubic adjacent cells of austenite a tetragonal cell that has the semidiagonal of the cube face as edge of the square base (with empty spaces to host the carbon in the martensite that coincide with those of the austenite). Under these conditions, the relation "c/a" for this tetragonal cell observed inside of the austenite would be $2^{1/2}$; greater than the real relation—variable with the carbon percentage $C_1$—between both parameters of the martensite:

$$\frac{c_M}{a_M} = \frac{2.861 + 0.116 \cdot C_1}{2.861 - 0.013 \cdot C_1} = 1 + 0.045 \cdot C_1 \tag{1.132}$$

The idea of Bain arises from this point. It is possible to form martensite from the tetragonal cell localized between two cells of austenite, by mechanical compression of the austenite according to a parallel direction to the vertical axis of the cell, in such a way that the iron atoms could get closer ones to others and, thus, reducing the tetragonality.

This situation would justify the transformation of the austenite into martensite without diffusion and with hardly movement of the carbon atoms. This formation of martensite would be almost instantaneous, as it does not require the participation of the variable time as happens in the reactions that take place through the mechanism of nucleation and growth. This question also explains the mechanism industrially

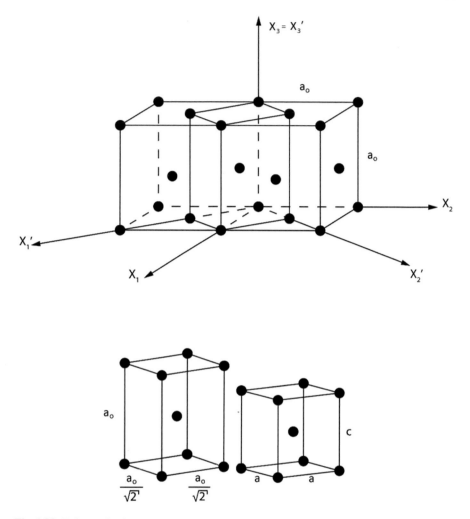

**Fig. 1.38** Bain mechanism

used to obtain martensite by simple mechanical deformation of the austenite at a temperature lower than $M_d$ (Pero-Sanz et al. 2017, 2018, 2019).

However, the mechanism proposed by Bain (1924) and Bain and Paxton (1966), even when it has the merit of being the first explanation that was given to explain the formation of martensite, is not truth. It is assumed in this mechanism a parallelism between the planes $\{100\}_\gamma$ of the original austenite and $\{100\}_m$ of the martensite, as well as between the directions $<110>_\gamma$ and $<100>_m$, which is not what happens in the reality. Kurdjumow and Sachs (1930) established the relations of epitaxy that exist between the martensite and the austenite in a steel of carbon content <1.4%, and they tried to explain the transformation of one cell into the other by a mechanism of simple shearing.

They have demonstrated that the closest planes {100} of the tetragonal martensite are almost parallel to the closest planes {111} of the austenite, while in the directions <110> of the austenite they are nearly parallel to the directions <111> of the martensite, which, considering the symmetry of the lattice, involves the existence of 24 different orientations of the martensite with respect to the austenite. The martensitic transformation is carried out by shearing on the planes {111} of the austenite according to the direction <211> followed by a second shearing according to the direction <101> of the austenite (that is to say, <111> of the martensite) and, finally, some slight modifications of the positions of the atoms to exactly obtain the dimension of the martensite cell.

In any case, it is convenient to mention that the border—called habitus plane—between the austenite and the non-transformed martensite is the plane $\{111\}_\gamma$ for %C < 0.4. On the contrary, it is the $\{225\}_\gamma$ for 0.4 < %C < 1.4% or the plane $\{259\}_\gamma$ for %C > 1.4. It is precisely the habitus plane the responsible of the micrographic morphology of the martensite: in lamellae when %C < 0.4, in lamellae and needles when 0.4 < %C < 1.4 and needles when %C > 1.4.

The transformation of austenite into martensite is always accompanied by volume increase, as it is possible to check, for example, in the steel with 0.6% C: the lattice parameter of the austenite is 3.571 Å and the lattice parameters of the martensite of 0.6% C are 2.853 Å—for $a_M$—and 2.93 Å—for $c_M$. As the austenite has four atoms of iron per each unit cell, and the martensite two per cell, the transformation of X austenite cells originates 2·X cells of martensite, which involves—if we calculate the volume of X cells of austenite and 2·X of martensite—a volume increase of 4.7%.

This volumetric expansion is much greater than that produced in the transformation of the austenite into ferrite and pearlite (around 1%) (Pero-Sanz et al. 2017). Moreover, the formation of martensite takes place at temperatures much lower than those required to obtain ferritic-pearlitic structures, and even bainitic. Consequently, this volume increase caused by the martensitic transformation is more dangerous than others if we consider stresses, deformations, and cracks that can be generated during the cooling.

If we focus on the crystallographic mechanisms, the thermodynamic possibility of the austenite → martensite transformation, at a temperature $T_1$, is given as first approximation by the thermal gradient $(\theta_\gamma - T_1)$ that makes possible the reduction of the free energy required for the transformation of the austenite into martensite (where $\theta_\gamma$ is the austenitization temperature of the steel). This reduction of the free energy involves a balance between the diminution of energy for the transformation $\gamma \rightarrow \alpha$ and an increase of energy resulted from the distortion of the $\alpha$ crystalline structure (as all the carbon of the austenite and all the alloying elements can be kept solved in the $\alpha$—tetragonal—martensitic). Therefore, it seems reasonable to allow that the formation of martensite will involve a great thermal difference the greater the carbon content and the greater the content in other alloying elements previously solubilized in the austenite.

## 1.8.2 Ms Temperature

Obtaining 1% of martensite (see Fig. 1.34) requires reaching by cooling a temperature Ms (martensite starting temperature) that depends, almost exclusively, on the chemical composition. The Ms does not depend, for example, on the cooling rate (if the cooling rate is greater than a critical value to impede the formation of pearlite or bainite, martensite would be obtained).

This Ms temperature is, for that reason, lower the greater the carbon and alloying elements contents the austenite had in solid solution. All the elements decrease— although in different manner—the Ms temperature (except the cobalt and the aluminum). Carbon, followed by the manganese, is the element that reduces more the Ms temperature.

Knowing the Ms temperature of each steel has industrial interest. It provides an orientation about the susceptibility of the steel to deform and to crack during the quenching (the more brittle the lower the Ms). The Ms temperature can be experimentally calculated but it is very useful to estimate it as a function of the chemical composition. We indicate in the following paragraphs the equations proposed by Steven and Haynes (1956), Nehrenberg (1946) and Hollomon and Haffe (1945) to calculate the Ms in the case of low alloy and medium alloy steel. The existence of several equations indicates the lack of accurateness that they offer.

Probably, the most precise equation is that proposed by Steven and Haynes (1956) (corrected by Irving). This equation allows calculating the Ms with an approximation of ± 2 °C when the composition of the steel is: 0.10–0.55% C, Ni < 5%, 0.10–0.35% Si, Cr < 3.5%, 0.30–1.70% Mn and Mo < 1%:

$$Ms(°C) = 561 - 474 \cdot \%C - 33 \cdot \%Mn - 17 \cdot \%Ni - 17$$
$$\cdot \%Cr - 21 \cdot \%Mo - 11 \cdot \%W - 11 \cdot \%Si \qquad (1.133)$$

The equation proposed by Nehrenberg (1946) is

$$Ms(°C) = 500 - 350 \cdot \%C - 40 \cdot \%Mn - 22$$
$$\cdot \%Cr - 17 \cdot \%Ni - 11 \cdot \%Si - 11 \cdot \%Mo \qquad (1.134)$$

and that proposed by Hollomon and Haffe (1945) is

$$Ms(°C) = 500 - 350 \cdot \%C - 40 \cdot \%Mn - 35 \cdot \%V - 20 \cdot \%Cr - 17 \cdot \%Ni - 10$$
$$\cdot \%Cu - 10 \cdot \%Mo - 5 \cdot \%W + 15 \cdot \%Co + 30 \cdot \%Al \qquad (1.135)$$

These equations do not consider the austenitic grain size of the steel and are useful for grain sizes of around 7 ASTM. The numerous grain boundaries seem to constitute an obstacle for the formation of martensite, because of their influence in the elastic and plastic dislocations that are involved in the energy of mechanical type. The Ms temperature generally decreases when the grain is very fine.

If we consider that the Ms is the temperature at which only 1% martensite appears, it seems logical to admit that, to obtain greater quantity of martensite, it is necessary to have a thermal difference greater than $(\theta_\gamma - Ms)$, where $\theta_\gamma$ is the austenitization temperature.

In fact, the thermal gradient must also compensate the increase of the energy of mechanical type because the formation of martensite involves:

– an increase of volume due to the formation of the martensite,
– plastic deformation of the austenite contiguous to the martensite,
– and, creation of a field of elastic stresses in the austenite contiguous to the plastically deformed austenite.

If only 1% of martensite is formed, this mechanical energy, which should be compensated with greater thermal difference from $\theta_\gamma$, is irrelevant. It is necessary to obtain greater martensite percentages to compensate this mechanical energy and, consequently, greater thermal differences $(\theta_\gamma - T)$ would be required.


## 1.8.3   Thermal Difference ($\theta_\gamma - T$) and Residual Austenite

We remind that to obtain martensite in a hypoeutectoid steel—to quench it—, it is possible to cool from a temperature $\theta_\gamma$ (called quenching-austenitization temperature) greater than the critical temperature $A_3$ of this steel. But it is convenient that $\theta_\gamma$ exceeds the critical temperature $A_3$. Among other reasons because a high $\theta_\gamma$ temperature involves almost always a reduction of the Ms temperature, and as a result, an increase of the quantity of residual austenite in this steel at room temperature.

This effect, clearly appreciated in high carbon and alloyed steels, seems to be consequence of several factors. First, a more complete redissolution of the possible precipitated phases in the austenite (carbides and carbonitrides). This redissolution would increase the carbon and alloying elements solubilized in the austenite, which would involve a reduction of the Ms temperature.

Nevertheless, increasing a lot the $\theta_\gamma$ temperature above $A_3$—practice that is not recommendable due to the overheating risks—not always involve a reduction of the Ms temperature. For example, non-redissolved carbides—if they were very dispersed and fines in the austenite—would contribute to maintain fine the austenitic grain size (and, as it was indicated, this would contribute to reduce the Ms temperature), but this influence would disappear when they redissolve in the austenite at very high temperature. Thus, the increase of the Ms due to the growth of the austenitic grain size would compensate the descent of Ms resulted from being richer the austenite in dissolved elements.

Numerous authors have tried to find a theoretical correlation between the percentual quantity of formed martensite and this final temperature, T, of cooling (always lower than Ms). Koistinen and Marburger (1959) proposed an equation that allows obtaining in an approximate manner the volume of residual austenite—or non-transformed into martensite:

$$V_a = 100 \cdot \exp[-0.011 \cdot (Ms - T)] \tag{1.136}$$

Therefore, a certain steel—whose Ms was known—will have less residual austenite the lower the temperature reached by cooling. Analogously, if we cool down to the temperature T, the steel will have a greater quantity of residual austenite the greater the content in carbon and alloying elements (because Ms will be lower).

The exponential character of the above-mentioned equation involves that to achieve the complete transformation of the austenite into martensite (to obtain 0% of residual austenite), it is necessary to cool down to a T equal to -∞ °C. This is equivalent to accept, theoretically, that there is not a temperature—industrially known as $M_f$—where 100% of martensite is obtained. However, it is not the real situation. For instance, a certain steel with 0.25% C—whose Ms temperature was 375 °C—, if we use the Eq. (1.136), it will theoretically contain 2% martensite at 25 °C, 1.6% at 0 °C and 1.2% at −25 °C. On the contrary, this steel quenched at 25 °C would really have 1% of residual austenite, which indicates that this temperature is practically the $M_f$ of the steel. It is habitually considered the martensite transformation completed when 99% martensite is obtained.

Many other steels are also fully martensitic at room temperature (Leslie 1982). Figure 1.39 (Speich and Leslie 1972) shows the carbon influence on the Ms temperature, the percentage of martensite at room temperature and, consequently, on the percentage of residual austenite. In fact, the percentage of residual austenite at room temperature is only relevant in steels with more than 0.5% C: for 0.5% C there is approximately 2%, 6% when 0.6%C and > 30% for steels with 1.25% C.

Even with caution, it is possible to estimate with sufficient approximation in the case of highly alloyed steels, by application of the Koistinen and Marburger (1959) equation, that the $M_f$ temperature is equal to Ms minus 300 °C (in almost all the steels). In fact, we would obtain replacing in the Eq. (1.136) the value of T by $M_f$ (that is to say, by Ms-300 °C):

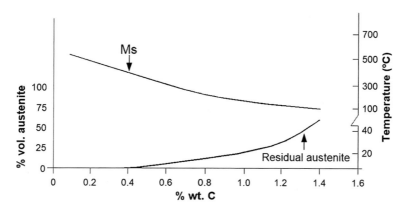

**Fig. 1.39** Correlation between %C, Ms temperature and quantity of residual austenite

$$V_a = 100 \cdot \exp[-0.011 \cdot 300] \simeq 3.7\% \qquad (1.137)$$

## 1.8.4 Stabilization of the Austenite by Interruption in the Cooling at T < Ms

All the above-mentioned, referred to the thermal difference $\theta_\gamma$ –T, implies—as it was indicated at the beginning of Sect. 1.8—that the continuous cooling rate from the austenitic state should exceed a certain rate, $V_c$, critical for each steel, with the purpose of avoiding that pearlite or bainite could be formed in this steel before reaching, by this cooling, the Ms temperature. On the other hand, considering this, the cooling rate below the Ms until reaching the final temperature T has also influence in the proportion of residual austenite. It is checked that, during the cooling, all the interruptions below the Ms temperature stabilize the austenite and increase the percentage of residual austenite in the steel.

We assume, for instance, a certain steel whose Ms temperature was 350 °C, and we also consider that 200 °C should be reached to obtain 80% of martensite by continuous cooling from the austenitic state. This same steel, cooled from the austenitic state down to 275 °C, and keeping constant at this temperature, would require continuing the cooling down to 180 °C if we want to obtain 80% of martensite.

It was also experimentally checked that the stabilization of the austenite at constant temperature, intermediate between Ms and $M_f$, increases with the time; it was also checked—as a consequence of the previously indicated—for continuous cooling that the stabilization of the austenite increases when the cooling rate decreases (always and when this does not result lower than $V_c$).

Several researchers assigned this stabilization of the austenite to the fact that the formation of martensite lamellae produces always, by plastic deformation, an accommodation of the surrounding austenite and creates a surface with great concentration of dislocations. The permanence at an intermediate temperature between Ms and $M_f$ will facilitate, with the time, the movement of dislocations and the pinning of the interfacial dislocations with carbon atoms.

This would also explain that the stabilization would be non-appreciable when there is only a small quantity of martensite in the matrix (that is to say, if the temperature of holding was barely lower than Ms). This would also explain that the stabilization is greater the lower (far from the Ms) the holding temperature.

The fundamentals expressed in Sect. 1.8.4 can be summarized in that the interruptions in the cooling below Ms, or the slower continuous cooling rate, decrease the $M_f$ temperature typical of a cooling theoretically instantaneous at this temperature, and that—for other reason, by decreasing Ms—, the quantity of residual austenite increases if the austenitization temperature grows.

# 1.9 Complementary Considerations About the Ferritic-Pearlitic Transformations

We have referred in previous pages to TTT curve to forecast the structures that would be obtained by continuous cooling of the steel from the austenitic state if they overlap the cooling curve. If the cooling rate is very slow, pearlitic structures are obtained. If the cooling rate was greater, bainitic structures would be obtained. If the cooling rate was sufficient to not intersect the TTT curve in any point, martensite would be obtained.

However, if we are rigorous, CCT curves (Continuous Cooling Transformation) must be used for continuous cooling: obtained by dilatometric measurements—and with the support of the metallography—for different cooling rates of the steel from the austenitic state. We show in Fig. 1.40 an example of a CCT curve.

**Exercise 1.12** Calculate the structures obtained in the dilatometer device with different cooling rates (controlled cooling) of the samples.

*Solution starts:*
We consider the curves represented in Fig. 1.40. The different structures are.
Curve 1: martensite (99%), ferrite (1%).
Curve 2: martensite (97%), ferrite (3%).
Curve 3: martensite (95%), ferrite (5%).
Curve 4: martensite (65%), ferrite (35%).
Curve 5: martensite (50%), ferrite (50%).

**Fig. 1.40** TTT curve of continuous cooling of a steel of 0.12% C, 0.79% Mn, 1.23% Si, 0.014% S, 0.0011% P, 0.43% Ni, 1.22% Cr, 0.54% Mo, 0.24% Cu, 0.053% As, 0.016% Ti. Austenitization at 950 °C for 30 min. Austenitic grain size: 9 ASTM

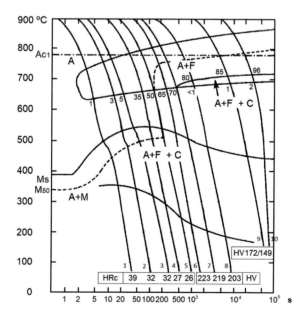

Curve 6: martensite (35%), ferrite (65%).
Curve 7: martensite (30%), ferrite (70%).
Curve 8: martensite (20%), ferrite (80%), pearlite (<1%).
Curve 9: martensite (14%), ferrite (85%), pearlite (1%).
Curve 10: martensite (2%), ferrite (96%), pearlite (2%).

*Solution ends.*

### 1.9.1  Ferritic-Pearlitic Transformations of the Austenite by Continuous Cooling

If the cooling rate from the austenitic state increases, the temperature $A_{3r}$ decreases for the same hypoeutectoid steel of $C_1$ carbon percentage. Figure 1.41 illustrates that the temperature for the beginning of ferrite formation—$A_{3r}$—is $T_1$ for the rate $V_1$, $T_2$ for the rate $V_2$, etc.

Complementarily, Fig. 1.41 reminds us that the critical size of the ferrite nucleus decreases with $A_{3r}$. This is coherent: it is necessary for the participation of less iron atoms to form an alpha nucleus, being this one more stable (that is to say, it is not disaggregated by thermal fluctuations), the thermal vibration is reduced with the temperature and, therefore, bond energy between atoms grows. Thermodynamically, it is also possible to demonstrate that the lower the $A_{3r}$, the greater the number of nuclei formed at this temperature; nuclei that, when they grow, will give greater number of proeutectoid ferrite grains (and, for that reason,—for the same quantity of ferrite—grains will be finer).

**Fig. 1.41** Transformation of the austenite by continuous cooling for a binary steel $C_1$ < 0.77% C

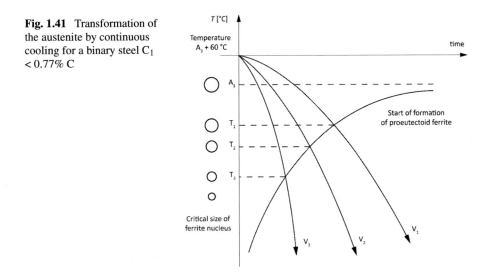

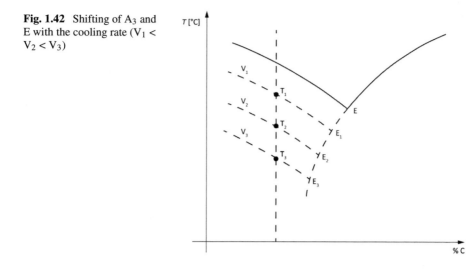

**Fig. 1.42** Shifting of $A_3$ and E with the cooling rate ($V_1 < V_2 < V_3$)

Furthermore, the reduction of $A_{3r}$ for faster cooling rates is translated into the fact that the transformation lines of the Fe–C metastable equilibrium diagram are modified as it is indicated in Fig. 1.42.

Likewise, the cooling rate hardly has influence in the curve of the carbon solubility in the austenite ($A_{cm}$). This modifies the coordinates of the eutectoid point, as for equilibrium cooling, the coordinates of the eutectoid point are 0.77% C and 727 °C. On the contrary, for the cooling rate $V_1$, the eutectoid will correspond to the temperature and composition of the point $E_1$. Analogously, for the rate $V_2$, the eutectoid would be $E_2$, for $V_3$ would be $E_3$, etc. Therefore, when the cooling rate increases, the eutectoid temperature and the carbon content in the pearlite decrease (diluted pearlite).

On the other hand, the proportion of cementite in the pearlite (which with equilibrium cooling is approximately 0.77/6.67≃12%) decreases with the cooling rate ($e_1/6.67$ for $V_1$, $e_2/6.67$ for $V_2$, $e_3/6.67$ for $V_3$, etc.). Reciprocally, the ferrite that exists in the pearlite (complement to 100 of the quantity of cementite) increases. For that reason, it is possible to say that these pearlitic structures with carbon percentages $e_1, e_2, e_3$, etc.—depending on the cooling rates $V_1, V_2, V_3$, etc.—are "diluted pearlite" (because they have more ferrite than the pearlite of 0.77% C). Thus, diluted pearlite of a eutectoid steel of 0.4% C would have only 6% of cementite, approximately.

It is convenient to mention that diluted pearlite, because of being formed at temperature <727 °C, has a separation between lamellae of cementite, $S_0$, smaller than the interlamellar distance of the coarse pearlite of 0.77% C (0.25–0.50 μm). Therefore, even with less carbon than 0.77%, diluted pearlites—as a result of being finer—are usually more resistant than this coarse pearlite whose ultimate tensile strength is estimated in 800 MPa, as said in Sect. 1.5.1.

Summarizing, and considering again the steel with $C_1$%, its pearlite fraction increases with the cooling rate. In fact, for equilibrium cooling, pearlite fraction is nearly $C_1/0.77$; on the contrary, this relation is $C_1/e_1$ for $V_1$, $C_2/e_2$ for $V_2$, $C_3/e_3$ for $V_3$, etc. Definitely, the increase of the cooling rate is translated into an increase of the tensile strength, $\sigma_u$, with respect to the value calculated for this ferritic-pearlitic steel with equilibrium cooling.

This steel of $C_1\%$ can be even 100% pearlitic for a certain rate $V_m$, although for greater rates than that indicated, it is not adequate to consider the equilibrium diagram because non-equilibrium structures would be obtained, as mentioned above.

## 1.9.2  Ferritic-Pearlitic Transformations in Low Alloy Steels

The information collected in the previous section is deduced from the decrease of the temperature $A_3$ when the cooling rate is increased from the austenitic state. However, for the same cooling rate, those metallographic consequences are potentiated with other intrinsic factors of the steel, for example, with content in gammagenous elements. These elements, which for the equilibrium cooling decrease the $A_3$ temperature of the carbon of this steel, will exhibit to a greater extent the diminution of the $A_{3r}$ when the cooling rate is increased, with the subsequent consequences, not only microstructural but also mechanic.

This way, for example, if we compare a binary steel of 0.2% C with other of 0.2% C–1.5% Mn, both air cooled, the ultimate tensile strength ($\sigma_u$) is greater in the alloyed steel. In fact, the value of $\sigma_u$ for a hypoeutectoid steel of ferritic-pearlitic microstructure—alloyed or not—can be approximately estimated equal to the addition of two terms: percentage of proeutectoid ferrite multiplied by the ultimate tensile strength of the alloyed ferrite, and percentage of pearlite (equal to 100 minus the percentage of proeutectoid ferrite) multiplied by the ultimate tensile strength of the pearlite. For the same cooling rate, the quantity of pearlite is greater in the steel with 0.2% C-1.5% Mn than in the steel with only 0.2% C. On the other hand, the effect of the grain refinement of the ferritic grain that intrinsically the manganese produces by decreasing of the $A_3$ also adds the weak hardening that produces in the ferrite by solid solution, which involves that the ultimate tensile strength of C-1.5% Mn steels rather exceeds that of carbon steel.

The modification of the point $A_3$ by gammagenous elements has other industrial consequences: for the same carbon percentage, the growing contents in manganese allow forming the steel in gamma state—without transformation into alpha—until temperatures lower than those corresponding to the plain carbon steel. The quenching of steels with manganese can be also carried out at lower temperatures. However, on the contrary, if we consider the same austenitization temperature (for forming or quenching) for two steels with the same carbon content, the plain carbon steel is less prone to overheating than its bearing manganese one.

# 1.10   Designation and Normalization of Steels

## 1.10.1   Different Methods to Designate Steels

It is not easy to give a single name to each type of steel. They are usually grouped into different designations depending on the chemical composition, microstructure, heat o thermal-chemical treatments, mechanical resistance, or other descriptors of quality. In other occasions, they are classified also as a result of the obtaining method, deoxidation practice, characteristics of the final forming of the process, etc.

### 1.10.1.1   Due to the Chemical Composition

Considering the chemical composition, steels are usually classified into *carbon steels*—unalloyed—, *low alloy steels*—the addition of the alloying content quantities does not exceed 5%—, *medium alloy steels*—when any alloying element exceeds 5% and the addition of all the alloying contents does not exceed 8% in weight—and *highly alloyed steels*—alloying elements exceed 8%.

Complementarily, steels are also designated as *hypoeutectoid*, *eutectoid*, or *hypereutectoid* if the carbon content is smaller, equal, or greater than that of the eutectoid austenite (0.77% C for binary steels—only iron and carbon—but lower carbon content if there are other alloying elements).

On the other hand, some highly alloyed steels, which reduce the maximum carbon soluble in the austenite, for high carbon contents (although smaller than 2.11% C), give a eutectic matrix of austenite and carbides when solidify. These steels are called ledeburitic steels.

The characterization of any steel usually begins with the determination of its chemical composition, followed by a metallographic analysis—optical microscope and/or scanning electron microscope—and other ensemble of laboratory tests—mechanic, radiographic, thermal, etc.—that allow defining the structure and properties of the steel.

A very easy to carry out test is used—although, obviously, it does not replace the chemical analysis, and it is not an unequivocal method—to facilitate the fast classification of a steel according to its composition, the *spark test*. It consists in producing the friction of the steel with a grinding wheel at high velocity and seeing the sparks emitted. The important characteristics are color, volume, nature of the spark, and length. See in Fig. 1.43a and b the shape of the sparks depending on the type of steel. It is convenient to advise the reader that with a certain practice, using a grinding wheel that rotates at high velocity, using the same pressure—if it is excessive, the spark might lead to wrong conclusions about the carbon content—and comparing the spark of the steel with that of the reference steel, it is possible to have very satisfactory results with this qualitative test. A skilled operator can determine with significant approximation the carbon content of the steel and the characteristics of the alloying element (if it was available).

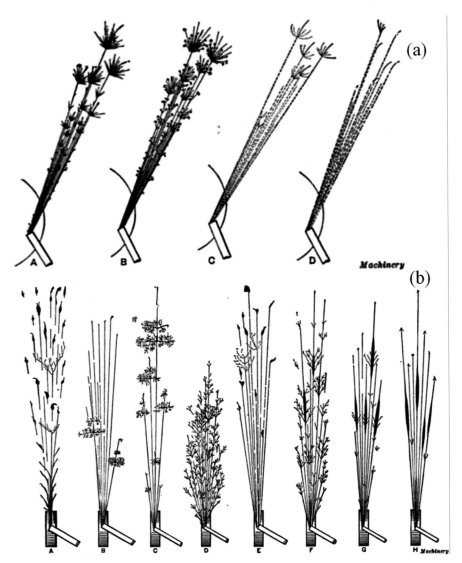

**Fig. 1.43 a** Spark for (A) High-carbon steel, (B) Manganese steel, (C) Tungsten steel, and (D) Molybdenum steel; **b** Spark for (A) Wrought iron, (B) Mild steel, (C) Steel with 0.50 to 0.85% C, (D) High-carbon tool steel, (E) High-speed steel, (F) Manganese steel, (G) Mushet steel and (H) Special magnet steel. *Source* Wikicommons

### 1.10.1.2   Other Designations

A same steel might have different metallographic constituents and, consequently, the properties would be different depending on the metallurgical history. However, it is also a common practice to designate steels—both alloyed and unalloyed—by their microstructure (Fig. 1.44).

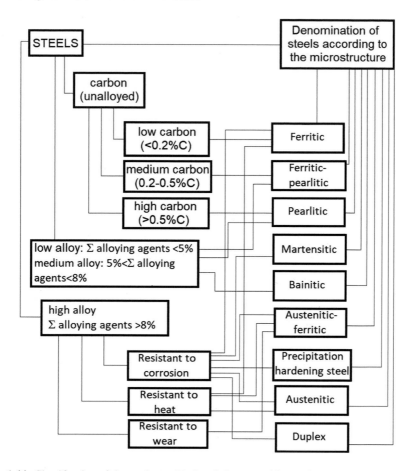

**Fig. 1.44** Classification of the steels considering their composition and structure

It is possible to distinguish between *ferritic* steels, *ferritic-pearlitic* steels, and *pearlitic* steels in carbon steels. Low and medium alloy steels can be *quenched and tempered*—their structure corresponds to that of the tempered martensite—, *bainitic, dual-phase*, TRIP (Transformation-Induced Plasticity steels), etc. The structural classification of highly alloyed steels includes ferritic, semiferritic, martensitic, austenitic, or austenitic-ferritic, as well.

Steels are usually designated by quality descriptors such as *stainless, refractories, with special properties*—magnetic, etc.—*corrosion resistant, sintered steels, high-speed, interstitial free, resistant to creep, tough at low temperatures, easily weldable*, and *easily machinable*. In this descriptive concept, we call ultra-high strength steels if their ultimate tensile strength exceeds 1500 MPa. Table 1.7 collects this property for steels and other materials. This mechanical quality—that for being considered in the case of steels—must be accompanied by the toughness: ability of materials to absorb energy by fast enough mechanical deformation before breaking.

**Table 1.7** Ultimate tensile strength ($\sigma_u$ in MPa) of some materials

| Material | $\sigma_u$ | Material | $\sigma_u$ |
|---|---|---|---|
| Muscle tissue | 0.1 | Hair | 192 |
| Artery tissue | 1.7 | Spiderweb wire | 240 |
| Cartilage | 3 | Ceramics | 34–350 |
| Polyurethane (foam) | 1 | Silver | 300 |
| Common brick | 5.5 | Silk | 350 |
| Low density polyethylene | 20 | Cotton fiber | 350 |
| Natural gum | 30 | Mild steel | 300–450 |
| Polypropylene | 33–36 | Linen | 700 |
| Leather | 41.1 | Aluminum and alloys | 300–700 |
| Polyurethane | 58 | Copper and alloys | 250–1000 |
| Tin and alloys | 14–60 | Artificial fibers | 350–1050 |
| Lead and alloys | 14–70 | Dual-phase and TRIP steels | 500–1000 |
| Tendons | 82 | High %C ferritic-pearlitic steels | 800–1300 |
| Cement | 4.1 | Titanium and alloys | 300–1400 |
| Hemp rope | 82 | Molybdenum and alloys | 665–1650 |
| Timber (lengthwise fiber) | 105 | Stainless steels | 400–1800 |
| Common glass | 35–175 | Nickel and alloys | 400–2000 |
| Glass fibered composites | 100–300 | Quenched and tempered steels | 680–2100 |
| Magnesium and alloys | 125–380 | Maraging steels | 1500–2500 |

   Considering the mechanical strength of structural materials, they are usually classified in the base of their yield strength ($\sigma_y$). Therefore, we have low strength steels if $\sigma_y < 250$ MPa, medium strength steels if $250$ MPa $< \sigma_y < 750$ MPa, high strength steels if $750$ MPa $< \sigma_y < 1500$ MPa, and very high strength if $\sigma_y > 1500$ MPa.

   In previous sections we have pointed out that the microstructural state of the steel is given by the thermal or thermal-chemical treatment to which the steel was subjected. Considering this matter, steels are also classified considering their state: annealed, normalized, quenched, tempered, case hardened, nitrided, carbonitrided, etc.

   On the other hand, it is possible to distinguish between *electric furnace* steels—all the alloyed steels but also many of the low alloy and even unalloyed steels—and *converter* steels depending on the obtaining method. Electric furnaces mainly use scrap and alloying elements to produce steels. In converters, on the contrary, steel is obtained by decarburizing, by means of oxygen injection in the molten pig iron—from liquid with approximately 4.3% C—coming from the blast furnace (which obtains the pig iron from iron ore, Verdeja et al. 2021). Converter steels are usually known with the name of integrated iron and steelmaking route steels as the Siemens procedure—acid process, open hearth furnace—has fallen in disuse.

Obtained the liquid steel, in electric furnace or in converter, and poured into ladles, the common practice is to cast in continuous, as the ingot casting is less used. In any case, depending on the practice of the deoxidation of the liquid steel (Verdeja et al. 2021), steel receives different names: *killed*, *semikilled*, *capped*, *rimmed*, or *vacuum casted* (degassed).

On the other hand, considering the characteristics of the final forming of the product, it is possible to distinguish between cast/molded steels and forged steels. We call cast/molded steels when the products or parts receive the final shape by casting in molds (sand molds, refractory molds, chill molds, etc.). On the contrary, we call forged steels (wrought steels), those where the shape is acquired by a thermal–mechanical forming—rolling, extrusion, die forging, etc.—, which starts from the solid obtained by continuous casting or ingot casting to obtain flat and long products such as plates, sheets, profiles, bars, rails, etc.

When until reaching its final shape this thermal–mechanical forming of the steel was carried out in gamma state (T > $A_3$), we can say that the steel was hot forged. If a mechanical forming at temperature lower than $A_1$ is possible—because the chemical composition allows it—, we can say that the steel was cold rolled—or wire drawn, stretched, deep drawn, etc.

## *1.10.2  International Standards for Steels*

Almost all countries have their own standards—commercial, professional, national—that include the classification, designation, chemical analysis, and properties of steels. We can include from all the international standards: AISI (American Iron and Steel Institute), SAE (Society for Automotive Engineers), DIN (Deutsches Institut für Normung), AFNOR (Association Française de Normalization), BSI (British Standards Institution), ISO (International Organization for Standardization), EN-CEN (European Committee for Standardization), ASTM (American Society for Testing Materials), ASME (American Society of Mechanical Engineers), and AWS (American Welding Society).

Steels are classified in Spain considering the UNE-AENOR (Una Norma Española-Asociación Española de Normalización) standards. Nevertheless, it is frequent the use of the North American designation AISI—used in almost all countries of America, even those of Spanish language—because facilitates the international exchanges, commercial and technical, with these countries. It is also common the utilization of the DIN standards and the European standards.

There has been interest in unifying the steels designations both in North America and Europe, because of obvious advantages. Moreover, it is frequently appreciated when comparing different specifications—that designate the steels using letters, chemical symbols, or number—, that a same alloy has a different name depending on the standard that is used, or the same naming is used for two different alloys.

This objective of unifying standards has led North American professional and commercial associations to try to designate the alloys—including steels—according

**Table 1.8** Primary UNS series

| UNS series | Metallic group |
|---|---|
| Axxxxx | Aluminum and aluminum alloys |
| Cxxxxx | Copper and alloys (brasses and bronzes) |
| Dxxxxx | Specified mechanical property steels |
| Exxxxx | Rare earth and rare earthlike metals and alloys |
| Fxxxxx | Cast irons |
| Gxxxxx | AISI and SAE carbon and alloy steels (except tool steels) |
| Hxxxxx | AISI and SAE H-steels |
| Jxxxxx | Cast steels (except tool steels) |
| Kxxxxx | Miscellaneous of steels and ferrous alloys |
| Lxxxxx | Low-melting metals and alloys |
| Mxxxxx | Miscellaneous non-ferrous metals and alloys |
| Nxxxxx | Nickel and nickel alloys |
| Pxxxxx | Precious metals and alloys |
| Rxxxxx | Refractory metals and alloys |
| Sxxxxx | Heat and corrosion resistant (stainless) steels |
| Txxxxx | Tool steels, wrought, and cast |
| Wxxxxx | Welding filler metals |
| Zxxxxx | Zinc and zinc alloys |

to the UNS specification (Unified Numbering System), to avoid the multiple designations resulted from the AISI, SAE, ASTM, AWS, etc. associations. In fact, almost all the technical literature in North America uses nowadays the equivalence UNS to the AISI, SAE, etc. designation to refer to a steel. It seems convenient to mention the future importance of the UNS whether we consider the North American influence in the industrial sector.

The UNS designation started with the publication in 1975 of the first edition of the UNS Handbook. The families of alloys group in UNS series designated by a letter, prefix, and five numerical digits (see Table 1.8), which identify the alloy. For example, UNS T30108 corresponds to the AISI A-8 tool steel, S30452 to the AISI 304 austenitic stainless steel, K93600 to the Invar, C26200 to the cartridge brass, etc.

# References

Andrews KW (1965) Empirical formulae for the calculation of some transformation temperatures. J Iron Steel Inst 203:721–725

Attwood B (1994) New technologies: Making steel the material of choice for the 21st century, AISI General Meeting, May 18, 1994

Bain EC (1924) The nature of martensite. Trans AIME 70:25

Bain EC, Paxton HW (1966) Alloying elements in steel. American Society for Metals, Metals Park, Ohio, USA

Calvo-Rodés R (1956) El acero, su elección y selección, INTA, Madrid, Spain

Flinn RA, Trojan PK (1979) Materiales de ingeniería y sus aplicaciones. McGraw-Hill, México

Hollomon JH, Jaffe LD (1945) Time-temperatures relations in tempering steel. Trans AIME 162:223–249

Koistinen DP, Marburger RE (1959) A general equation prescribing the extent of the austenite-martensite transformation in pure iron-carbon and plain carbon steels. Acta Metall 7:59–60

Kurdjumow G, Sachs G (1930) Über den Mechanismus der Stahlhärtung. Z Phys 64:325–343

Leslie WC (1982) The physical metallurgy of steels, 1st edn. McGraw-Hill International Book Company, London, UK

Massalski TB, Murray JL, Bennett LH, Bakker H (1986) Binary alloy phase diagrams. American Society for Metals, Metals Park, Ohio, USA

Nehrenberger AE (1946) The temperature range of martensite formation. Trans AIME 167:494–501

Paxton, HW (1979) The changing scene in steel. Metall Trans A 1815–1829

Pero-Sanz JA, Quintana MJ, Verdeja LF (2017) Solidification and solid-state transformations of metals and alloys, 1st edn. Elsevier, Boston, USA

Pero-Sanz JA, Fernández-González D, Verdeja LF (2018) Physical metallurgy of cast irons. Springer International Publishing, Cham, Switzerland

Pero-Sanz JA, Fernández-González D, Verdeja LF (2019) Structural materials: properties and selection. Springer International Publishing, Cham, Switzerland

Pickering FB, Gladman T (1963) Metallurgical developments in carbon steels. ISI Special Report 81:10

Porter DA, Easterling KE (1981) Phase transformations in metals and alloys, 1st edn, Vam Nonstrand Reinhold (International) Co. Ltd., Wokingham, UK

Speich GR, Leslie WC (1972) Tempering of steel. Metall Trans 3:1043–1054

Steven W, Haynes AG (1956) The temperature of formation of martensite and bainite in low-alloy steels. J Iron Steel Inst 183(8):349–359

Verdeja JI, Fernández-González D, Verdeja LF (2020) Operations and basic processes in ironmaking. Springer International Publishing, Cham, Switzerland

Verdeja LF, Fernández-González D, Verdeja JI (2021) Operations and basic processes in steelmaking. Springer International Publishing, Cham, Switzerland

# Chapter 2
# Heat Treatment of Steels

## 2.1 Introduction

The ensemble of controlled heating and cooling processes that can be applied to a solid alloy with the objective of modifying its microstructure—and consequently its properties—without changing the chemical composition of the alloy is called *heat treatment*.

We call thermochemical treatments to those that additionally involve a chemical modification in the periphery of the alloy (case hardening).

Finally, we call thermomechanical treatments to those that involve structural modifications of the alloy by heating and, simultaneously, mechanical forging of the solid alloy before the cooling.

## 2.2 Austenitization

Several heat and thermomechanical treatments involve, first, achieving that the steel had austenitic structure to, from these high temperatures, transform later by cooling that structure into the desired constituents.

This is the case, for instance, of several heat treatments that are indicated below—for instance, full annealing, normalizing, quenching, isothermal annealing, etc.—and also of some thermochemical treatments (for instance, case-hardening) or thermomechanical treatments (for instance, recrystallization in gamma phase, forming of the austenite, etc.).

## 2.2.1  Heating to Austenitize

A steel, it does not matter the initial structure—tempered martensite, bainite, pearlite, ferritic-pearlitic, cementitic-pearlitic, etc.—experiences above $A_3$ the following transformation of the structure:

$$Fe_\alpha + Fe_3C \rightarrow Fe_\gamma \qquad (2.1)$$

as the structure is always of ferrite and carbide at temperature close and lower than $A_e$.

The previous reaction takes place by a mechanism of nucleation in the ferrite and cementite interface: the formation of austenite by allotropic transformation of the ferrite starts there. Contiguous cementite easily disaggregates its contours to give ferrite and carbon and, consequently, the ferrite of the interface—already transformed into austenite—absorbs this carbon. The transformation from ferrite into austenite takes place faster than the decomposition of carbides.

For the complete austenitization, it is necessary a prolonged holding at austenitization temperature $\theta_1$ greater than $A_{3c}$. The driving force for the austenitization reaction is the temperature difference $\theta_1$–$A_{3c}$. It is necessary for a time of permanence at temperature $\theta_1$ to not only transform all $Fe_\alpha$ into $Fe_\gamma$ and disaggregate the cementite but also to achieve—by diffusion in the volume of carbon—a homogeneous austenite.

The greater the ferrite/cementite-free surface, the shorter times are required for this austenitization at temperature $\theta_1$. For that reason, the austenitization of lamellar cementite is faster than that of the globular cementite. And, for the same morphology (lamellar or globular), the austenitization is performed in shorter times the smaller the cementite size.

The time of permanence at temperature $\theta_1$ to achieve not only full austenitization but also chemical uniformity of all the austenite is of 30 min in industrial muffle furnaces each inch in thickness when the heating process up to $\theta_1$ has lasted 1 h per inch. The holding when the heating has lasted half an hour per inch in thickness is usually, on the contrary, of 1 h per inch. The total duration of the treatment is shorter when the heating is carried out in molten salts furnaces.

Exercise 2.1 The steel in Fig. 1.34 (C–Cr–Ni–Mo) is austenitized before the quenching from a temperature of 880 °C. It has 0.1% of the microalloying element vanadium to inhibit the austenitic grain growth. Would it be effective for this purpose?

Data: Assume that $N_f$, free nitrogen in the steel, is 0.010%. Temperature of $C_3V_4$ precipitation start is given by

$$\log[\%C] \cdot [\%V]^{4/3} = -\frac{10800}{T} + 7.06 \qquad (2.2)$$

Temperature of precipitation of the NV:

$$\log[\%\text{N}] \cdot [\%\text{V}] = -\frac{7733}{\text{T}} + 2.99 \qquad (2.3)$$

*Solution starts*

- Temperature at the start of precipitation, in the cooling, or end of redissolution (ideal) in the heating of the $C_3V_4$:

$$\log[0.5] \cdot [0.1]^{4/3} = -\frac{10800}{\text{T}} + 7.06 \rightarrow \text{T} = 1243 \text{ K } (970\,^\circ\text{C}) \qquad (2.4)$$

- Temperature at the start of precipitation, in the cooling, or end solving of redissolution (ideal) in the heating of the NV:

$$\log[\%\text{N}] \cdot [\%\text{V}] = -\frac{7733}{\text{T}} + 2.99 \rightarrow \text{T} = 1291 \text{ K } (1018\,^\circ\text{C}) \qquad (2.5)$$

As the temperatures of redissolution are equivalents and greater by around 90–140 °C than the temperature of treatment, both intermetallics will be efficient inhibitors to avoid the growth of the austenite. It would be better than the NV, although the quantity of free nitrogen in steel can be critical regarding the toughness (0.01% of free nitrogen increases the transformation temperature to 70 °C, see Pickering's equation, Eq. 1.68).

*Solution ends.*


### 2.2.1.1  Overheating and Burning

Austenite grains, which are formed by nucleation and growth at the temperature $\theta_1$, impinge after some time, and some of them grow at the expense of their neighbors to decrease the number of grain boundaries and, consequently, decrease the energy of the system. The average grain size is greater, the greater the temperature or larger the time of permanence, although significant increases in time are only equivalent to small rises in temperature, according to the relationship of semilogarithmic type ($\ln t = \text{Constant}/\text{T}$ or which is equal to $\text{T} \cdot \log t = \text{Constant}$).

The times required for the austenitization decrease for austenitization temperatures greater than $\theta_1$. However, it is not convenient to increase the temperature to reduce the time of carbides redissolution and uniformize the formed austenite in carbon; neither to prolongate more than necessary the time of permanence at the austenitization temperature. The reason is that the grain size might excessively grow to produce that is known as overheating of the steel, which involves a noticeably impairing of the steel properties.

For example, ferrite usually adopts an acicular shape for air cooling from the austenitic state—that is to say, at a rate slightly greater than that required for the $\text{Fe}_\gamma \rightarrow \text{Fe}_\alpha + \text{Fe}_3\text{C}$ equilibrium transformation. It was frequent to find these acicular

**Fig. 2.1** Widmanstäten
structure

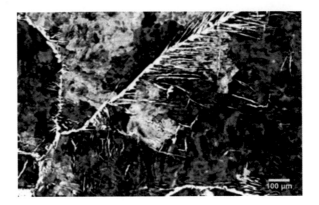

shapes of ferrite in as-molded, as-rolled, or as-forged steels and in weld beads.
The ferritic-pearlitic structure that resulted from cooling hypoeutectoid steels of
big austenitic grain size in air is called Widmanstätten structure (Fig. 2.1) due to
its analogy with the macrostructure observed by this researcher with naked eye in
meteorites in the nineteenth century.

The acicular ferrite of a Widmanstätten structure is unfavorable due to the low
mechanical properties that confer to the steel, particularly the low toughness, resulted
from the notch effect of the acicular ferrite, which would not be observed in the case
of the polyhedral ferrite.

Proeutectoid ferrite adopts a polyhedric shape only when the cooling rates are very
slow and the austenitic grain size is small. The growth of a ferrite crystal partially
depends on both the rate of latent heat release for the gamma-alpha transformation
and the rate of excess carbon diffusion rate from the formed ferrite.

On the other hand, when the austenitic grain has great size and the cooling rate
is high, the quantity of latent heat per unit of time that should be ceded through the
ferrite/austenite interface is notable. The growth of the ferrite crystals by the tips and
edges—in acicular shape, not polyhedric—facilitates that the latent heat could be
released in multiple directions and the gamma-alpha transformation is not stopped.

Besides, this morphology of the ferrite also facilitates the carbon expelling and
diffusion from the ferrite in multiple directions and reaches central zones of the
austenitic grain without requiring significant distances to be covered by the carbon,
by diffusion, as the carbon tends to accumulate in the contiguous austenite to the
formed ferrite.

When the chemical composition of the steel is such that gives fine precipitates—for
example, nitrides, carbides, carbonitrides—that remain unsolved at the austenitization temperature $\theta_1$, those phases oppose the mobility of the austenitic grain boundaries, difficult the grain growth—for that reason, they are called grain inhibitors—and
they reduce the overheating risks. The average grain size of the austenite, $d_y$, is related
to the volume fraction of grain inhibitor precipitates, $f_p$, and their average size, $d_p$,
by the following equation:

$$d_\gamma = \frac{d_p}{f_p} \qquad (2.6)$$

Thus, a 2‰ of precipitated grain inhibitor, of 200 Å (20 nm), would be able of maintaining an austenitic size, $d_\gamma$, of 10 $\mu$m.

Alphagenous elements or elements that stabilize the iron in the allotropic phase alpha—for instance, vanadium, molybdenum, etc.—are also favorable to avoid over-heating as the austenitic grain growth cannot start until the ferrite has completely transformed into austenite during the heating. It seems also convenient to mention that the elements that increase the solidus temperatures in high-carbon steels—see chromium, molybdenum, cobalt—equally reduce the overheating risks.

Very high austenitization temperatures favor—apart from the overheating—the formation of intergranular oxide in the steel and can even produce intergranular melting if the temperature reaches the solidus temperature. In both cases, we can say that the burning of the steel takes place. And, as opposed to the overheating—that, as it will be later mentioned, can be corrected by normalizing heat treatment—it is not possible to regenerate the burned steel structure and, therefore, the material is to be rejected (scrapped).

Finally, it is convenient to advise that the overheating of the steel is also unfavorable when the cooling of the structure is not ferritic-pearlitic, of Widmanstätten type. For example, the resulting structure is of martensite when severe cooling, and this appears as lamellae or needles of big size, which exhibit lower toughness than fine martensite obtained from austenitic grains of small diameter.

The grain is habitually designed by a number, N, defined by the American Society for Testing Materials (ASTM) as follows:

$$n = 2^{N-1} \qquad (2.7)$$

where n is the number of grains that exist per square inch when observed with a microscope at 100x. We indicate in Table 2.1 the correlation between the number N, the mean linear intercept (assuming that all they are tetradodecahedrons of the same dimensions), and the number of grains of the material per $mm^2$.

**Exercise 2.2**  One important characteristic of a microstructure is the size of its cells (grains). There are several ways to measure and characterize the grain size. One is in terms of the grains per area. The ASTM grain size number, N, is defined by the relation (Equation 2.8):

$$n = 2^{N-1} \qquad (2.8)$$

where n measures the number of grains per square inch on the photomicrograph taken at 100x. The simplest way to find n is to count the grains in a representative rectangular field as

**Table 2.1** Grain size

| Number N ASTM | Grains/mm$^2$ | Average length in mm |
|---|---|---|
| −3.5 | 0.685 | 1.051 |
| −3.0 | 0.969 | 0.884 |
| −2.5 | 1.370 | 0.743 |
| −2.0 | 1.938 | 0.625 |
| −1.5 | 2.740 | 0.526 |
| −1.0 | 3.875 | 0.442 |
| −0.5 | 5.480 | 0.372 |
| 0 | 7.750 | 0.313 |
| 0.5 | 10.96 | 0.262 |
| 1.0 | 15.50 | 0.221 |
| 1.5 | 21.92 | 0.186 |
| 2.0 | 31.00 | 0.156 |
| 2.5 | 43.84 | 0.131 |
| 3.0 | 62.00 | 0.110 |
| 3.5 | 87.68 | 0.0929 |
| 4.0 | 124.0 | 0.0781 |
| 4.5 | 175.4 | 0.0657 |
| 5.0 | 248.0 | 0.0552 |
| 5.5 | 350.7 | 0.0465 |
| 6.0 | 496.0 | 0.0391 |
| 6.5 | 701.4 | 0.0328 |
| 7.0 | 992.0 | 0.0276 |
| 7.5 | 1403 | 0.0232 |
| 8.0 | 1984 | 0.0195 |
| 8.5 | 2806 | 0.0164 |
| 9.0 | 3968 | 0.0138 |
| 9.5 | 5612 | 0.0116 |
| 10.0 | 7936 | 0.00977 |
| 10.5 | 11,220 | 0.00821 |
| 11.0 | 15,870 | 0.00691 |
| 11.5 | 22,450 | 0.00581 |
| 12.0 | 31,740 | 0.00488 |
| 13.0 | 63,500 | 0.00352 |
| 14.0 | 127,000 | 0.00249 |

$$n = n_i + \frac{n_e}{2} + \frac{n_c}{4} \tag{2.9}$$

where $n_i$ is the number of grains entirely within the rectangular field, $n_e$ is the number of grains which are cut by an edge of the field and $n_c$ (that is equal to 4) is the number of grains on the corners. $A_{100x}$ is the area (in$^2$) of the field at 100x (or corrected to $100 \times$ for other magnifications).

*Linear intercept grain size*

Another simple measure of grain size is the mean linear intercept, $\overline{L}$, of lines drawn randomly with respect to the microstructure. The quantity $\overline{L}$ is a measure of the grain size and is conventionally given in mm (in real space). The system is to lay a large length of line (or lines) randomly on the microstructure and count the number of intersections per length $N_L$. The average linear intercept is

$$\overline{L} = \frac{1}{N_L} \tag{2.10}$$

If the microstructure is random, the analysis lines may all be in the same direction (parallel), but its grains are elongated (band and cold-worked structures), and the analysis lines must be randomly oriented. For random microstructures $\overline{L}$ and ASTM, grain size is related as can be seen in the resolution of Exercise 2.2.

Calculate the number of real grains per mm$^2$ in steel with 7 ASTM grain size.

*Solution starts*

By definition:

$$n = 2^{N-1} \tag{2.11}$$

where n is the number of grains per square inch in the micrograph at 100x. Therefore:

$$n = 2^{7-1} = 64 \frac{\text{grains}}{\text{in}^2} \tag{2.12}$$

Thus, the number of grains per square millimeter will be equal to

$$a = \frac{64}{(25.4/100)^2} = 992 \frac{\text{grains}}{\text{mm}^2} \tag{2.13}$$

In the case of tetrakaidecahedrons of Thomson, the intersections by a plane will be hexagons of area equal to

$$\frac{3 \cdot \sqrt{3}}{2} \cdot L^2 \tag{2.14}$$

where L is the side of the hexagon. Therefore,

$$\frac{3 \cdot \sqrt{3}}{2} \cdot L^2 \cdot 992 \text{ grains} = 1\text{mm}^2 \tag{2.15}$$

It is possible to demonstrate by quantitative metallurgy that the average intersected length (grain size) is equal to

$$\sqrt{2} \cdot L \rightarrow \overline{L} = 27.9 \ \mu m \tag{2.16}$$

See Table 2.1.

We are going to calculate now the ASTM grain size from the micrograph in Fig. 2.2. We determine the number of complete grains and the number of partial grains from the micrograph in Fig. 2.2. There are 8 full grains and 15 partial grains (they count 1/2):

$$\text{Total number of grains} = 8 + 15 \cdot 0.5 = 15.5 \text{ grains} \tag{2.17}$$

According to the American Society for Testing Materials (ASTM),

$$n = 2^{N-1} \text{ (at 100x)} \tag{2.18}$$

where n is the number of grains in a square inch at $100\times$ and N is the ASTM grain number.

The size of the micrograph is 77 mm $\times$ 101 mm.

We calculate n by

**Fig. 2.2** Micrograph of the steel (100x)

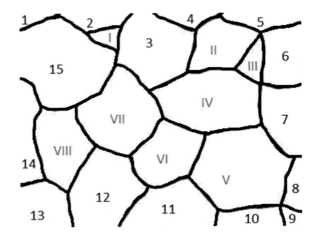

$$n = \frac{\text{Total number of grains}}{\text{size of the micrograph}/25.4^2} = \frac{15.5 \text{ grains}}{\left(77 \cdot 101/25.4^2\right)\text{inch}^2} = 1.3 \text{ grains/inch}^2$$

$$(2.19)$$

Using Eq. 2.18,

$$n = 2^{N-1} \to 1.3 = 2^{N-1} \to \log 1.3 = (N-1) \cdot \log 2 \to N = \frac{\log 1.3}{\log 2} + 1 = 1.4$$

$$\to N \simeq 1 \text{ ASTM}$$

$$(2.20)$$

We determine the grain size assuming squared grains:

$$\frac{\text{number of grains}}{\text{mm}^2}(\text{at } 1x) = \frac{15.5 \text{ grains}}{77 \cdot 101/100 \cdot 100} = 20.3\frac{\text{grains}}{\text{mm}^2} \qquad (2.21)$$

$$\frac{\text{number of grains}}{\text{mm}^2} \cdot \text{Area of the grain} = 1 \text{ mm}^2 \to 20.3\frac{\text{grains}}{\text{mm}^2} \cdot \left(\overline{L}\right)^2 \text{mm}^2 = 1 \text{ mm}^2$$

$$(2.22)$$

$$\left(\overline{L}\right)^2 = \frac{1}{20.3\frac{\text{grains}}{\text{mm}^2}} \to \overline{L} = \sqrt{\frac{1}{20.3\frac{\text{grains}}{\text{mm}^2}}} = 0.222 \text{ mm} \qquad (2.23)$$

We determine the grain size assuming hexagonal grains:

$$\frac{\text{number of grains}}{\text{mm}^2}(\text{at } 1x) = \frac{15.5 \text{ grains}}{77 \cdot 101/100 \cdot 100} = 20.3\frac{\text{grains}}{\text{mm}^2} \qquad (2.24)$$

$$\text{Area of the grain} = \frac{\text{perimeter} \cdot \text{apothem}}{2} = \frac{3 \cdot \left(\overline{L}\right)^2 \cdot \sqrt{3}}{2} \qquad (2.25)$$

$$\frac{\text{number of grains}}{\text{mm}^2} \cdot \text{Area of the grain} = 1 \text{ mm}^2$$

$$(2.26)$$

$$\to 20.3\frac{\text{grains}}{\text{mm}^2} \cdot \frac{3 \cdot \left(\overline{L}\right)^2 \cdot \sqrt{3}}{2} \text{ mm}^2 = 1 \text{ mm}^2$$

$$\left(\overline{L}\right)^2 = \frac{2}{20.3\frac{\text{grains}}{\text{mm}^2} \cdot 3 \cdot \sqrt{3}} \to \overline{L} = \sqrt{\frac{2}{20.3\frac{\text{grains}}{\text{mm}^2} \cdot 3 \cdot \sqrt{3}}} = 0.138 \text{ mm} \to d$$

$$= \overline{L} \cdot \sqrt{2} = 0.138 \text{ mm} \cdot \sqrt{2} = 0.195 \text{ mm}$$

$$(2.27)$$

Now we calculate the grain size using the intersection method. We draw the diagonals of the micrograph (Fig. 2.3). Then, we count the intersections grain-grain boundary, in this case, six in one diagonal and nine in the other diagonal.

We calculate the number of grains per millimeter:

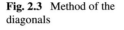

**Fig. 2.3** Method of the diagonals

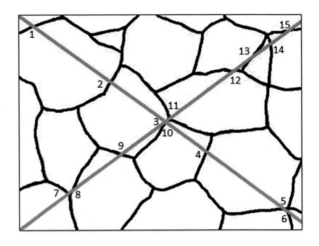

$$N_L = \frac{15}{2 \cdot 125/100} = 6\frac{\text{grains}}{\text{mm}} \qquad (2.28)$$

And, therefore, the grain size:

$$L_3 = \frac{1}{6} = 0.167 \text{ mm} \qquad (2.29)$$

There are some scatters in the obtained results. They will be diminished, improving the number of micrographs available for the same specimen.

*Solution ends.*

### 2.2.1.2  Dimensional Variations During the Heating

Dimensional variations during the heating until the austenitization are due to the allotropic transformation alpha-gamma that takes place with volume contraction. But before reaching temperature $A_e$, where the allotropic transformation starts, there are also risks of stresses and deformations if we have parts or profiles with heavy thickness.

It is convenient in these big parts that the heating was slow, and the most uniform possible before reaching $A_e$ temperature (that in binary steels is 727 °C) to avoid temperature differences between the nucleus and the periphery of the part with the aim of preventing risks of internal stresses that could produce deformations, and even cracks, in parts. It seems very reasonable the recommendation of Apraiz (1985): "do not put cold parts directly in the furnace—whose temperature was greater than 350 °C—if those exceed 200 mm of diameter" because the cold steel, with low plasticity, could break due to the stresses created between the periphery and the nucleus of the part.

However, the risks increase when temperature $A_e$ exceeds that is when the steel starts to contract because of the alpha-gamma transformation ($\simeq -1\%$ by volume). It is very convenient holding at a temperature close to and lower than $A_e$ with the aim of uniformizing the temperature of periphery and nucleus before the alpha-gamma transformation. On the contrary, in big parts, when the temperature of the periphery reaches $A_e$, they begin to contract, while the nucleus, as it has still not reached this temperature, is dilating in alpha phase and, therefore, stresses are even greater. And, when the periphery has exceeded $A_{3c}$, and therefore, it is expanding, the nucleus—because it did not reach such temperature—is still contracting.

It is clear that in parts of very small thickness, and simple geometry, temperatures are almost the same in the periphery and nucleus when heating. For that reason, dilatations before reaching $A_e$, contractions between this $A_e$ temperature and the temperature $A_{3c}$, and subsequent dilatations above this temperature take place at the same time and equally in all the part: the part homothetically dilates and/or contracts, which involves few risks of deformation and breaking.

Slow heating rates, on the contrary, are convenient when the greater size of the parts or the more complex its geometry is and the more polished or mechanized the part surface is, because of the difference in temperatures between the periphery and the nucleus of the part, for each time $t_1$, is greater, in this case, than when the surface is wrinkled or oxidized. Therefore, it is recommended for the heating to last two or three times more when the surface is "brightly" than when it is not.

In steels with alloying contents that decrease the thermal conductivity of the steel—as in the case of steels with chromium content greater than 12%—it is convenient that the duration of the heating until the austenitization was twice that of the low alloy steels.

In general orientation, whether the recommendation of Apraiz (1985) is considered, it is convenient that the difference of temperatures between two points of the same radius situated at 22 mm of distance in cross sections of parts does not exceed the 20 °C. To achieve it, it is recommended that the duration of the heating from the room temperature up to 850 °C was greater than 30 min per inch of diameter, and, preferentially, that the duration of the heating was of 60 min per inch of diameter.

These stresses, deformations, and cracks during heating until the austenitization are common to full annealing, quenching, normalizing, rolling, and forging in gamma state, etc. If the cracks are produced in gamma state, the tips of the crack usually decarburize; this allows ruling—frequent discussion between blacksmiths and those operators of heat treatments—when a crack has arisen at high temperature, or if, on the contrary, the crack has appeared in the cooling by quenching (in this case the tip cracks will not be decarburized).

### 2.2.2 Cooling from the Austenitic State

The cooling can be continuous and performed in different manners (cooling inside the furnace, or in air, or in water, or in oil, etc.) from the austenitic state. In other cases, the cooling is isothermal, and the heat treatments are called isothermal when the cooling is achieved, for example, by immersion in a salts bath or in a molten lead bath.

The microstructure of steel at the end of the heat treatment will be the result of the interaction between the cooling curves—of the surface and the nucleus of the part—and the TTT curve. If the cooling from the austenitic state was more or less fast, the same part of steel could have in a certain point inside of the part martensitic, bainitic, fine pearlite, or coarse pearlite microstructures.

The factors that conditionate the microstructure of a part at the end of a heat treatment with previous austenitization can be summarized as follows:

– the heat transfer factor of the coolant (the ability to absorb heat from the surface of the part),
– size of the part to be treated,
– TTT curve of the considered steel.

### 2.2.2.1  Heat Transfer Factor of the Coolant

The cooling capacity of a coolant can be quantified not only qualitatively (water has more heat transfer factor than oil and this last one than air) but also quantitatively. For this, we can be warned about how the steel surface cools.

The heat flow rate through the metal-coolant interface follows the law of Newton:

$$q = A \cdot (T_S - T_M) \qquad (2.30)$$

where $T_S$ is the temperature of the part surface, $T_M$ is the temperature of the coolant, and A is the thermal conductivity of the interface.

The coefficient A can vary during the cooling of the part surface. This happens, for example, if the coolant is rest water. In fact—as indicated in Fig. 2.4—in the first moment, when the contact between water and steel surface is established (steel that is at the austenitization temperature), water vaporizes and the vapor bubbles difficulty the heat transfer (the value of A is small).

In the second stage, the heat transfer is fast as the bubbles detach and the convection mechanism starts (A grows). Finally, in the third stage of direct contact between rest water and the surface, the temperature descent in the steel surface is slow (A decreases). In the considered case—three different values of the heat transfer coefficient A—we take an average value M of the three values.

We call heat transfer factor, H, of the coolant to the ratio M/2·k, where k is the thermal conductivity of the steel (that can be considered constant for all steels, except for those with high chromium contents).

The value of H is greater than the capacity of the coolant to absorb heat. Therefore, H is greater for water than for air and, as it is logical, the value of H for the same coolant increases if the coolant is stirred. We indicate in Table 2.2 the heat transfer factors, H, for different coolants and different degrees of stirring.

**Fig. 2.4** Cooling of the steel surface

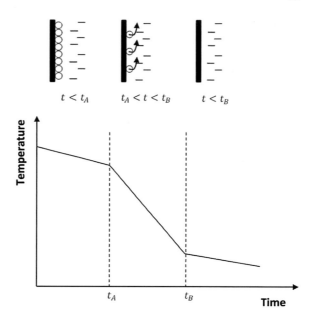

**Table 2.2** Cooling heat transfer factors, H, for different coolants and degrees of stirring

| Stirring | Air | Salts | Oil | Water | Salt water (brine quenched) |
|---|---|---|---|---|---|
| Without | 0.02 | 0.25–0.30 | 0.25–0.30 | 0.90–1.00 | 2.00 |
| Weak | – | 0.30–0.35 | 0.30–0.35 | 1.00–1.10 | 2.00–2.20 |
| Moderate | – | 0.35–0.40 | 0.35–0.40 | 1.20–1.30 | – |
| Medium | – | 0.40–0.50 | 0.40–0.50 | 1.40–1.50 | – |
| Strong | – | 0.50–0.80 | 0.50–0.80 | 1.60–2.00 | – |
| Very strong | 0.08 | – | 0.80–1.00 | 4.00 | 5.00 |

We could say about a coolant, ideal, that it has an infinite heat transfer factor, H, if the periphery of the part instantaneously reaches the temperature of the coolant when the part is submerged in it. We remind that this does not involve the same with the temperatures of the nucleus of the part, as it would progressively reduce the temperature as the heat flows from inside of the part to the surface in contact with the ideal medium.

## 2.2.2.2 Size Factor

Dimensions of the steel part to be treated are other factors that have an influence on the microstructure obtained at the end of heat treatment with cooling from the austenitic state.

We assume a cylinder of diameter $D_1$ and so long that the heat release would be only radial (not through the extremes of the cylinder) when cooling in a coolant of heat transfer factor $H_1$ (for instance, no-stirring in still water, see Fig. 2.5a). A gradient of temperatures is established between the periphery and the axis of the cylinder when cooling. This gradient changes with time because the calories (the heat stored inside of the cylinder), which must be released, are proportional to the volume. On the contrary, heat that is released to the coolant is proportional to the lateral surface. The ratio of volume to surface is equal to $D_1/4$. For that reason, the relation between the stored heat and the heat that can be released increases with increasing diameter.

This way, for the same coolant of heat transfer factor $H_1$ (for instance, no-stirring in still water, quiet water), when comparing the cooling curves of periphery and nucleus of two steel cylinders whose diameters were $D_1$ and $D_2$ ($D_1 < D_2$), we can see (Fig. 2.5a, b) that for the round bar of diameter, $D_2$, both the nucleus and the periphery cool slower than they do in the round bar of diameter $D_1$. After a time $t_1$, not only is the temperature of the periphery greater in the round bar of diameter $D_2$ but also its gradient of temperature between the periphery and the nucleus is greater than that in the round bar of diameter $D_1$.

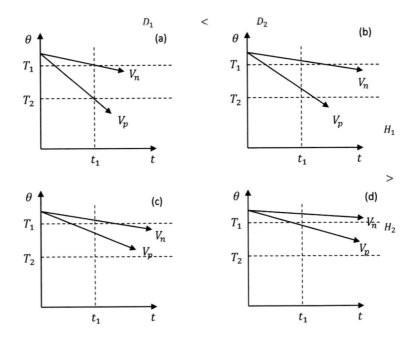

**Fig. 2.5** Cooling of the surface ($V_p$) and the nucleus ($V_n$) when the rod size is changed ($D_1 < D_2$) holding the heat transfer factor as a constant: **a** round bar of diameter $D_1$ cooled in water; **b** round bar of diameter $D_2$ cooled in water; **c** round bar of diameter $D_1$ air-cooled; **d** round bar of diameter $D_2$ air-cooled

On the other hand, if a coolant with less heat transfer factor than $H_1$ (e.g., air) is used, these gradients would be smaller in both cases. Compare Fig. 2.5a, c, which correspond to round bar $D_1$. Or compare Fig. 2.5b, d, which correspond to round bar $D_2$.

The microstructure of the steel at the end of the heat treatment will be the result of the interaction between the cooling curves—of the surface and the nucleus of the part (which depends on the heat transfer factor of the coolant and the size of the part)—and the TTT curves (which only depend on the chemical composition of the steel and the austenite grain size).

That is to say, the same steel part might have a martensitic structure in a point inside of the part, or any other structure (v. gr. bainitic, fine pearlite, or coarse pearlite), depending on the cooling rate from the austenitic state.

**Exercise 2.3** Two round bars of the same steel of equivalent diameters $D_1$ and $D_2$, after being austenized at the same temperature, have been quenched in oil. The penetration of the quenching (depth of hardening) is 3 mm in the periphery of the round bar $D_1$ and 7 mm in the round bar $D_2$. Answer the following questions, with adequate justification of the answers:

1. What of the round bars will have a bigger diameter?

   *Question*

   The round bar of diameter $D_1$ because the bigger the diameter, the lower the surface cooling rate, for the same heat transfer factor of the quenching.
2. Is it possible to say that a round bar of diameter $D_2$ has more hardenability than the round bar of diameter $D_1$?

   *Question*

   No, the hardenability is a property of the steel (in this case it is the same steel) and not of its diameter: if its size is bigger than $D_{cr}$, as defined, its structure in the nucleus will not be martensitic (hardened).
3. If the quenching would have been performed in water, the depth of the quenching would have been: the same, bigger, or smaller than in the case of the quenching in oil?

   *Question*

   The quenching in water is more severe so: a faster cooling rate and more depth of hardening are available.
4. Would the proportion of retained austenite change due to quenching in water and quenching in oil?

   *Question*

   Theoretically not (see the equation of Koistinen and Marburger in Pero-Sanz et al. 2018a), because the proportion of retained austenite only depends on the $M_s$, for equality of temperature of the coolant, $T_m$. In the practice, using a less severe cooling medium increases the proportion of retained austenite but also diminishes the susceptibility to quench cracking.

5. Modifying any extrinsic factor, could be possible achieving a depth of hardening bigger than 3 mm, for oil quenching, in the round bar of diameter $D_1$?

*Question*

Yes, increasing the austenitization temperature, because more hardenability of the steel shifts to the right of the TTT curves (CCT) from the origin of times.

## 2.3  Annealing to Soften the Steel

The aim of the annealing is to soften the steel to machine or cold form it. This is possible because ferritic-pearlitic structures would have been obtained in the cooling, which are softer than the martensitic or bainitic structures.

There are two types of annealing that we call supercritical because they require the previous austenitization of the steel above the critical temperature $A_e$: the full annealing and the intercritical annealing.

### 2.3.1  Full Annealing

The steel is heated up to temperatures greater than its $A_{3C}$ and held at this temperature until obtaining an austenitic structure in all points of the steel to perform full annealing. From this temperature, cooling must be sufficiently slow (Fig. 2.6) with the purpose of achieving the equilibrium constituents of the $Fe$–$Fe_3C$ diagram. The cooling must be slower (few degrees per minute) than in the normalizing treatment (few degrees per second).

Fast cooling, from $A_{3r}$ to $A_e$ with the objective of reducing the time elapsed on the treatment, is convenient. However, this fast cooling implies that fine grains of proeutectoid ferrite are obtained, and, for that reason, maximum softening is not obtained.

Later, cooling should be sufficiently slow from this temperature slightly greater than $A_e$ to obtain coarse lamellar pearlite. Once the full ferritic-pearlitic structure is attained, the cooling rate does not matter, and it is usually performed in air. The ultimate tensile strength of a $C_1\%$ carbon-content steel like this cooled, as primary approximation, is $\sigma_u$ (MPa) $= 300$-$650 \cdot C_1$. Its Brinell hardness has an approximate value—considering the correlation between strength and hardness—of HB$\simeq 3 \cdot (30 + 65 \cdot C_1)$, see Chap. 1

Due to technical issues, full annealing is only used in hypoeutectoid steels when lamellar pearlite is required, for instance, for subsequent machining or boring. Other annealing treatments are preferred for economic reasons, e.g., intercritical annealing

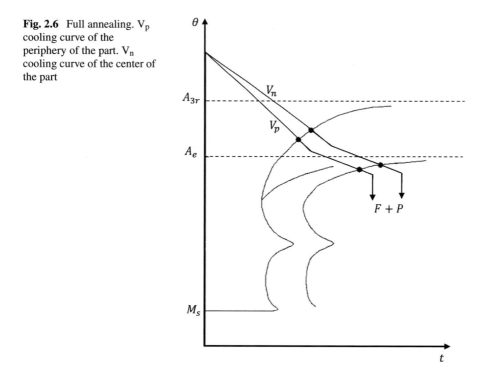

**Fig. 2.6** Full annealing. $V_p$ cooling curve of the periphery of the part. $V_n$ cooling curve of the center of the part

or even, as we will mention later, subcritical annealing. Full annealing is not used in hypereutectoid steels because it involves overheating and/or burning risks (the temperature interval between $A_{cm}$ and $T_S$ decreases with increasing carbon content).

## 2.3.2 Intercritical Annealing

The steel part is heated to a temperature $T_1$ comprised within $A_e$ and $A_{3C}$ with the purpose of performing intercritical annealing treatment. The part is held at this temperature for a time shorter than that required to fully achieve the equilibrium phases, which are ferrite and austenite. At this temperature, there will be cementite apart from ferrite and austenite. This cementite comes from the pearlite (or from the bainite or the martensite) that was previously in the steel and did not fully decompose into iron and carbon atoms during the heating up to $T_1$.

Next, the steel part is slowly cooled from this temperature $T_1$ (Fig. 2.7). A structure of ferrite and an aggregated similar to the pearlite, called "globulite" (see Fig. 2.8), is obtained at the end of the transformation. Its morphology is formed by globules of coarse $Fe_3C$ dispersed in a matrix of ferrite.

**Fig. 2.7** Intercritical
annealing

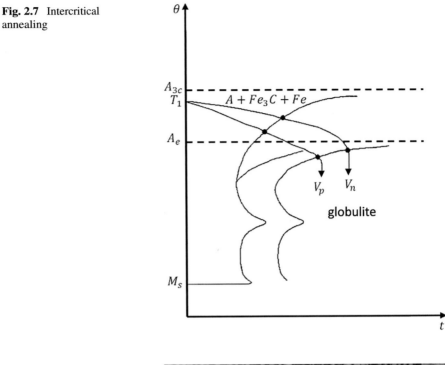

**Fig. 2.8** Globular cementite

The structure of the steel after the intercritical annealing is softer than that obtained in the full annealing, and, for that reason, more advantageous to perform the operations of cold forming of the steel. Or of lathing in the case of steels that are not too soft, as happens in the case of very low-carbon steels.

Definitely, to soften steel, it is almost always advantageous the intercritical annealing if compared with full annealing. Moreover, this annealing that requires

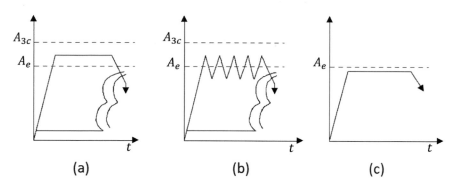

**Fig. 2.9**  Cycles a, b, and c used to achieve the structure of globular cementite

greater austenitization temperatures than $A_3$, consumes more energy. On another note, the intercritical annealing does not present risks of overheating and can be also used in hypereutectoid steels.

Intercritical annealing is usually also called "spheroidizing annealing" due to the non-lamellar shape that the pearlite adopts. Alternation of temperatures before cooling favors the globular morphology and the softening of the steel (see Fig. 2.9b). These thermal fluctuations, as facilitate the heterogeneous nucleation of the cementite, allow faster transformation of the eutectoid austenite into "globulite".

We have assumed in previous paragraphs that cooling, both in the full annealing and intercritical annealing, is continuously performed, e.g., programming the cooling rate of the furnace.

### 2.3.3  Isothermal Annealing

Steel can be also cooled from full or partial austenitization temperatures by immersion in a molten-salt bath at a constant temperature. This temperature must be lower (and closer) to the temperature $A_e$. In this case, the treatment is known as "isothermal annealing" (Fig. 2.10).

We can share several reasons to use isothermal annealing in a certain steel, either if the heating has been of full or incomplete austenitization. On another note, this constant temperature can be fixed, and the transformation of austenite would be uniform at any point of the part. This allows controlling the interlamellar spacing of pearlite, $S_0$, and the homogeneity of this pearlite both in massive as well as thinner sections of the part.

Furthermore, the temperature of the molten salts that could allow the full transformation of the austenite into diluted pearlite (without the appearance of proeutectoid ferrite) can be chosen and this also allows hiding the banded structure, which is typical in forged steels (see Verdeja et al. 2021).

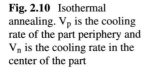

**Fig. 2.10** Isothermal annealing. $V_p$ is the cooling rate of the part periphery and $V_n$ is the cooling rate in the center of the part

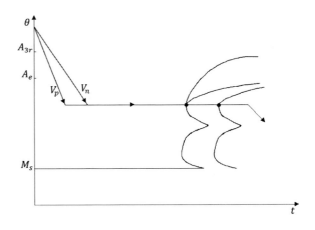

Isothermal annealing is particularly required when the steel has high hardenability, as happens in the case of many self-hardening steels used for tools. In these cases, slow continuous cooling between $A_e$ and $A_e$–50 °C is necessary to transform the austenite into pearlite, which can take tens of hours. On the contrary, if slow continuous cooling is not used, the final structures would be bainitic or martensitic. However, with isothermal cooling, the isothermal transformation of austenite into pearlite can be performed rapidly due to the great capacity to absorb heat from the molten salts if we compare it with air- or furnace-cooling.

This also justifies that, for steels of great hardenability, it is advisable—after forming a part in gamma state—to submerge the part immediately into molten salts, the temperature of which has been decreased closer to $A_e$. On the contrary, if this part was cooled in air after forging, a hard structure of bainite and martensite would be obtained, which would not be adequate for those complementary operations. With isothermal cooling after forging, a pearlitic structure, soft and suitable for machining and borehole operations (that the part could subsequently need), is directly obtained. We should remember that the mechanical strength, $\sigma_u$, of pearlite is inversely proportional to the square root of the interlamellar spacing, $S_0$.

## *2.3.4  Subcritical Annealing*

Subcritical annealing is another type of heat treatment used to soften steels. This treatment is performed by heating the steel part to a temperature close to but lower than the $A_e$ and holding it for a time at that temperature (Fig. 2.9c). The cooling rate after subcritical annealing is not important because there will not be any allotropic transformations. The structure obtained at the end of holding at this constant temperature is of globular cementite distributed in a ferrite matrix. Obtaining this structure softens the steel no matter what the initial structure is, whether it was martensite, bainite, or pearlite.

The softening, after being at the subcritical temperature of the treatment, is enough for the purpose required in parts to be machined (Verdeja et al. 2020).

The softening rate, as a function of the holding time at that temperature, is fast at the beginning and decreases asymptotically. A certain time later, the duration of the treatment can be extended, but the softening is not increased.

This type of annealing is also usually known as "spheroidizing annealing," due to the shape that the cementite adopts. But this name is usually reserved for intercritical annealing, as said above (Sect. 2.3.2).

The softening achieved using a subcritical annealing is not enough for some alloyed self-hardening and tool steels and full or intercritical annealing heat treatments are required.

With the limitations indicated in the previous paragraph, subcritical annealing offers several advantages with respect to the annealing treatments described in this chapter. First, energy is saved because temperatures greater than $A_e$ are not required. Second, deformation and/or cracking risks derived from allotropic transformations are avoided.

## 2.4   Normalizing

The heat treatment of normalizing is based on austenitizing a steel part and then, from these temperatures, cooling it in air. The austenitization temperature used for the normalizing should exceed the $A_3$ of the steel between 50–70 °C. This way, the total duration of the treatment is shortened (less time is required to fully austenitize the structure the greater the austenitization temperature). However, recommended temperatures should not be exceeded with the objective of avoiding overheating and burning risks.

If hypoeutectoid steel is not self-hardening, that is to say, if martensite is not produced by simple air cooling from the austenitic state, the resulting structure in the normalizing is usually of ferrite and lamellar pearlite, both in the periphery and the nucleus of the normalized part (Fig. 2.11). If the steel was self-hardening—because of being very alloyed (TTT curve is very far from the origin of times)—the air cooling from the gamma state would be a simple air hardening and, in this case, the utilization of the word "normalizing" would not be adequate to designate this heat treatment.

The air cooling from the gamma state is faster than the equilibrium cooling corresponding to the $Fe-Fe_3C$ diagram. The cooling that would take place, from the gamma state of the steel, very slowly, inside of the furnace with a closed door, as it is habitually performed in the case of the full annealing (a few °C/hour), is usually considered of equilibrium.

Therefore, if we compare the results of a normalizing with those of full annealing in the same hypoeutectoid steel with $C_1$%, the normalized steel will exhibit a greater proportion of pearlite, smaller proeutectoid ferrite grain diameter, smaller separation, $S_0$, between cementite lamellae, and a greater proportion of ferrite inside of the pearlite. To conclude, the yield strength, the ultimate tensile strength, the hardness,

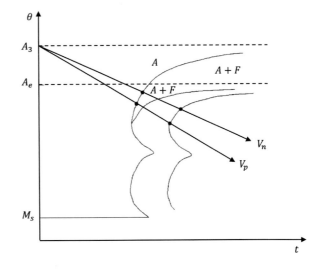

**Fig. 2.11** Normalizing by air cooling. $V_p$ is the cooling rate of the part periphery and $V_n$ is the cooling rate of the center of the part

and, almost always, the toughness of the normalized steel will be greater than that of the annealed steel.

Generally, in almost all the materials, when the hardness increases, the toughness decreases. However, in the case of normalized steel, the toughness will not be necessarily lower than that of the same steel in the annealed state and, therefore, softer. We should consider that the pearlitic cementite lamellae will be thinner after the normalizing than in the annealing. These thin lamellae can, therefore, be less brittle than the coarse lamellae as they admit a certain plastic deformation in service, and, if they break during the mechanical deformation, the generated microcracks can be retained inside of the ferrite that surrounds the broken cementite lamellae. Thus, depending on the dilution grade of the pearlite, also the toughness can be improved by a normalizing.

Habitually, the objective of the normalizing treatment is obtaining a ferritic-pearlitic structure "normal" of the steel, that is to say, a ferritic grain size equal to or smaller than the value 7 ASTM (see Table 2.1) and fine lamellar pearlite. This treatment has usually the finality of correcting a previous deficient structure—as the Widmanstätten—derived from an austenitic big size. It is usually employed to refine the grain of steels in as-cast state, in as-rolled state, or overheated during an austenitization. Definitely, it is possible to say that normalizing is a correcting treatment.

On other occasions, the normalizing is employed as utilization treatment as an alternative to quenching and tempering. In this case, the objective is simply achieving a lamellar pearlitic structure—diluted and fine—distributed into a proeutectoid ferrite matrix of very fine grains. The ultimate tensile strength, $\sigma_u$, of the steel with this structure and its toughness, as well, can be sufficient for many of the requirements in service, without requiring quenching and tempering the steel. Thus, a great part of the steels used in rail transport—for instance, to be used in wheels and boogies—are used in normalized state. In the case of the railways, a special hardening to the head is sometimes applied, equivalent to a normalizing, with the objective of refining the pearlite and increasing its wear resistance.

In any case, we should consider the size factor of the part to be treated. If the normalized part is big, the structure of the pearlite in the periphery of the part would be finer than in the center, because the peripheral cooling was faster. Therefore, the ultimate tensile strength—and the other mechanical properties—of a specimen removed from the center of this part of big dimensions will be lower than that of another specimen removed from a zone close to the surface.

## 2.5 Quenching

Quenching a steel part is achieving that its structure was fully martensitic after cooling. For this, it is necessary to austenitize the part and cold it in such a way that the cooling rate in the center of the part was greater than a certain critical value $V_c$ with the objective of achieving that the austenite could transform into martensite without previous transformations in the bainitic or pearlitic zones.

The critical rate of quenching depends on the chemical composition. Smithells (1984) points out that, for low alloy steels—$\%Ni < 3$, $\%Cr < 1.5$, $\%Mo < 0.3$–, the critical rate $V_c$ (in °C/s) to achieve a 99% martensitic structure (in parts of small diameter) can be calculated using the following equation (where C, Mn, Cr, Ni, Mo are the percentages of these elements in the steel composition):

$$\log V_s = 4.3 - 3.27 \cdot \%C - \frac{\%Mn + \%Cr + \%Ni + \%Mo}{1.6} \tag{2.31}$$

For parts of small diameter, v. gr. wires of 3 mm, the cooling curves of the periphery and nucleus of the part almost coincide and, therefore, a cooling that would allow obtaining martensite in the periphery of the part would also achieve that the structure of the nucleus was also martensitic. However, when the part size is big, it is also possible that, when cooling in a coolant with heat transfer factor $H_1$, martensite is obtained in the periphery of the part, but not inside (incomplete quenching).

The austenitization temperature before the quenching, which is called quenching temperature—we must not mistake with the $M_S$ temperature—cannot be excessively high with the purpose of avoiding overheating (which originates coarse martensite, few tough) and/or burning the steel. On the contrary, an incomplete austenitization of these steels will produce, when quenching, a mixture of ferrite (soft) and martensite (hard) and, therefore, lower resistance than in the case that the structure was fully martensitic. We consider as quenching temperature for hypoeutectoid steels:

$$\theta_\gamma = A_{3c} + (\text{between } 40\,°C \text{ and } 60\,°C) \tag{2.32}$$

On the contrary, hypereutectoid steels must be partially austenitized, 100% martensite is not obtained when quenching but a dispersion of proeutectoid cementite in a matrix of martensite (cementite is also a hard constituent). The reason for quenching the hypereutectoid steels in this way—the complete austenitization would require exceeding the $A_{cm}$ temperature—is to avoid the overheating and/or burning of the steel. Hypereutectoid steels are very prone to overheating due to the closeness of $A_{cm}$ to the solidus line of the Fe-Fe$_3$C diagram. We advise that, for instance, for a steel with 2% C, the complete austenitization would require reaching temperatures close to the melting point (a binary steel of 2.11% C would start to melt when 1148 °C is reached).

### 2.5.1   Hardenability

The hardenability of steel is the easiness to be quenched (Davenport and Bain, 1930). Thus, hardenability, which relates to the minimum quenching rate that will produce martensite and will permit the steel to escape from the transformation into softer constituents, controls the size of the section that may be uniformly heat-treated to initial high hardness and controls also the depth of hardening in larger sections of homogeneous chemical composition (Bain and Paxton 1961). It is not the same concept hardenability as the depth of the quenching. We call the depth of the quenching to the distance, from the periphery, where the part has a structure fully martensitic. The depth of the quenching is given by the intersection of the cooling curves of the different points of the part—function of the part size and the heat transfer factor of the part—and the TTT curve of the steel. The depth of the quenching can be increased, for instance, using a more severe coolant.

Hardenability, on the contrary, is something characteristic of the steel, and independent of the part size and the heat transfer factor of the used cooling. The hardenability is defined by the TTT curve. It is possible to say that a steel is more hardenable, the farer from the origin of times the TTT curve is.

Therefore, the hardenability exclusively depends on the chemical composition of the steel—TTT curves far from the origin of times the more alloyed (because the dissolved elements in the austenite delay the pearlitic and bainitic transformations)—and the austenitic grain size (greater the higher the quenching temperature).

Sometimes, to increase the hardenability of few hardenable steels, we try to increase the austenitic grain size, quenching from high temperatures; or increasing the holding time at the austenitization temperature $\theta_y$. However, this practice is not recommendable because the martensite formed in the quenching from an austenitic grain size very big is few tough even after a tempering.

It is not necessary to determine the TTT curve to know the hardenability of steel. There are other tests such as Jominy's or the calculation of the ideal critical diameter that determine the hardenability of the steel.

**Fig. 2.12** Center hardness after quenching the same steels in oil or water of bars of different diameters

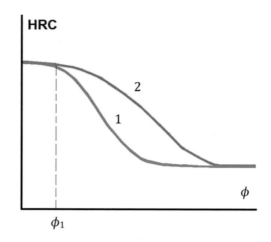

Exercise 2.4 Figure 2.12 indicates in y-axis the hardness obtained, after quenching, in the center of round bars of different diameter of the same steel S. Both curves were obtained by quenching the round bars in oil (ones) and in water (the others). The following questions are proposed:

(a)   Indicate whether it is or not correct the shape of the curves.
(b)   What curve corresponds to the quenching performed in oil?
(c)   How the curves would be in the case of a steel F with more hardenability than the steel S?

*Solution*

(a) and (b) Curves (1) and (2) are correct. The steel has sufficient hardenability to quench in both mediums, water ($H \simeq 1$, curve 2) and oil ($H \simeq 0.5$, curve 1). In both mediums, the steel quenches for small diameters ($\phi_1 = \phi_2$) where both curves HRC-$\phi$ are tangent. Probably for big diameters, $\phi$, the hardness after quenching 2 would be slightly greater than the hardness after the quenching 1 (where the ferritic-pearlitic structures would be coarser).

(c) If the TTT curves would correspond to a steel F of greater hardenability than the S, they will be farer from the origin of times. Curves 1 and 2 will shift toward the right and the value of the critical diameter, $\phi$, for the quenching in both mediums will increase.

*Solution ends.*

### 2.5.1.1   Jominy's Curves

We have indicated in the previous paragraph that the hardenability is given by the TTT curve, and the depth of the quenching by the interaction of this curve with those of the cooling. Therefore, if we compare two results of quenching a part of the same size using equal cooling heat transfer factor, the result of the quenching will exclusively depend on the steel. This is the basis of Jominy test.

The test consists in cooling by water stream at 25 °C the lower base of a cylindrical specimen of 25 mm in diameter and 100 mm in length, previously austenitized. The test—details are given by the UNE-EN ISO 642:2000 ERRATUM:2011—assumes first a heating up to a temperature greater than the $A_{3c}$ of the considered steel, in the suitable furnace, for 30 min, to even ensure that the nucleus of the part was austenitized. The specimen is later transferred to a cooling device (see scheme in Fig. 2.13) constructed considering the standard. The time elapsed from the moment when the specimen is removed from the furnace until the cooling should not exceed 5 s.

The quenching device has a fixing and centered element for the specimen, located in the vertical of the hole through which the water flows with the aim of achieving that the cooling could exclusively happen on the lower end (the device must be protected from air currents during the test). The specimen like this quenched has the cooling rates in all its points (therefore also those situate in the later surfaces, it does not matter if they are more or less far from the base) fixed and perfectly known. We indicate in Table 2.3 the cooling rates as a function of the distance from each point of the generatrix to the quenched end.

Once completely cooled the specimen, two generatrixes are drawn by grinding, located at 180° one from the other, until a depth of 0.4 mm (to eliminate the possible surface diametral decarburizing and to measure the hardness of the specimen from the quenched extreme). It is convenient to indicate that the grinding must be carried

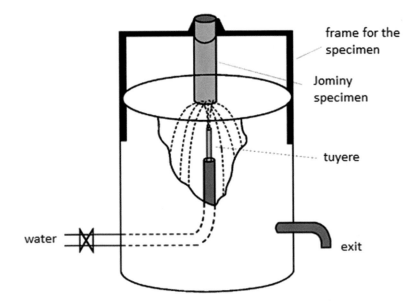

**Fig. 2.13** Device for the quenching of Jominy's specimen

**Table 2.3** Cooling rates in Jominy's test at different distances from the quenched end of the specimen

| Distance to the quenched extreme (in inches) | Cooling rate in °C/s at 704 °C | Distance to the quenched extreme (in inches) | Cooling rate in °C/s at 704 °C |
|---|---|---|---|
| 1/16 | 271.7 | 11/16 | 10.8 |
| 2/16 | 170.6 | 12/16 | 9.0 |
| 3/16 | 108.3 | 13/16 | 8.0 |
| 4/16 | 68.8 | 14/16 | 6.9 |
| 5/16 | 42.9 | 15/16 | 6.4 |
| 6/16 | 31.3 | 1 | 5.6 |
| 7/16 | 23.3 | 20/16 | 3.9 |
| 8/16 | 17.9 | 24/16 | 2.8 |
| 9/16 | 13.9 | 28/16 | 2.2 |
| 10/19 | 11.9 | 2 | 1.9 |

out in such a manner that any heating is avoided because this heating could involve tempering risks.

We compare two steels in Fig. 2.14: one with low hardenability—TTT curve close to the origin of times—and the other with greater hardenability. If we consider that the cooling rates are equal in both cases (same bar diameter), the results of the quenching exclusively depend on the TTT curve of the steel and, therefore, provide a

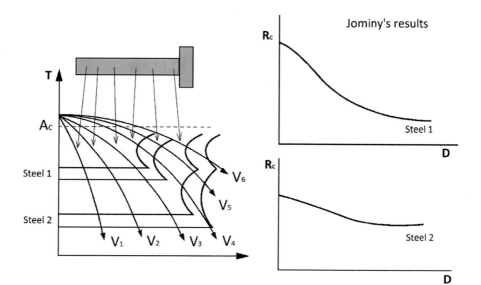

**Fig. 2.14** Comparison of results of Jominy's test in two steels

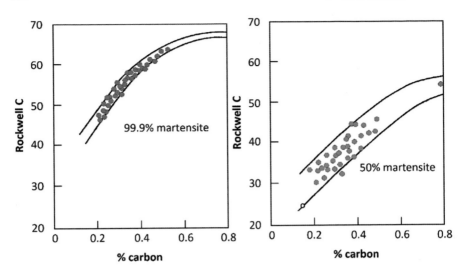

**Fig. 2.15** Hardness of structures 99.9% and 50% martensitic as a function of the carbon content

qualitative measurement of the hardenability: the flatter Jominy's curve of hardness of the steel is, the more hardenable the steel.

The $J_{99}$ measurement can be used to estimate—quantitatively—the hardenability of steels whose Jominy's curves were known. This is Jominy's distance to the quenched end of the specimen where the structure of each steel is 99% martensitic; or which is equal, the distance to the quenched end where the hardness corresponds to 99% martensitic microstructure in this steel. And the hardness of a structure 99% martensitic of each steel is well known as, it practically depends only on the carbon percentage (Fig. 2.15). In many practical cases, instead of the $J_{99}$, the $J_{50}$ measurement is accepted: distance from the quenched end where the hardness corresponds to 50% martensitic microstructure.

### 2.5.1.2    Real and Ideal Critical Diameters

The hardenability of a steel can be also determined by its critical real diameter when the same coolant is used. The comparison between real critical diameters of several quenched steels in this coolant medium allows comparing their hardenabilities.

Figure 2.16 illustrates the result of quenching, in the same coolant—of heat transfer factor $H_1$—several round bars of two steels:

(a)    a carbon steel,
(b)    a low alloy (Cr-Ni-Mo) steel,

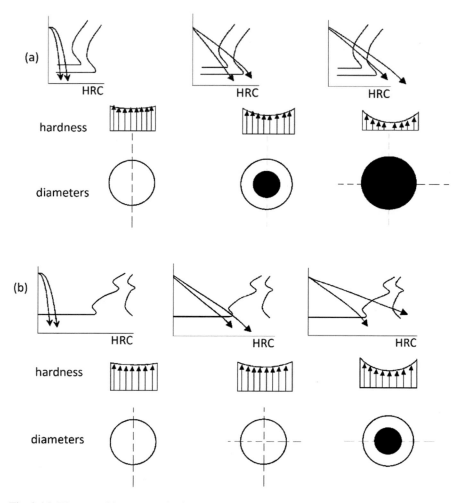

**Fig. 2.16** The quenching penetration/depth for the same heat transfer factor $H_1$ depends on the TTT curve: **a** carbon steel; **b** low alloy steel

whose TTT curves are indicated in Fig. 2.16. As the coolant is the same, the results of the quenching—and, for that reason, their differences regarding the hardenability—will exclusively depend on the TTT curves of one and the other steel if the round bar used in the quenching have the same diameter in both cases.

We should advise, in Fig. 2.16, that for a round bar of small diameter both the periphery and the nucleus of one and the other steel are quenched.

However, for a greater diameter, the first steel (a) has martensite in the periphery but not in the center. On the contrary, for the same diameter, the second steel (b) has martensite both in the periphery and the nucleus of the round bar. Therefore, although it is not same the hardenability as the depth of the quenching, it is possible to ensure that the second steel has greater hardenability than the first one.

When quenching a steel in a certain coolant—of heat transfer factor $H_1$—we call real critical diameter of the steel, $D_C$, for heat transfer factor $H_1$, to the bigger round bar of this steel that quenched in this coolant has in its center 99% martensite. The real critical diameter of each steel, for this coolant of heat transfer factor $H_1$, can be determined by measuring the depth of the quenching in the cross section of round bars of growing size, as is schematized in Fig. 2.17.

We advise that the real critical diameter of a certain steel will be greater the more severe the quenching coolant is. Thus, we indicate in Fig. 2.18, for this steel, the real critical diameter for several growing heat transfer factors. Experimentally, we also advise that the curve behaves asymptotically at a certain value for infinite heat transfer factor of the coolant.

We call ideal critical diameter of a steel, $D_{Ci}$, to the real critical diameter for infinite heat transfer factor, that is to say, to the biggest round bar of this steel,

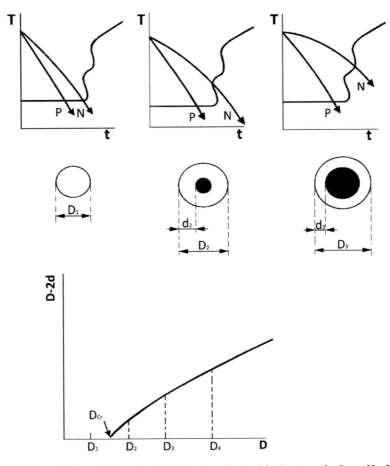

**Fig. 2.17** Determination of the real critical diameter of a steel for heat transfer factor $H_1$. D is the diameter of the round bar and d is the depth penetration of the quenching

**Fig. 2.18** Real critical diameter of a certain steel for several quenching heat transfer factors. $D_{cr} \simeq D_{ci}$ for $H \to \infty$

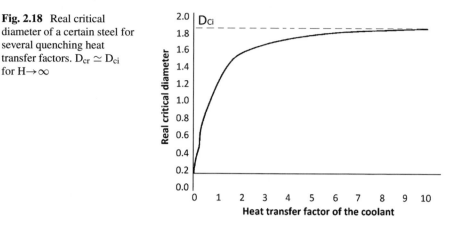

which quenched in a coolant of infinite heat transfer factor, has in its center 99% martensite. We call ideal critical diameter because implies that the coolant has an ideal capacity of heat absorption in such a way that the periphery of the round bar instantaneously acquires the temperature of the cooling agent. That is to say, the superficial heat transfer coefficient M is infinite (although the interior of the round bar will not instantaneously acquire the temperature of the quenching medium).

If we calculate the ideal critical diameters of several steels—the heat transfer factor of the quenching is fixed, infinite, in all the cases—the steel with the greatest ideal critical diameter will have the greatest hardenability. However, as the ideal critical diameter of the steel does not depend on the real coolant where the steel is quenched, $D_{Ci}$ is an objective measurement of the hardenability of each steel: the most far from the origin of times the TTT curve of the steel is, greater the ideal critical diameter will be. It is reasonable to admit that—without requiring to experimentally calculate it—we can calculate the ideal critical diameter of a steel if we know its TTT curve; or which is the same, it is determined as a function of both the chemical composition of the steel and the austenitic grain size.

Equally, considering that Jominy's test is also a hardenability test, it is possible to assume that there will be a certain correlation between the results of Jominy's test of a steel and its ideal critical diameter. Effectively, if we know the distance $d_j$ (99) of a steel from the experimental Jominy's test curve, it is possible to calculate the ideal critical diameter of this steel with the support of the Fig. 2.19 (also determined by experimental methods).

Knowing the ideal critical diameter of a steel is very useful to predict the results of the quenching of this steel in any coolant of heat transfer factor $H_1$ (remember Table 2.2). It is possible to calculate the real critical diameter of certain steel with the support of Grossmann's graphic (Fig. 2.20a, b, Grossmann and Bain 1972) as a function of the ideal critical diameter of a steel and the heat transfer factor of the coolant where it will be quenched.

**Fig. 2.19** Equivalences between the ideal critical diameter of a steel and its Jominy's distance with 99% martensite

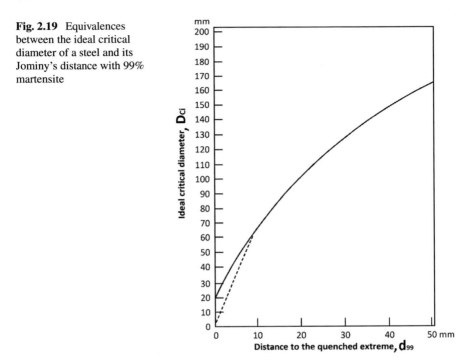

The first step for the selection in materials engineering of the most convenient steel to obtain the martensite structure at any point of a certain part, when cooling this part in a coolant $H_1$, consists in previously calculating its equivalent round bar. That is to say, finding the diameter of a cylinder—of great length—whose nucleus could cold at the same rate as the point of the slowest cooling rate of the part, having martensitic structure.

Once determined for $H_1$ the equivalent round bar of the part, we will select a steel whose real critical diameter for quenching with heat transfer factor $H_1$ was equal to that of the equivalent round bar. This way, it is possible to ensure that martensite will be obtained in the center (inner) of the round bar and, therefore, also at the point of the slowest cooling rate of the part (with greater reason, we will have a martensitic structure in any other point of the part). Definitely, the part of this steel, cooled in the coolant of heat transfer factor $H_1$, will be completely quenched.

Nevertheless, it is sufficient in many cases to choose a steel whose real critical diameter at 50% martensite for quenching in $H_1$ will coincide with the diameter of the equivalent round bar. That is to say, a steel that when quenched in this medium would have 50% martensite in the center of a cylinder of diameter $D_{Cr50}$.

**Exercise 2.5** How do you determine the real diameter $D_{cr}$ of certain steel for oil quenching if we know Jominy's curve? Indicate tables/data that would be necessary to know considering the abacus and diagrams included in the text.

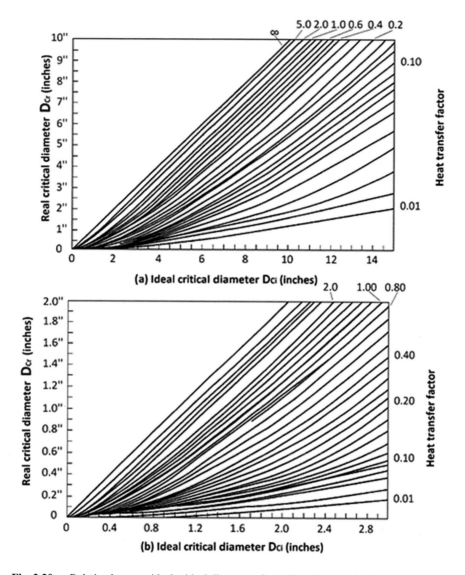

**Fig. 2.20 a** Relation between ideal critical diameter of a steel and their real critical diameters for several heat transfer factors, **b** detail of the previous one

**Necessary steps:**

(a) Knowing the chemical composition of the steel, assuming its carbon content practically expresses by itself the hardness of the martensite.

(b) Calculate the HRC as a function of the carbon content for hardness in the nucleus of 50 and 99% martensite, Fig. 2.15.

**Table 2.4**  Distances to the quenched extreme and results of Jominy test for two steels A and B

| $d_j$ (1/16 inch) | Steel A (HRC) | Steel B (HRC) |
|---|---|---|
| 2 | 50 | 50 |
| 4 | 32 | 48 |
| 6 | 24 | 47 |
| 8 | 19 | 46 |

(c)  Using the values HRC/$d_j$ in Jominy's curve and calculating the distances corresponding to structures 50% and 99% martensitic: $d_j$ (50) and $d_j$ (99). Obviously: $d_j$ (99) < $d_j$ (50).

(d)  Calculate $D_{Ci}$ (critical diameter) for $d_j$ (99), Fig. 2.19. There is an analogous abacus (it approximately coincides with the previous figure, to calculate $d_j$ (50). Obviously: $d_j$ (50) > $d_j$ (99).

(e)  Using the abacus in Fig. 2.19 to calculate $D_{cr}$ (in inches) as a function of the $D_{Ci}$ (in inches) for the oil quenching (H≃0.5). Obviously: $D_{cr}$ (99) < $D_{cr}$ (50).

(f)  The same procedure should be followed for any steel if we know its carbon content (chemical composition), hardness of the martensitic structure (50 or 99%), Jominy's curve of the steel, $D_{Ci,}$ and $D_{cr}$ for the chosen heat transfer factor, H.

**Application**:

The results of Jominy's test of two steels A and B are the following, as a function of the distances to the quenched end (Table 2.4).

(a)  If carbon content is similar in both steels, justify what is the most alloyed?

(b)  What of the two steels would have greater $D_{Ci}$? Assume a hardenability 50% martensitic.

(c)  For a diameter of 12 mm (≃8/16"), one of the two steels requires to be quenched in salt water with stirring (1.6 < H < 2), Table 2.2. What of the two steels? Why?

(d)  The other steel can be oil quenched (H≃0.5) if the rounds are smaller than 40 mm in diameter (≃1.6"). Justify the obtained result when quenching in salt water (brine quenched), with fast cooling, a part of this steel whose equivalent round bar was 40 mm in diameter, if the part has great differences of mass factor.

*Solution starts*

(a)  It will be more alloyed that steel where the HRC decreases more slowly with $d_j$, that is to say, the steel B has more hardenability.

(b)  For hardenability 50% martensitic in the nucleus:

$$(\%C)_A \simeq (\%C)_B \simeq 0.45\% \to HRC \simeq 45 \qquad (2.33)$$

Thus, for the steel A:

$$d_j < \frac{4''}{16} \tag{2.34}$$

for the steel B:

$$d_j > \frac{8''}{16} \tag{2.35}$$

(c)  The value of the critical diameter is, in one and other steel:

$$D_{Ci}(A) \simeq 50 \text{ mm } (2'') \tag{2.36}$$

$$D_{Ci}(B) \simeq 75 \text{ mm } (3'') \tag{2.37}$$

(d)  Steel A:

$$D_{Ci}(A) \simeq 2'' \text{ and } H \simeq \frac{1.6}{2} \rightarrow D_{cr} = 1.4'' > 0.5'' \text{ (12 mm)} \tag{2.38}$$

Steel B:

$$D_{Ci}(B) \simeq 3'' \text{ and } H \simeq 0.5 \rightarrow D_{cr} = 1.6'' \text{ } (\simeq 40 \text{ mm}) \tag{2.39}$$

For the salt quenching with fast cooling ($H\simeq1.6/2$, the $D_{cr}$ would be greater than 2"). It would not be recommendable because the difference in mass factor would produce internal cracks between the zones of different thicknesses with the subsequent risk of the appearance of deformations and cracks.

*Solution ends.*

## 2.5.2   Susceptibility to Cracking Due to Quenching

It is convenient to make here reference to the cracking risks of steel parts during the quenching cooling. These risks are particularly presented when the coolant has a great heat transfer factor and medium or high alloy steels are considered (whose $M_S$ temperature was smaller than 300 °C).

We should remember (Fig. 2.5a) that when the coolant has a great heat transfer factor during the cooling, for instance, stirred water, and the diameter of the equivalent round bar is big: there is a noticeable difference between the temperature of the periphery and that of the nucleus during the cooling. This difference grows as time progresses. Thus, assuming that the cooling could allow avoiding the pearlitic and bainitic curves—both in the periphery and the nucleus of the part—the transformation of the austenite into martensite is not simultaneous in the periphery and the nucleus. When martensite is formed in the periphery—and the volume increases in this zone—the nucleus is still austenitic.

In this case, stresses to which the nucleus of the part is subjected as a result of the outer expansion are absorbed as plastic deformation of the nucleus. However, when the austenite of the nucleus is transformed into martensite, the external zones—already martensitic—are at lower temperatures and have few plasticity. Therefore, when the nucleus is transformed into martensite, also with volume increase, the peripheric zones, martensitic and with few plasticity, act as a hoop around the nucleus. The expansion of the nucleus can even produce, by breaking off this peripheral hoop, the appearance of cracks (along the generatrix of the round if we are considering a cylindrical part).

The lower the $M_S$ temperature of the steel, the lower the plasticity of the peripheral martensite at the moment when the nucleus transforms into martensite, and a volume increase is produced. This low plasticity of the martensitic periphery will favor the possibility of the crack appearance at this moment. Thus, it is usually said that the more alloyed the steel is, and therefore with lower $M_S$ temperature, the more susceptible to cracking in coolants with more heat transfer factor the steel is than in the case of few alloyed steels.

So, highly alloyed steels have this risk when the quenching involves a great heat transfer factor. However, if a coolant with low heat transfer factor is used, for instance, air (self-hardening steels), when the great hardenability of the steel allows it, the difference in temperature between the periphery and the nucleus of the part during the cooling is small. In this case, the transformation of austenite into martensite is almost simultaneous in both the periphery and the nucleus. There is no hoop effect and neither the subsequent cracking risks. There are hardly any plastic deformations due to temperature differences between zones of different size factors of the part (non-deformable steels in the quenching). Although it is true that the quenching in coolants with low heat transfer factor, as the cooling is slower both in the periphery and the nucleus, involves a greater quantity of residual or retained austenite.

### 2.5.3 Surface Quenching

It consists in obtaining hard peripheric layers by means of a surface quenching of the steel. Only the periphery of the part is austenitized to achieve the surface quenching—by oxyacetylene burner or by heating through high-frequency induction currents—or even more recently, concentrated solar energy was used for this purpose (Fernández-González et al. 2018). There are three zones of temperature from the periphery to the center of the part: external zone ($\theta > A_3$), intermediate zone ($A_1 < \theta < A_3$), and central zone that is not affected by the heating ($\theta < A_1$). The time required for the austenitization depends on the depth that is desired (it is sufficient with around 2 min in the case of thicknesses from 1 to 2 mm). It is convenient that the quenched layer, and also the extension of the intermediate zone ($A_1 < \theta < A_3$), were small and, for that reason, a high difference in temperatures is required.

The outer cooling by aspersion, or immersion in water, or in oil, will transform into martensite in the outer zone but the structure of the nucleus of the part is not modified. The inalterability of this zone creates a state of compression stresses in the periphery, which is advantageous for the behavior against wear and fatigue and contributes to the fact that the periphery is harder than in the habitual quenching. A tempering at around 150–180 °C against residual stresses is given after the quenching (greater temperatures are not adequate because they will destroy the favorable state of outer compression stresses) (stress-relieving treatment, Pero-Sanz et al. 2018a).

Unalloyed steels (plain carbon steels) are used for the surface quenching—considering that a special hardenability for the steel is not required—which is advantageous not only due to its lower price but also because it gives less quantity of residual austenite after quenching. The surface quenching is not usually employed in steels with more than 0.6% C due to the risk of scaling the hard layer.

It is convenient that the starting structure—before making the surface quenching—was of lamellar pearlite. This allows a faster austenitization than when the structure of the cementite is globular (less relation surface/volume). Increasing the temperature would be required in the case of globular structures for a soon redissolution of the cementite, and this involves an increase in the austenitic grain and a reduction of the martensite toughness. Therefore, it is usually recommended, before performing the surface quenching, to make a previous normalizing treatment of the part.

## 2.6 Tempering

### 2.6.1 General Questions

In the surface quenching, and in some other cases—for instance when we want to achieve a hard-cutting edge in tools like knives, shears, etc.—the aim of the quenching is to harden the steel. However, another objective is habitually pursued: obtaining, by heating (tempering) the martensite produced in the quenching, a fine structure of cementite dispersed in a ferrite matrix. Thus, as it will be later mentioned, this simultaneously improves the resistance $\sigma_u$ of the steel and its toughness.

#### 2.6.1.1 Changes in the Martensite During the Heating

The parameters "a" and "c" of the martensite, as it is illustrated in Fig. 1.36, linearly vary with the carbon content which already advises about the metastable character of the martensite and its evolution with the temperature.

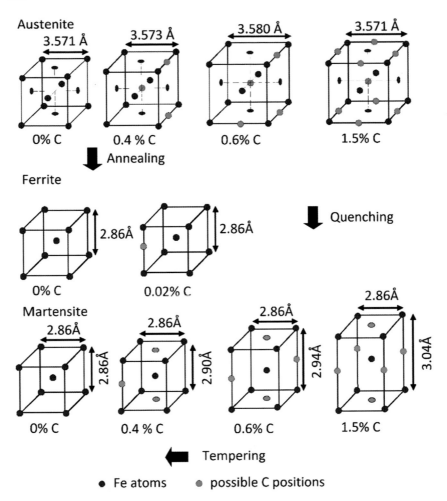

**Fig. 2.21** Scheme of quenching and tempering. Fe atoms (red), possible C interstitial positions (green)

In fact, if martensite, whose carbon content $C_1$%, is subjected to a prolonged increase in temperature, the diffusion of the carbon atoms located in the interatomic spaces is facilitated. Some of them left these positions and react with iron atoms to produce carbides. Proceeding like that, it is lower the $C_1$ percentage of carbon dissolved atoms—in interstitial solid solution—in the crystalline structure. Therefore, the tetragonality of the martensite will decrease as the inserted carbon diminishes and the lattice parameters are modified: the parameter "c" decreases and the parameter "a" slightly increases.

Thus, the carbon percentage in solid solution can be nearly zero by means of suitable heating of the martensite. This leads to a martensite structure completely cubic (ferrite); that is to say, of parameters "c" and "a" equal to 2.86 Å (Fig. 2.21). Therefore, if the heating temperature and the holding time are adequate, the martensite of $C_1\%$ of carbon fully transforms into an aggregate of carbides and ferrite-tempered martensite.

As it is obvious, the tempering temperature must be always lower than the eutectoid $A_e$. If this temperature is exceeded, the $\alpha \to \gamma$ allotropic transformation starts and, the austenitization of the steel would start. The following inconvenience is identified in this case: a structure of tempered martensite would not be only obtained when cooling from the tempering temperature, but a mixture of tempered martensite and the constituents that the austenite would give by transformation during the cooling.

The $A_e$ temperature is determined by the alloying elements that the steel has. If the character of the elements dissolved in the austenite is mainly gammagenous—v. gr. low and medium alloyed steels with Mn, Ni, Co—the $A_e$ temperature will be lower than 727 °C (which is the eutectoid temperature of a plain carbon binary steel).

On the contrary, if the chemical composition of the steel mainly consists of alphagenous elements—v. gr. Mo, W, V, Si (always when their contents were lower than those that could make even disappearing the $\delta \to \gamma$ transformation)— the $A_e$ temperature grows and, therefore, makes possible the tempering at greater temperatures.

Tempering produces a softening of the martensite. This is due to the loss of tetragonality of the iron crystal structure and the decrease in the number of dislocations, by recovery, and even recrystallization, of the ferritic structure that is produced by the tempering of the martensite (Leslie 1981, VII-13 p. 231).

The hardness of the biphasic aggregate that is finally obtained—of cementite dispersed in a ferrite matrix—is always smaller than the hardness of the quenching martensite, and this is even observed when the size of the precipitates if they are very fine, nanometric, could produce a real structural hardening of the ferrite.

The steel with a structure of tempered martensite—less hard and less resistant in tension than in the quenched state—is harder and more resistant than in the case where its structure was ferritic-pearlitic (i. e. the habitual of a normalizing treatment). Moreover, its toughness—determined by the CVN resilience (Charpy V-notch impact test), or by the critical stress intensity factor or fracture toughness, $K_{IC}$—also results greater than in the case of having ferritic-pearlitic structure. For example, a steel with 0.35% C in normalized state usually has at room temperature a value of $\sigma_u$ equal to 715 MPa and a resilience of 14.7 J/cm², while after quenching and tempering, values can reach 1250 MPa and 98.1 J/cm² for $\sigma_u$ and resilience, respectively.

The ensemble of heat treatments of full quenching and tempering of the structure improves the characteristics of the steel if compared with those of the normalized state. Quenching and tempering are interesting for many applications. This is the case, for instance, of the transport industry (shaft, steering shaft, crankshaft, gears, etc.): apart from an increase in the security due to the improvement of the steel toughness, the increase in $\sigma_u$ by quenching and tempering allows using smaller sections in parts and, therefore, reducing the weight (and cost). Quenched and tempered structures are also interesting in pressure vessels and, in general, in important-critical parts subjected to alternating stresses. The quenching and tempering of the steel, together with an increase in the toughness and mechanical resistance, improves its resistance to fatigue of a great number of cycles (high cycle fatigue). Cementite nanometric precipitates divert the trajectory of the crack and are favorable to, during the fatigue, increase the stage of the crack propagation.

### 2.6.1.2  Evolution of the Bainite and Pearlite

We understand by tempering of the bainite and pearlite the ensemble of alterations suffered by non-martensitic structures by heating that, as the martensite tempering, does not involve the $\alpha \rightarrow \gamma$ allotropic transformation, that is to say, heating until temperatures lower than $A_e$, or subcritical heating.

The lower bainite—formed by needles of ferrite with carbides inside—experiences during the heating a coalescence of the carbides and finally provides an analogous structure to that of the tempered martensite.

Although, as opposed to the tempered martensite, the tempered bainite is less uniform—it is started from a single constituent in the aging of the martensite—and, consequently, the toughness is lower.

Regarding the pearlite, the subcritical heating facilitates the breaking up of the cementite lamellae and the globulization of these fragments (tends to adopt a more stable morphology: less surface/volume). The structure softens with this procedure: the morphology corresponds to ferrite and dispersed carbides (of greater size than those of the tempered bainite). The degree of globulization depends on the temperature and the heating time. It is necessary to heat up to high temperatures (although always lower than $A_e$) to achieve the tempering of the pearlite. Therefore, the range of temperatures for this tempering of the pearlite is very small. In this process—in fact, a subcritical annealing—the main factor for the softening is the holding time.

The heating hardly has an influence on the proeutectoid products coming from the austenite decomposition. That is to say, if the steel structure is ferritic-pearlitic, the proeutectoid ferrite remains invariable and, if the steel was hypereutectoid—cementite-pearlite structures—the only effect of the heating would be the spheroidization (globular cementite would be formed).

This possibility of tempering the bainite and pearlite remarks other aspects: the complementary interest of the tempering in the case of incomplete quenching. The tempering softens the structural heterogeneities and the hardness differences in the cross section of the part (apart from transforming the peripheric martensite into ferrite and cementite and facilitates the carbides' coalescence into both bainite and pearlite in the most internal zones). Therefore, structures in different points of the cross section will be more similar and the hardness too.

### 2.6.1.3  Colors of the Tempering

A small film of oxide, whose thickness varies with the reached temperature, is formed when heating a steel in an oxidizing environment, on clean and polished surface. The

**Table 2.5** Colors according to the tempering temperature

| Color | Carbon steels and low alloy steels | Stainless steels |
|---|---|---|
| Pale yellow | 220 °C | – |
| Straw yellow | 230 °C | 280 °C |
| Gold yellow | 245 °C | 320 °C |
| Brown-yellow | 255 °C | 350 °C |
| Dark violet | 265 °C | 400 °C |
| Purple violet | 275 °C | 470 °C |
| Light blue | 290 °C | 510 °C |
| Brown-blue | 297 °C | 550 °C |
| Greenish blue | 330 °C | 640 °C |
| Grey black | 400 °C | 725 °C |

surface color of the steel, due to phenomena of interference between the reflected light by the metal and the light reflected by the oxide, also varies with the temperature.

These colors—when the thickness of the oxide layer is uniform—allow roughly knowing the temperature at which the steel was heated (Table 2.5).

The formation of oxide—and therefore the color—does not depend on the original structure either martensitic or ferritic-pearlitic and, in this way, the naming of "tempering colors" is not appropriate. Thus, when we observe the chips of a ferritic-pearlitic part removed by turning, it is possible to observe these tempering colors, which vary from a pale yellow for slow machining rate, to greenish-blue if the chipless rates are faster and the chip reaches greater oxidation temperatures.

Normally, the color of the oxide is equal to any steel except in the case of high chromium steels. For example, thinner oxide layers are produced in stainless steels (more resistant to oxidation than other steels) than in carbon steels or low alloy steels when they have been heated at the same temperature. Thus, the tempering colors of the stainless steels appear at higher temperatures.

## 2.6.2 Stages of the Martensite Tempering

The transformation by continuous heating of the martensite takes place by diffusion, nucleation, and growth to finally obtain a structure of precipitated cementite finely dispersed into ferrite. And, as it will be later more detailed, temperature and time are complemented between them: an increase in the temperature produces similar effects to longer holding at a lower temperature, as well.

### 2.6.2.1   Transformation of the Tetragonal Martensite

The first stage of the tempering—segregation of the carbon and precipitation of the carbide ε—starts at room temperature in steel with carbon content comprised between 0.3 and 1.5% C (Schrader and Rose 1966). Carbon atoms diffuse from their insertion positions toward the numerous dislocations that exist in the martensite and precipitate there (Fig. 2.22) in the form of hexagonal ε carbides ($Fe_{2.4}C$). The diffusion and precipitation are faster and more complex at greater temperatures (v. gr. 200 °C).

In this stage, martensite, losing carbon, finishes with less than 0.25% C in solid solution. The parameters "a" and "c" for this inserted carbon content are almost equal and, for that reason, this martensite is called "cubic martensite". Cubic martensite can be metallographically distinguished from the tetragonal martensite by means of etching with Nital that darkens the cubic martensite but does not alter the tetragonal martensite.

**Fig. 2.22** Evolution of the microstructure during the tempering, precipitation of the carbide ε. Relief replica. The bigger martensite needles have a granular structure of varying intensity, which is caused by non-uniform precipitation of ε- carbide. Between these martensite needles, small smooth untempered martensite needles remain in the areas which were originally residual austenite. It cannot be decided whether they have formed during the cooling after tempering or during the preparation of the microsection (5000x)

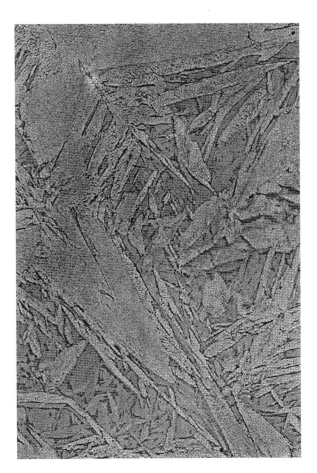

It is also convenient to advise that this cubic martensite—or ferrite supersaturated in carbon—has also a great number of crystalline dislocations, and it is, for that reason, very work-hardened and susceptible of being much more softened by recovery, or recrystallization, at higher temperatures.

### 2.6.2.2 Transformation of the Residual Austenite

If the steel has more than 0.4% C, there is still a certain proportion of residual austenite (also known as retained austenite) in the quenching, the greater the carbon content (see Fig. 1.38).

Heating above the $M_S$ favors the transformation of this residual austenite into bainite with volume and hardness increase. Effectively, if the steel has a high-carbon content, during the tempering of the martensite—v. gr. 1 h at 230–250 °C—it is possible to appreciate a peak or increase in the hardness due to that transformation. In low-carbon steels, this hardness increase is non-appreciable, the hardness of the martensite when tempered continuously falls with the temperature for equal times of tempering.

### 2.6.2.3 Brittleness During the Tempering

The progressive loss of martensite tetragonality during the stage of carbide ε precipitation is accompanied by an increase in the toughness.

We indicate in Fig. 2.23a the resilience after tempering of 1 h at different temperatures for a certain steel. We should also observe that the toughness of the tempered martensite at 350 °C is almost zero. This fall of the toughness, especially observed in the range of 250–350 °C, is usually called low tempering brittleness (Schrader and Rose 1966).

The low toughness is the consequence of a redissolution of the carbides ε that is accompanied by a reprecipitation of cementite with rod-like morphology (Fig. 2.23b). Due to the small size, of around nanometers, these precipitates are not appreciated when using the optical microscope, they are only visible, at great magnification, when using the electron microscope. The precipitation of these rod-like cementites is produced in the zones of the material that have a great number of dislocations, especially—as almost continuous constituent—at the grain boundaries of the former austenite (from which it was obtained, by quenching, the martensite). The rod-like cementite precipitates also in the interfaces between lamellae or needles of martensite, and even inside of the martensite (in the dislocations), with defined crystallographic orientations, analogous to that of Widmanstätten. The morphology of these elongated cylindrical precipitates—remember the notch effect—and the character of almost continuous framework are the reason for the low toughness—or brittleness—that is observed.

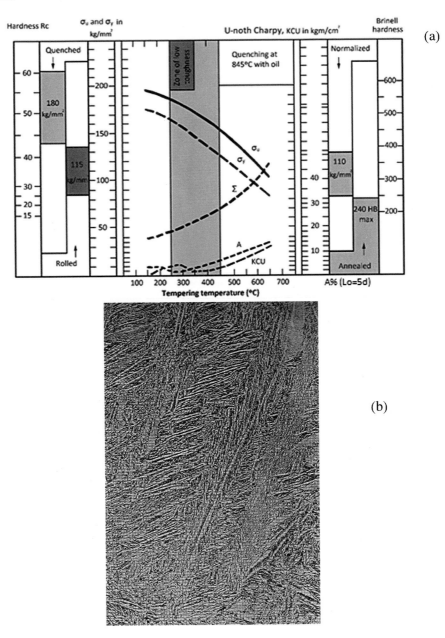

**Fig. 2.23** **a** Evolution of the mechanical characteristics during the tempering of a steel with 0.37–0.43%C, 0.70–1.00%Mn, 0.15–0.40%Si, 0.030%max. P, 0.035% max. S, 0.40–0.70%Ni, 0.15–0.25%Mo. **b** Fine rods of cementite (brittleness of low tempering). Relief replica. The precipitates can be seen in many of the former martensite needles, as fine rods. Sometimes coarser precipitates occur at the martensite needle boundaries (2000x)

This brittleness of the tempering at low temperature increases with the carbon content in the steel, which seems logical if we consider that, the greater the carbon content, the greater is the framework of cylindrical precipitates.

If the temperature of the embrittled steel is increased, for instance, heating later at 400 °C, the globulization of the cementite is favored as well as the discontinuity of the framework, and a noticeable improvement in the toughness is achieved. This suggests that, to avoid the brittleness of the low tempering, it is advantageous all that contributes to delay the temperature and time at which the precipitation of the rod-like cementite takes place. For instance, by making the tempering of the martensite at a temperature greater than 400 °C with fast heating up to this temperature. Thus, the ferrite produced by tempering of the martensite is recovered ferrite and, as it has less dislocations—fewer active centers for the carbide precipitation—the cementite adopts globular forms of greater size, although non-visible when using the optical microscope, instead of nanometric cylindrical bars that would form the continuous framework that leads to the low tempering brittleness.

It is also advantageous the presence of silicon in the steel. This element, because of being graphitizing and anti-carbide forming element, delays the precipitation of cementite to higher temperatures, and as a consequence, the loss of toughness typical of the low tempering is not produced. Due to other reasons, molybdenum and tungsten are also beneficial as they have greater affinity for the carbon than the iron and they delay the cementite precipitation in the critical stage.

On the contrary, phosphorus, antimony, arsenic, tin, and nitrogen are unfavorable for the brittleness of the low tempering. The main reason is probably that they segregate toward the austenite grain boundaries and, even when they do not precipitate, they weaken the adhesion of the cementite to the matrix and intergranular fractures might start there.

The weakening of the grain boundaries by local enrichment of atoms seems also responsible for other brittleness that, sometimes, appears in some steels tempered at high temperatures. It is called Krupp brittleness because it was observed for the first time during the war of 1914 in some barrel cannons manufactured by this German company. It is identified in several low/medium alloy steels (%Ni > 3.5, %Cr > 1.5, %Mn > 0.8) when they contain phosphorus (500 ppm), antimony (800 ppm), arsenic (500 ppm), tin (500 ppm) impurities. If the temperature of the martensite is increased, the alloying agents and the impurities diffuse toward the ancient austenitic grain boundaries. This enrichment, although it did not produce precipitates, leads to a weakening of the grain boundaries and produces an embrittlement that is observed when these steels have been tempered at 450–550 °C or if after the quenching a tempering at high temperatures was carried out, but the cooling between 600 and 350 °C was rather slow, or when tempering at higher temperatures—and it does not matter the cooling rate—the steel is in service in a range of temperatures close to 500 °C. The phenomenon is reversible: if the steel affected by this embrittlement is heated again at 600 °C and is rapidly cooled from this temperature, the brittleness disappears. Therefore, the atoms' diffusion has an influence although the precipitates do not appear. It is embrittlement produced by nucleation and growth (C curves with a vertical maximum of 450–550 °C).

The effects of the embrittlement are observed in the resilience test. Steels affected by the Krupp embrittlement break at low values of CVN, generally with intergranular fracture through the ancient austenite grain boundaries. This brittleness—that is not a consequence of the precipitate's formation—is not usually observed in the tensile test. Even when, intergranular corrosion is produced in NaOH environment, under stresses, through the austenitic grain boundaries enriched in atoms segregated in these zones (Stress Corrosion Cracking, SCC).

A remedy to avoid the Krupp embrittlement consists, as was already indicated, in the fast cooling after the tempering to avoid the vertical maximum of the C curve of brittleness. It is also possible to improve the behavior of the steels that are susceptible to this type of embrittlement (chromium steels, Cr-Ni steels, Cr-Mn steels, etc.) using as an alloying agent, also, molybdenum. This element delays the recovery of the ferrite (tempered martensite) and, therefore, there exist a great number of dislocations and vacancies inside of the tempered martensite. These crystalline defects, acting like scavengers for the atoms during their movement by diffusion, limit the arrival of impurities to the grain boundaries.

### 2.6.2.4   Recovery-Recrystallization of the Ferrite and Spheroidization of the Cementite

Above 400 °C and up to 600 °C, it is observed a progressive loss of hardness of the ferrite (tempered martensite, almost without carbon content) due to the mobility and disappearance of the linear crystalline defects or dislocations by means of a process similar to the recovery that takes place by heating of the work-hardened structures (climbing and elimination of dislocations of opposite sign).

Globulization of the cementite takes place while the ferrite recovers as the cementite is energetically more stable since the surface/volume relation diminishes. The formation of cementite spheroids does not matter its size—although they were still not visible using the optical microscope—is sufficient to break the continuity of the framework of nanometric rod-bars of cementite that could have precipitated at lower temperatures (a framework that is responsible for the brittleness at low-temperature tempering). Said in another manner, the brittleness of low-temperature tempering disappears above 400 °C.

At very high temperatures—it is necessary to reach temperatures of around 700 °C (although in any case lower than $A_e$)—true recrystallization of the ferrite takes place. Instead of boundaries between lamellae or needles of ferrite, inherited from the martensite, ferrite grain boundaries, more or less equiaxed, appear. This recrystallization is faster in low-carbon steels. It is observed in high-carbon steels that the numerous carbides inhibit the recrystallization process as they obstacle (pinning) the movement of the boundary nucleus/work-hardened matrix.

At these high temperatures, apart from the ferrite recrystallization, the process of coalescence and globulization of the cementite continues and, therefore, these particles can reach a size that allows their observation with the optical microscope.

The progressive reduction of the hardness that is experienced in this stage of the tempering at high temperature is the cooperative result of the recovery-recrystallization of the ferrite and the globulization of carbides. This reduction of hardness is additionally accompanied by an increase in the toughness of the tempered steel. In any case, the toughness, hardness, and $\sigma_u$ are greater than that of this steel with ferritic-pearlitic structure.

### 2.6.2.5   Secondary Hardening

Hardened steels destined to operate at temperatures of around 600 °C—v. gr. steels for dies of hot working, high-speed steels, etc.—usually have carbide-forming elements such as vanadium, molybdenum, tungsten, etc., in their chemical composition.

If before the quenching of the steel, it contains these elements dissolved in the austenite—in the form of solid solution with the austenite—they will be after the quenching in substitutional solid solution inside of the martensite, not only in the martensite of the quenching but also in the martensite that is still being tempered. As the martensite is transforming into ferrite, the elements still occupy, in the cell of the tempered martensite, substitutional positions.

When temperatures of around 600 °C are reached, these carbide-forming elements migrate by diffusion toward the dislocations of the ferrite and there, they react with the carbon and precipitate as nanometric particles to produce a structural hardening of the ferrite called secondary hardness.

Any fine dispersion of precipitates in a metallic matrix contributes to increase the hardness of any alloy. However, it is necessary for this that the size of the precipitates—and the distance between them—was so small that could make difficult the advance of the dislocations during the plastic deformation. The size of the precipitates, and its dispersion, must be around $10^{-9}$ m (nanometric), which is not appreciable when using the optical microscope. The precipitates that are observable with the optical microscope—such as inclusions or others—with sizes of around $10^{-6}$ m (1 μm), do not interfere with the dislocations and, for that reason, do not produce structural hardening.

The secondary hardening or structural hardening is only observed in certain alloys, both ferrous and non-ferrous (v. gr. duralumin), whose precipitation process from a solid solution involves a sequence of transition phases—or pre-precipitation states—that favor this final precipitation at nanometric size. This hardening is proportional to the quotient:

$$\frac{G \cdot b}{L_p} \tag{2.40}$$

where G is the shearing modulus of the matrix (iron in this case), b is the Burgers vector of the dislocation, and $L_p$ is the tangent distance between precipitates.

For certain chemical compositions—i. e. steels alloyed with vanadium, molybdenum, tungsten, etc. quenched—and after a decrease in the martensitic hardness by tempering in the first stages, the subsequent structural hardening produced by the carbides reach levels of hardness similar to those of the as-quenched martensite with greater toughness.

Not any carbide-forming element has this behavior. They must be markedly carbide-forming and thermodynamically more stable than the $Fe_3C$. We have in order of greater or smaller affinity by the carbon elements as vanadium, molybdenum, tungsten, and—with lower influence—chromium. However, despite the great affinity of these elements by the carbon, they will not form their carbides until the temperatures reach the range of 500–600 °C, as the precipitation of cementite is already produced from lower temperatures.

It seems logical this previous precipitation of the cementite. It is sufficient to consider that, for the formation of those carbides, it is necessary a migration toward the crystalline defects not only of the carbon atoms—as it would be the case of the $Fe_3C$—but the atoms of vanadium, molybdenum, tungsten, or chromium should also migrate toward these defects (they are in substitutional solution). Considering that these atoms replace those of the iron in the crystalline lattice, its diffusion is slower than the diffusion of the carbon atoms (they are in interstitial solid solution).

The ferrite matrix in the range of 500–600 °C—with carbide forming elements in substitutional solid solution—has practically no carbon as it has already migrated at lower temperatures to form cementite. Thus, the precipitation of carbides—v. gr. of vanadium, molybdenum, tungsten, etc.—requires a previous redissolution of the cementite by decomposition of their constituent atoms—iron and carbon—and later precipitation of the carbides by nucleation and growth (Ostwald ripening effect, Pero- Sanz et al. 2017, p. 222). The carbides of these elements that, ultrafine, produce secondary hardening in steels with 0.3–0.4% C are $V_4C_3$, $Mo_2C$, $W_2C$, $Cr_7C_3$. The temperatures at which the maximum secondary hardening takes place are 600–625 °C for the $V_4C_3$, 575 °C for the $Mo_2C$, and 500 °C for the $Cr_7C_3$.

Vanadium, even for contents as low as 0.1%, gives a noticeable peak or increase in hardening when nanometric precipitates of $V_4C_3$ are formed. These maintain its fine dispersion—it is very resistant to the coalescence—until temperatures of around 700 °C, which makes the vanadium being an alloying element interesting for hardened steels that should be used in service at high temperatures. Additionally, vanadium, due to its alphagenous character, increases the eutectoid temperature $A_e$ and, therefore, increases the limit of stability before the austenitization. Anyway, vanadium content in ultra-resistant hardened steels does not usually exceed the 0.5% V because the austenitization before the quenching would require very high temperatures to decompose and dissolve the vanadium carbides in the austenite. These high temperatures of austenitization involve risks of an austenitic grain size excessively coarse (overheating).

Molybdenum percentage in the steel to achieve sufficient structural hardening should be greater than 1% (other carbides—orthorhombic and of complex cubic lattice—different from the $Mo_2C$ (intermediate between $Mo_2C$ and $Mo_6C$) precipitate for lower contents). The optimal content of molybdenum should be between 2 and 2.5%. For greater contents, a prolonged permanence at high temperature, v. gr. 25 h at 700 °C, favors that the carbide $Mo_2C$ could evolve to give $Mo_6C$, which does not produce structural hardening. When the steel, apart from molybdenum (2–2.5% Mo), contains vanadium, this does not give vanadium carbides but dissolves in $Mo_2C$, stabilizes it, and impedes the over-aging.

Tungsten precipitates as $W_2C$—hexagonal isomorphic carbide with the $Mo_2C$—which also produces structural hardening. For the same atomic percentage as molybdenum, tungsten hardens but less than molybdenum because its precipitates have greater size. The upper limits of weight percentage, analogous to that of the molybdenum, have also as objective avoiding that the carbide $W_2C$ could evolve to give $W_6C$.

Chromium, despite being less carbide-forming than the above-mentioned elements—diffuses in the ferrite at a greater rate and, therefore, for the same time, precipitates as carbide at lower temperatures—500 °C—than the other considered elements. Additionally, it coalesces more than the other carbides and, because of being less fine, its precipitates harden less than the other carbide-forming elements.

The greatest descent of hardness in the tempering curve of the martensite with chromium in solid solution is produced at 500 °C and this alloying agent is less effective if the tempering temperature is 700 °C due to its easy coalescence. On another note, it is convenient to additionally take into account that when there are other elements that are more carbide-forming, for instance, 2% of molybdenum together with 5% of chromium precipitates more $Mo_2C$ with preference to the $Cr_7C_3$.

The chromium carbide that produces the structural hardening is the $Cr_7C_3$ ($K_2$ carbide, see Pero-Sanz et al. (2018a, pp. 168–177, b, pp. 272–275,). Its genesis logically depends on the chromium content of the steel. Thus, a peak of secondary hardness is not usually appreciated for %Cr < 7%. On the contrary, in martensitic stainless steels (%Cr > 12), there is this peak of structural hardening with precipitation of $Cr_7C_3$, although in ferrous alloys of high chromium content, $Cr_{23}C_6$ can also precipitate, and this phase grows at the expense of the $Cr_7C_3$, which can even disappear. Some elements as the tungsten favor the disappearance of $Cr_7C_3$, while others—v. gr. the vanadium—favor the stability of the $Cr_7C_3$.

It is possible to assess as a summary of this section that the improvement in the mechanical resistance of hardened steels can be achieved in different manners: increasing the carbon content (substitutional resistance to tempering), improving the intensity of the structural hardening, achieving more resistant precipitates to the coalescence, etc. However, increasing the carbon content of the steel to have a harder martensite resulted from the quenching and also tempered martensite with greater hardness and resistance (in comparison with that of other steel with lower carbon content), is not a suitable method: it involves drawbacks as the reduction of the toughness and weldability and an increase in the propension to the carbides coarsening. It is preferable to use lower carbon contents (substitutional alloys that improve hardenability and increase tempering resistance) and improve the intensity of the structural hardening, as well.

### 2.6.3   $T_1$ Temperature and Time t in the Tempering

#### 2.6.3.1   Hardness After 1 h of Tempering at Temperature $T_1$

We indicate in Fig. 2.24 the tempering curves of several steels for treatments of 1 h: the variation of hardness as a function of the tempering temperature. We have only considered steels of carbon content lower than 0.4% that, for that reason, do not have residual austenite in the quenching, but they produce 99% martensite in the quenching. Similar structural changes are observed in all these steels depending on the temperature range, for example, carbon segregation and carbide ε precipitation (until 200 °C); precipitation of cementite as rod-like of around 15 nm in diameter and less than 200 nm in length (between around 250 and 350 °C); recovery of the ferrite (between 400 and 600 °C); recrystallization of the ferrite (between close to 700 °C and $A_e$).

We observe that the hardness fall is faster in high-carbon steels (and, for that reason, it is difficult to control its hardness by tempering). This relative greater softening, with respect to its hardness in the quenching, in high-carbon steels, shows that the driving force for the tempering of a steel increases with the carbon content in the steel, or which is the same, with the difference of hardness between the martensite of quenching and that of tempering.

We should advise, additionally, that when we work with high tempering temperatures—v. gr. at 700 °C –the hardness differences are less relevant between the tempered martensite of different initial carbon percentages. This is explained because the particles of precipitated cementite—although more numerous—have greater size for greater carbon content in the steel, and consequently the distance between precipitates—which is the question that mainly determines the hardness increase—hardly changes.

**Fig. 2.24** Hardness of binary martensite of Fe and C tempered at different temperatures for 1 h

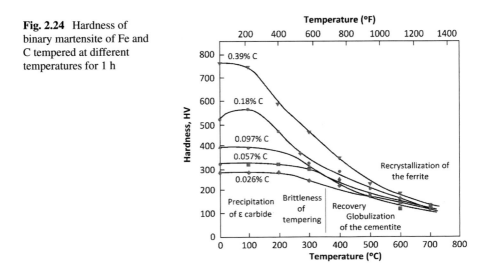

**Fig. 2.25** Tempering curves of two steels of the same carbon content, alloyed one of them (1) and not the other (2)

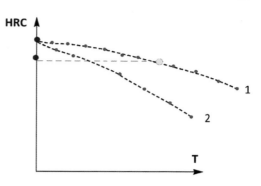

The alloying elements that were solubilized in the austenite before the quenching pass to be in solid solution in the martensite of quenching, but they do not significantly increase its hardness. On the contrary, they delay, some more than others, the softening of the martensite during the tempering. This is explained because these elements continue in solid solution in the ferrite (they are there since the tempering of the martensite), and this involves a greater hardness than that of a ferrite without elements in solution. For example, we compare in Fig. 2.25 the softening curves by tempering two martensites with identical carbon content, $C_1\%$, one alloyed (1) and not the other (2).

Nickel has little effect on the hardness of the tempered martensite in medium and low alloy steels, for quenching and tempering, and its effect is almost equal at any tempering temperature. With respect to the silicon, it is advised that its effect is greater at 315 °C than at other tempering temperatures due to its graphitizing character—anti-carbide-forming—inhibits the conversion of the ε carbide into cementite (which, as it was already indicated, is favorable to avoid the brittleness of low tempering). Regarding chromium, the increase in the hardness due to this element is usually maximum at 425 °C (by precipitation as carbides), and lower at greater temperatures due to the coalescence of carbides. The effects of molybdenum and vanadium are noticeable for high tempering temperatures, v. gr. 600 °C, because of secondary hardness, as was already mentioned.

### 2.6.3.2 Equivalence Temperature–Time in the Tempering

The successive transformations of the martensite during the tempering take place by diffusion and, therefore, temperature and time are complementary. Figure 2.26 illustrates, for a certain steel, the times required to achieve that its quenching hardness to decrease—by tempering at temperature $T_1$—until obtaining a hardness $H_1$ or until a hardness $H_2$ or until finally achieving the minimum hardness corresponding to 100% tempered martensite (cementite finely dispersed in a matrix of recrystallized ferrite).

Diffusivity enhances with the temperature and, therefore, using higher temperatures, the time required to temper the martensite until a certain temperature is lower.

**Fig. 2.26** Effect of
temperature T and time t, of
tempering, over the hardness
of tempered martensite for
the same steel. **a** Curves 1, 2,
3, and 4 correspond to the
combinations of T and t for
the respective hardness of
$H_1$, $H_2$, $H_3$, and $H_4$ ($H_1 > H_2$
$> H_3 > H_4$). **b** curves 1 and 3
(represented in coordinates
$Y = \ln(1/t)$ and $X = 1/T$)

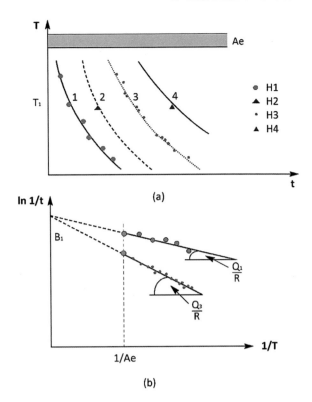

Tempering temperature should never exceed the $A_e$ temperature because austenitization would start. Tempering times between 30 and 90 min are habitually used in
industry.

We assume in Fig. 2.26a a pair of values of time (t) and temperature (T) that allow
obtaining tempered martensite of hardness equal to $H_1$ by heating this quenched
steel. If we consider as y-axis the values of $\ln 1/t$ and as x-axis the values of $1/T$, it is
possible to experimentally assess that this ensemble of values is over a straight line
of slope $-Q_1/R$ (Fig. 2.26b), where R is the Boltzmann constant.

For the same steel, the pairs of values of t and T experimentally leading to other
tempering hardness $H_3$ lower than $H_1$ would appear over another line of the same
origin ordinate, $B_1$, but of slope $(-Q_3/R)$. The value of Q depends only on the final
hardness chosen for the tempered martensite of this steel.

The correlation observed between temperatures and times of tempering to obtain in
this steel a hardness $H_1$ points out—as $\ln 1/t = -Q_1/R \cdot T + \ln B_1$—that the softening
rate of its martensite of quenching until reaching a hardness equal to $H_1$ follows
the Arrhenius law. The rate (inverse value of the time to achieve this softening) is
obtained as follows (see Verdeja et al. 2021):

$$\frac{1}{t} = B_1 \cdot \exp\left(-\frac{Q_1}{R \cdot T}\right) \qquad (2.41)$$

In the previous equation, R is the Boltzmann constant, as said, and $Q_1$ would be the difference between the addition of energies of the quenching martensite and the activation energy required to make possible its tempering minus the energy of the tempered martensite of hardness $H_1$.

That equality can be expressed as follows:

$$\log(B_1 \cdot t) = \frac{Q_1}{R \cdot T} \cdot \log e = \frac{L_1}{T} \qquad (2.42)$$

This value $L_1$, which receives the name of the parameter of Hollomon-Jaffe,—equal to $T \cdot \log(B_1 \cdot t)$—illustrates that the same hardness $H_1$ can be achieved with this steel by means of several combinations of temperature and time defined by the equation $T \cdot \log(B_1 \cdot t)$ or also by $T \cdot (c_1 + \log t)$ where $c_1$—equal to $\log B_1$—is a characteristic value of this steel.

It would be sufficient to determine the value of $B_1$—or that of $c_1$—experimentally finding two combinations of T and t that, for this steel, would give equal hardness of the tempered martensite. Both cases would answer to the same value of the activation energy Q and, for that reason, equal value of L; or which is the same, the following relations should be verified:

$$T_1 \cdot (c_1 + \log t_1) = T_2 \cdot (c_1 + \log t_2) \qquad (2.43)$$

and the following relation would be obtained:

$$T_1 \cdot c_1 + T_1 \cdot \log t_1 = T_2 \cdot c_1 + T_2 \cdot \log t_2 \rightarrow c_1 \cdot (T_1 - T_2)$$

$$= (T_1 \cdot \log t_1 - T_2 \cdot \log t_2) \rightarrow c_1 = \frac{(T_2 \cdot \log t_2 - T_1 \cdot \log t_1)}{T_2 - T_1}$$

$$(2.44)$$

It is possible to conclude from various experimental calculations that the value of the coefficient $c_1$, for time in hours, varies for steels of medium and high-carbon (0.25 < %C < 0.40) and tool steels (between 0.90 and 1.20%C) within the values $c \simeq 15$ and $c \simeq 20$.

What is before, we have assumed that the hardness of the martensite decreases always during the tempering. Thus, the values calculated for c cannot be simply applied to steels that have secondary hardness during the tempering.

Precisely the constancy of the value $c_1 \simeq 20$ allows to use, for any medium and low alloy hardened carbon steel (0.25 < %C < 0.40)—and always and when it has no secondary hardness—an interesting abacus that relates temperatures and times of tempering with parameters of Hollomon-Jaffe.

We have already mentioned the equation:

$$\log(B_1 \cdot t) = \frac{L_1}{T} \qquad (2.45)$$

and if we take logarithms in both sides of the equality, we obtain:

$$\log(\log(B_1 \cdot t)) = \log L_1 - \log T \tag{2.46}$$

that for different values of $L_1$, it can be represented by a bound of parallel straight lines ($Y = X - \log T$) of y-axis $\log(\log(B_1 \cdot t))$ and x-axis the logarithms of the parameters of Hollomon (we should remember that, for the same steel, each hardness of the tempered martensite, $H_1$, corresponds a different parameter of Hollomon, $L_1$).

We should advise that as $B_1$ is almost equal in all these steels (because $c_1$ is approximately equal to 20), the y-axis depends only on the time. Therefore, for a certain steel quenched and tempered until a certain hardness $H_1$ (that is to say for the same value of $L_1$), it is possible to determine with the support of the abacus in Fig. 2.27—where we have considered that $B_1$ is equal to $10^{20}$ when fixing in hours the y-axis—the pair of values of temperature and time that provide this same hardness $H_1$ of its tempered martensite. It would be sufficient with drawing through

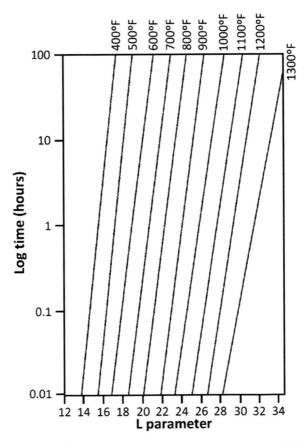

**Fig. 2.27** L parameter for different temperatures and times of tempering ($c \simeq 20$)

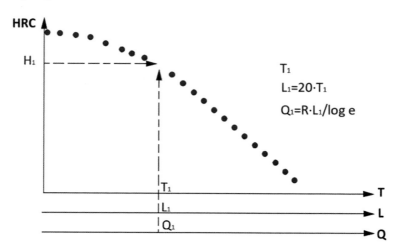

**Fig. 2.28** Hardness after tempering for one hour a certain previously quenched steel as a function of the tempering temperature T, the parameter L, and the energy Q

$L_1$ a perpendicular line to the x-axis and calculate the intersections with the bound of parallel lines.

Consequently, if we experimentally determine, in one of these steels, its curve of variation of hardness (HRC-$T_x$) for tempering of 1 h at different temperatures $T_x$, it is possible to know the effects that the tempering produces in this steel for any other combination of temperature and time. It is sufficient to consider that the curve of tempering, experimentally determined for each steel (Fig. 2.28), allows drawing the curve (HRC-L) that relates the real hardness of the tempered martensite and its corresponding parameter L. It is the same curve that was already determined (HRC-$T_x$), where we simply change the x-axis values: to each HRC hardness will correspond in the x-axis, instead of a value of $T_x$, a value of $L_x = T_x \cdot (20 + \log 1)$, that is to say, $L_x = 20 \cdot T_x$.

This experimental curve of tempering of the steel for 1 h, HRC-$T_x$, and therefore the curve HRC-L, also allows—if we change the values of the x-axis—knowing the energy $Q_1$ corresponding to each hardness $H_1$ of the tempering of this steel. In fact, if $T_1$ is the tempering temperature and $L_1$ is the parameter to obtain this hardness $H_1$, the energy $Q_1$ will be equal to $L_1 \cdot R/\log e$.

### 2.6.3.3 Potential Hardness and Softening Coefficient

For steels with carbon content in the range 0.31–1.15% C, unalloyed, or few alloyed (with maximum content in carbide forming elements of 5% Mo and 5% Cr), tempered (within 100 °C and 710 °C) for times between 6 s and 1000 h, Hollomon and Jaffe, (1945) concluded—as result of a wide experimentation—that the real hardness $H_r$ after tempering is determined by the following equation:

$$H_r = H_p - 0.00457 \cdot L \tag{2.47}$$

where L is the Hollomon parameter that is equal to $T \cdot (c + \log t)$, and $H_p$ is a value that chiefly depends on the chemical composition of the steel called "characteristic hardness" or potential hardness of the steel. Potential hardness has not a physical meaning.

Potential hardness of this steel can be experimentally determined by measuring the real hardness $H_1$ after tempering at temperature $T_1$ for a time $t_1$. The potential hardness is determined by the real hardness and the parameter $L_1$.

The potential hardness can be calculated, with sufficient approximation, simply as a function of the chemical composition of the steel.

Already known the potential hardness of the steel, it is possible to calculate the real hardness after quenching and tempering for any temperature T and time t of tempering (for any L) in medium and low alloy steels with 0.25–0.40%C (where $c \simeq 20$).

The reader may be surprised by the possibility of calculating the real hardness of a tempered steel by simply subtracting two magnitudes without apparent physical meaning: the potential hardness $H_p$, which depends only on the chemical composition of the steel, and a softening coefficient of this potential hardness, which only depends on the temperature and tempering time.

But in reality, $H_p$ and $H_r$, as also $0.00457 \cdot L$, have energy meaning. In fact, $H_p$—although expressed in values of Rockwell C—is an energy $Q_C$, which results by the addition of the energy stored in the steel by quenching (harder martensite the greater the carbon content in the steel) and another term, the activation energy required (that could be quantified, in hardness, by subtracting from $H_p$ the quenching hardness) to later transform this martensite into tempered martensite of $H_r$. The quenching martensite will reach a state energetically more stable by means of this tempering, of lower energy than that of the quenching, expressed also quantitatively by a hardness, its $H_r$ hardness. The difference between the energies expressed by $H_p$ and $H_r$, which in terms of hardness is $0.00457 \cdot L$, is the energy ceded by softening of the potential hardness. In fact, it is possible to appreciate that, by its equal $0.00457 \cdot Q_1/R \cdot \log e$ is an energy value.

Regarding these considerations, it seems convenient to underline the difference between $Q_c$ and $Q_1$. The same real hardness of the steel after being tempered corresponds to each value of $Q_1$. Therefore, $Q_1$ radically depends on the difference between the potential hardness of the steel $H_p$ (that includes the hardness of the quenching and the activation energy for the tempering of the martensite) and its real hardness $H_r$, and this energy $Q_1$ can be quantified in unities of Rockwell with base in both hardness.

Regarding L—which is equal to $T \cdot (c + \log T)$—its value can be immediately known, for each combination of temperature T and time t, with the support of the abacus in Fig. 2.27 (determined for $c \simeq 20$). However, it is easier and faster to transform this abacus into other. Therefore, if we take logarithms on both sides of the equation:

$$T \cdot (c + \log t) = L \qquad (2.48)$$

and using the graphical representation—using in y-axis $Y = \log T$ and in x-axis $X = \log (c + \log t)$—we will obtain a bound of parallel straight lines whose origin ordinates are the values of $\log L$.

Figure 2.29 corresponds to this representation with time expressed in hours. We have taken $c \simeq 20$ for this caption and, it is useful, therefore, for steels with carbon content in the range 0.25–0.40% without alloying elements or weakly alloyed low in carbide-forming elements. It is sufficient to determine the real hardness of the steel by subtracting from its potential hardness the dimensioned value as Rockwell hardness (for each T and t of tempering) in Fig. 2.29. Value that is called softening coefficient as the origin ordinate of each line of the bound—$\log L$—is already multiplied by 0.00457.

We should use other abacus corresponding to $c \simeq 15$ in the case of steels whose carbon content was comprised within 0.90 and 1.20% C and—as in the steels of the previous paragraph—they do not contain great percentages of carbide-forming elements.

In the case of steels with vanadium, molybdenum, chromium—that can give secondary hardness—for tempering temperatures in the range 390–600 °C (v. gr. $T_x$ °C), the potential hardness requires to previously calculating two values of the hardness, one at 390 °C (we assume that is equal to $H_p$) and other at 600 °C (we

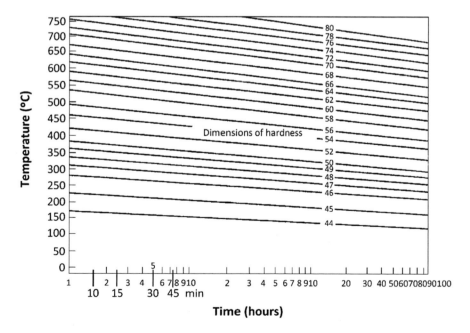

**Fig. 2.29** Values of the "softening coefficient" expressed in values of hardness HRC

assume that is equal to $H'_p$). We take as value of the hardness:

$$H_c = H_p + \frac{(T_x - 390) \cdot (H_p - H'_p)}{600 - 390} \tag{2.49}$$

Regarding equivalences between hardened steels, it is possible to mention that if two steels of different compositions, with a structure 99% martensite, have the same potential hardness, they will be equivalent in the tempering it does not matter the combination of T and t (as the tempering curves for 1 h will coincide).

**Exercise 2.6** The following values of HRC hardness are measured after tempering of 1 h carried out at 300, 400, 500, and 600 °C, respectively, in an unalloyed, medium carbon, steel, quenched in oil, see Table 2.6.

*Solution starts*

From the general equation:

$$HRC_{Q+T} = -a \cdot M + b \tag{2.50}$$

the parameter "a" transforms the softening coefficient into HRC hardness.
  Made this regression by plotting on the y-axis the values of hardness (quenching and tempering) and on the x-axis the values of M (softening coefficient), we have

$$HRC_{Q+T} = -0.0035 \cdot M + 80.2 \tag{2.51}$$

The potential hardness of the steel will be 80.2. For example, the hardness under conditions of hard tempering (200 °C) will be

$$HRC_{Q+T}(200°C) = -0.0035 \cdot (200 + 273) \cdot (15.3 + \log 3600) + 80.2 = 49 \tag{2.52}$$

Under conditions of tough tempering (550 °C),

$$HRC_{Q+T}(550°C) = -0.0035 \cdot (550 + 273) \cdot (15.3 + \log 3600) + 80.2 = 26 \tag{2.53}$$

| Table 2.6  Values of HRC hardness at different tempering temperatures | T (°C) | HRC (Q + T) hardened | M = T·(15.3 + logt), T in K and t in s |
|---|---|---|---|
| | 25 (quenching) | 56 | |
| | 300 | 42 | 10,807 |
| | 400 | 36 | 12,693 |
| | 500 | 30 | 14,579 |
| | 600 | 22 | 16,465 |

**Table 2.7** Rockwell C hardness obtained by tempering at indicated time and temperature

| Nominal Time of Tempering | 100 °C | 150 °C | 200 °C | 250 °C | 300 °C | 400 °C | 500 °C | 600 °C | 700 °C |
|---|---|---|---|---|---|---|---|---|---|
| 10 s | 47.5 | 48.0 | 49.0 | 49.0 | 46.5 | 45.5 | 35.5 | 26.5 | 21.5 |
| 90 s | 49.0 | 46.0 | 49.5 | 47.5 | 43.5 | 39.5 | 30.5 | 22.5 | 16.0 |
| 900 s | 49.0 | 47.5 | 48.0 | 46.0 | 42.5 | 35.5 | 25.5 | 18.5 | 13.5 |
| 9000 s | 47.5 | 47.0 | 46.5 | 44.0 | 39.5 | 30.5 | 19.5 | 16.0 | 7.0 |
| 86,400 s | 48.5 | 47.5 | 44.0 | 42.5 | 37.5 | 28.5 | 21.0 | 12.5 | 3.5 |

NOTE: This steel is used in the manufacture of axles, machinery elements, and transmission (gears) of vehicles of medium responsibility. In parts of small thickness (t < 10 mm), the quenching and tempering noticeably improve the properties of resistance and toughness.

End of the exercise.

**Exercise 2.7** Calculate the HRC and the Hollomon-Jaffe parameter curve, equation, and potential hardness for the steel 0.31% C-0.52% Mn-0.10% Si-0.026% S-0.007% P. Use the data in Table 2.7.

*Solution starts*

We calculate the Hollomon-Jaffe parameter, L, as follows:

$$T \cdot (15.9 + \log t) = L \tag{2.54}$$

Results are collected in Table 2.8.

We represent hardness versus time–temperature parameter for tempering fully quenched 0.31% carbon steel (Fig. 2.30).

We draw the trend line in the zone where HRC decreases with the Hollomon-Jaffe parameter (Fig. 2.31).

The potential hardness of the steel is 88.44.

NOTE: Indentation hardness is perhaps the most convenient quantitative measure of the tempered structure, or of what might be called the "degree of tempering". Undoubtedly, it is not an exact measurement, and in some cases (as when temper-brittleness occurs), pieces of the same hardened steel may have widely differing properties at the same hardness. However, hardness has the advantage of being simple to determine and is certainly the most widely used indication of the degree of tempering. Moreover, for fully quenched steels that are not temper-brittle and do not contain retained austenite when quenched, it appears that, in general, if the hardness of the steel is the same, the tensile and impact properties will also be identical, whether a high or a low tempering temperature is used.

*Solution ends.*

**Table 2.8**  Values of the Hollomon-Jaffe parameter

| Nominal time of tempering | 100 °C | 150 °C | 200 °C | 250 °C | 300 °C | 400 °C | 500 °C | 600 °C | 700 °C |
|---|---|---|---|---|---|---|---|---|---|
| 10 s | 6303.7 | 7148.7 | 7993.7 | 8838.7 | 9683.7 | 11,373.7 | 13,063.7 | 14,753.7 | 16,443.7 |
| 90 s | 6659.6 | 7552.3 | 8445.1 | 9337.8 | 10,230.5 | 12,015.9 | 13,801.3 | 15,586.8 | 17,372.2 |
| 900 s | 7032.6 | 7975.3 | 8918.1 | 9860.8 | 10,803.5 | 12,688.9 | 14,574.3 | 16,459.8 | 18,345.2 |
| 9000 s | 7405.6 | 8398.3 | 9391.1 | 10,383.8 | 11,376.5 | 13,361.9 | 15,347.3 | 17,332.8 | 19,318.2 |
| 86,400 s | 7772.0 | 8813.9 | 9855.7 | 10,897.7 | 11,939.3 | 14,022.9 | 16,106.6 | 18,190.3 | 20,273.9 |

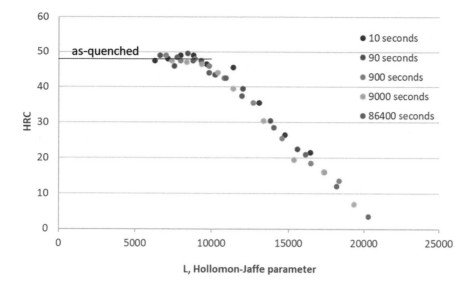

**Fig. 2.30** Hardness versus temperature parameter for tempering fully quenched 0.31% carbon steel

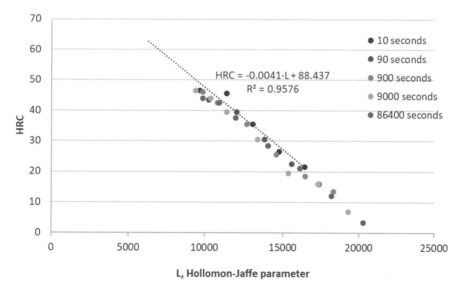

**Fig. 2.31** Hardness versus temperature parameter for tempering fully quenched 0.31% carbon steel (trend line)

**Exercise 2.8** It is usually said that "when two hardened steels of different chemical composition, have, on the contrary, equal their four "key" numbers, are equivalent and mutually replaceable it does not matter their temperatures and times of tempering". Reasoning about this sentence, if we know as key numbers the following ones: %C, $M_S$, ideal critical diameter ($D_{Ci}$), and potential hardness (Hp):

*Solution starts*

- If they have the same carbon content, their hardness (in carbon and low alloy steels) would be equivalent, the same in quenched state.
- If they have the same ideal critical diameter ($D_{Ci}$), their Jominy's distance ($d_j$ (99)), will be similar, Fig. 2.19. The same will happen with the real for any heat transfer factor of the quenching, H.
- If they have the same $M_S$ temperature, the temperature difference required from the austenitic state (T-$M_S$) will be the same, as well as the tendency to cracking (if $M_S$ < approximately 200 °C) and the proportion of retained austenite (Koistinen and Marburger, 1959).
- If they have the same potential hardness, Hp, for the same temperature and time of tempering, they will have the same hardness (mechanical resistance) as:

$$H_{Q+T} = H_p + M \qquad (2.55)$$

Figure 2.29, where M is the softening coefficient, equivalent to the Hollomon-Jaffe parameter

Therefore, both steels, regarding their quenching and tempering, are metallurgically and mechanically interchangeable.

The reader could ask about the reasons why steels of different carbon content are considered equivalent. In fact, it would seem adequate to use always the same steel (of sufficiently high-carbon content, i. e. 0.6% C) and after quenching, performing a tempering at a suitable temperature (always greater than 400 °C) to avoid the brittleness of the tempering with the aim of obtaining an adequate resistance $(\sigma_y)_1$. For instance, if a lower resistance, $(\sigma_y)_2$, was required, it would be sufficient to make the tempering of this steel at higher temperature: $(\sigma_y)_1 > (\sigma_y)_2$.

However, it is advantageous to choose from several steels (Pero-Sanz 2004, p. 349). The resistance $(\sigma_y)_2$, after tempering in the useful zone (at more than 400 °C), can be obtained with the steel of 0.6% C or with the steel of 0.4% C. In both cases, the mechanical resistance after quenching and tempering would be the same $(\sigma_y)_2$, but, on the contrary, the toughness would not be the same. The toughness of the 0.6% C steel would be smaller than that of the 0.4% C steel. In the second steel, the toughness would be greater because of having less quantity of carbides, considering the lower carbon content (Pero-Sanz 2004, p. 350). In the case of parts of size greater than 10 mm in unalloyed steels, it is necessary to use alloyed steels (low alloy) for this matter, that is to say, steels with greater or sufficient hardenability for resistance to fatigue.

End of the solution.

**Exercise 2.9** It seems that having a carbon steel, unalloyed, and with a carbon content sufficiently high—i. e., 0.6% C—it would be possible to obtain by quenching and tempering any mechanical resistance of those habitually required to manufacture parts resistant to the fatigue of low amplitude and great number of cycles, $800 < \sigma_y < 1200$ MPa, we ask:

(a) Why is it necessary to have a series of carbon steels, unalloyed, with different carbon percentages?

(b) Why are also required the alloyed steels? See Pero-Sanz et al. (2019, pp. 303–304).

*Solution starts*

(a) It is convenient to consider that the mechanical resistance, $\sigma_y$, is always associated to the toughness of the steel. Value that is necessary together with the first one to ensure the resistance to fatigue.

It is necessary to have a series of binary steels, for instance, of growing carbon contents, to achieve, when designing a part with a certain equivalent diameter that could ensure in the tough tempering, the same resistance avoiding the zones of tempering brittleness (Fig. 2.24). If %C grows, the hardness in quenching state increases, and the hardenability increases but inside of a narrow margin (see Eq. (2.31) of Smithells 1984), the cracking risk increases ($M_S$ decreases). The $D_{cr}$ (equivalent diameter) limits the maximum admissible carbon content.

(b) Returning to pages 303 and 304 in Pero-Sanz et al. (2019), we have four subgroups of low alloy steels as a function of the equivalent diameters 0/30, 30/60, 60/90, 90/120; we consider again key numbers mentioned in the previous exercise. Fundamentally, alloy elements increase the hardenability (for an interval %C 0.30–0.45%), Jominy's hardenability $d_j$ (50 or 99%) grows, noticeably increases $D_{Ci}$ and $D_{cr}$ quenched in oil, the $M_S$ does not noticeably decrease (avoiding the cracking tendency when using cooling mediums not necessarily severe (oil)) and increasing very particularly the softening resistance in the tempering (slightly increase Hp and significantly reduce M), which allow in tough tempering to obtain great resistances at temperatures between 550 °C and 650 °C (Figs. 2.24 and 2.25). Structures comprise recovered/recrystallized ferrite and sub-microscopic carbides: the mechanical resistance is given by the precipitation; the toughness, by the ferrite, although not exempt from the ductile–brittle transition.

*Solution ends.*

### 2.6.4 Multiple Tempering Treatments

Some alloyed steels, such as many of those used for tools and some of the self-hardening steels, require more than one tempering after the quenching.

These steels have low $M_S$ temperature and, therefore, they have a great proportion of highly alloyed residual austenite. This austenite is refractory to the transformation by heating—it would require very prolonged times—and, for that reason, it is not sufficient a single tempering for its transformation into bainite.

In these cases, the first tempering at a sufficiently high temperature (close to the 700 °C) destabilizes, or "conditions", this highly alloyed austenite. Carbides precipitate from the austenite at this temperature—it must be always lower than $A_e$, as was already mentioned. This austenite, already depleted in carbon and other alloying elements, is less refractory to the transformation (it has a $M_S$ temperature greater than that of the former austenite). This makes possible the transition into martensite by cooling from the temperature reached in this first tempering. As it is obvious, it is not indifferent to the cooling rate after the conditioning tempering as it would not be possible to obtain martensite but bainite. And a second tempering (double tempering) to temper later the martensite like this obtained is necessary.

Sometimes, it is not sufficient with a single tempering to "condition" the residual austenite after the quenching. Therefore, if once carried out the cooling that continues to the conditioning of the austenite, there is still certain percentage of austenite that is not transformed (because the $M_S$ temperature of the "conditioned" austenite is not sufficiently high) and a second tempering would be required to "condition" this austenite and make possible its transformation into martensite during the cooling. Logically, it is necessary for another tempering (triple tempering) to temper that martensite. Some tool steels require even four tempering treatments to make disappearing all the residual austenite.

It seems convenient to remember that the kinetics of the austenite transformation (into pearlite or bainite) is not the same for the cooling from the austenitic state and for the heating from the residual austenite (the C curves do not coincide). Therefore, the times required for the isothermal transformation of the residual austenite into bainite are not the same as those of the austempering. The factors that stabilize the austenite in the cooling, stabilize it also more to the tempering.

**Exercise 2.10** (Case Study: Hardening a wrought low alloy steel (SAE 4137)) A disc of 60 mm in diameter is taken from the axle of a loader that had experienced severe wear in the zone of contact with the bearings. We also include a specimen for dilatometry and a Jominy's specimen, taken both in the lengthwise direction of the axle by machining, as well as other little samples for heat treatment. The chemical composition of the steel, expressed in weight percentage, is collected in Table 2.9.

The axle is in quenched and tempered state that is expected to be studied.

The critical rate of quenching is calculated using the Eq. (2.31) for the composition indicated in Table 2.9.

$$
\begin{aligned}
\log V_S &= 4.3 - 3.27 \cdot \%C - \frac{\%Mn + \%Cr + \%Ni + \%Mo}{1.6} \\
&= 4.3 - 3.27 \cdot 0.36 - \frac{0.74 + 1.04 + 0.16}{1.6} = 1.91 \rightarrow V_S \\
&= 81.3\,°C/s
\end{aligned}
\tag{2.55}
$$

**Table 2.9** Chemical composition of the steel (SAE 4137)

| Element | wt. % |
|---|---|
| Carbon | 0.36 |
| Silicon | 0.30 |
| Manganese | 0.74 |
| Phosphorus | 0.025 |
| Sulphur | 0.017 |
| Chromium | 1.04 |
| Molybdenum | 0.16 |
| Aluminum | 0.063 |

And the critical diameter, $D_{cr}$, is

$$D_{cr} = \left( \frac{10000}{81.3} \right)^{1/1.8} \simeq 14.5 \text{ mm} \tag{2.56}$$

Experimental:

1. Determine the microstructure in the periphery and nucleus of the part. Calculate the curve of hardness from the periphery to the nucleus of the part.
2. Draw the curve of absolute dilatometry. Obtain the critical points of the steel.
3. Make Jominy's test of the steel and plot the corresponding curve.
4. Metallography of the steel in normalized state, full annealing, and globulizing annealing. Calculate the corresponding values of hardness.
5. Hardness of the steel in quenched state. Change of the hardness as a function of the tempering treatments performed for 1 h at 200, 300, 400, 500, and 600 °C, respectively.

**Questions**:

1. Justify the influence of the alloy elements (Cr, Ni, Mo) in the critical points of the steel. As a function of the data obtained in the dilatometry test, select the suitable temperatures for forging, quenching, annealing, and normalizing. Give the approximate value of $M_S$.
2. Ideal critical diameter of the steel for structures 50% and 99% martensitic, calculated from Jominy's test.
3. Evaluate, from both the chemical composition and the austenitic grain size of the steel, the bainitic and pearlitic hardenabilities at 50 and 99%, respectively. Calculate the critical diameters and compare them with those obtained in the question 2,
4. Calculate the potential hardness and the softening coefficients for tempering treatments performed for 1 h at 200, 300, 400, 500, and 600 °C, respectively. Compare the obtained hardness with experimental results. What treatment has experienced the steel of this case study?

5. This steel, is prone to decarburizing during the austenitization treatment before the quenching? What $CO/CO_2$ protecting atmosphere would you use?

6. The axle has broken in service by bending fatigue. The construction plane (drawings) indicates, for this part, an average hardness of 30 HRC. Do you think that the axle was subjected to the adequate heat treatment? How would you improve the fatigue resistance of this steel?

7. In what state do you supply the steel to facilitate the machining of the axle?

8. Determine the specification, according to the UNE standard (Una Norma Española, Spanish Standard), of this steel, as well as the corresponding specifications according to the AISI, AFNOR, and DIN standards. Properties and applications of the steel.

*Solution starts*

Experimental:

1. We determine first the microstructure in the periphery and nucleus of the part. After, we calculate the curve of hardness from the periphery to the nucleus of the part.

The observed microstructure, as it is shown in Figs. 2.32 and 2.33, micrographs at 100x and 600x, respectively, in the periphery of the axle, correspond to a quenched and tempered structure with tough tempering, as it is deduced from the intense degree of decomposition that the bainite exhibits. The microstructure in the nucleus is similar to that observed in the periphery, which is corroborated by the hardness values measured from the nucleus to the periphery of the axle (in two perpendicular directions).

The values of hardness measured from the periphery to the nucleus are collected in Table 2.10.

We plot in Fig. 2.34 the values of the hardness as a function of the distance.

2. Now we draw the curve of absolute dilatometry. We obtain the critical points of the steel.

**Fig. 2.32** Periphery of the axle. Quenched and tempered microstructure. HRC≃18 (100x)

**Fig. 2.33** Periphery of the axle. Quenched and tempered microstructure (600x)

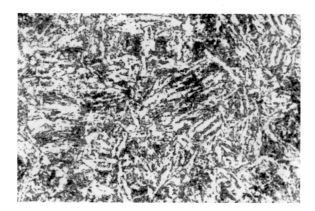

**Table 2.10** Hardness values measured from the periphery to the nucleus

|           | Distance (mm) | HRC  | HB  |
|-----------|---------------|------|-----|
| Periphery | 27            | 19   | 220 |
|           | 24            | 20   | 223 |
|           | 21            | 18   | 217 |
|           | 18            | 18   | 217 |
|           | 15            | 18.5 | 219 |
|           | 12            | 18   | 217 |
|           | 9             | 18   | 217 |
|           | 6             | 17   | 212 |
|           | 3             | 17.5 | 214 |
| Nucleus   | 0             | 18.5 | 219 |
|           | 3             | 17   | 212 |
|           | 6             | 17.5 | 214 |
|           | 9             | 17   | 212 |
| Periphery | 12            | 17.5 | 214 |
|           | 15            | 17   | 212 |
|           | 18            | 18   | 217 |
|           | 21            | 18   | 217 |
|           | 24            | 16.5 | 209 |
|           | 27            | 17   | 217 |

The length of the specimen is 50.0 mm.

Table 2.11 collects the values required to plot the dilatometric curve.

Now we plot the dilatometric curve obtained from the values of the test. Critical points are also indicated (Fig. 2.35). Critical points obtained for our steel using the equations of Andrews are

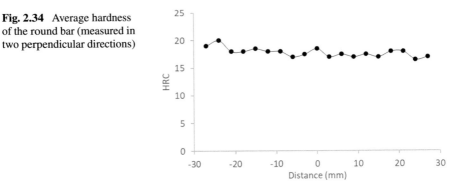

**Fig. 2.34** Average hardness of the round bar (measured in two perpendicular directions)

$$A_3 = 912 - 203 \cdot (\%C)^{0.5} - 30 \cdot (\%Mn) - 15.2 \cdot (\%Ni) - 11 \cdot (\%Cr)$$
$$- 20 \cdot (\%Cu) + 44.7 \cdot (\%Si) + 31.5 \cdot (\%Mo) + 13.1 \cdot (\%W)$$
$$+ 104 \cdot (\%V) + 120 \cdot (\%As) + 400 \cdot (\%Ti) + 400 \cdot (\%Al)$$
$$+ 700 \cdot (\%P) = 818\,°C \tag{2.58}$$

$$A_1 = 727 - 10.7 \cdot (\%Mn) - 16.9 \cdot (\%Ni) + 29.1 \cdot (\%Si) + 16.9 \cdot (\%Cr)$$
$$+ 290 \cdot (\%As) + 6.38 \cdot (\%W) = 745\,°C \tag{2.59}$$

As it is possible to check, there is a good approximation between the theoretical and experimental values.

We are going to calculate the average coefficient of expansion of the specimen in the range of 600–700 °C. The coefficient of linear expansion is given by the following equation:

$$\alpha = \frac{1}{L_1} \cdot \left[ \frac{(L_2 - L_1)}{(T_2 - T_1)} \right] \tag{2.60}$$

where $L_1$ represents the length of the specimen at the temperature $T_1$ and $L_2$ is the length at the temperature $T_2$. In our case:

$$T_1 = 600\,°C \rightarrow L_1 = 50.0 + 0.4125 = 50.4125 \text{ mm} \tag{2.61}$$

$$T_2 = 700\,°C \rightarrow L_2 = 50.0 + 0.4875 = 50.4875 \text{ mm} \tag{2.62}$$

since:

$$L_1 = L_0 + \Delta L \tag{2.63}$$

**Table 2.11**  Values to plot the dilatometric curve

| T (°C) | FEM (mV) | Dilation ($\cdot 10^{-2}$ mm) | |
|---|---|---|---|
| | | Heating | Cooling |
| 22 | 0 | 0 | −2 |
| 100 | 4.10 | 6.0 | |
| 160 | 6.53 | 10.0 | |
| 200 | 8.13 | 12.5 | |
| 240 | 9.75 | 15.5 | |
| 280 | 11.39 | 18.25 | 15 |
| 300 | 12.21 | 19.75 | 16.25 |
| 320 | 13.04 | | 17.5 |
| 340 | 13.88 | 22.5 | 18.75 |
| 360 | 14.71 | | 20 |
| 380 | 15.55 | 25.25 | 21.25 |
| 400 | 16.40 | 27 | 22.75 |
| 420 | 17.24 | | 24 |
| 440 | 18.09 | 30 | |
| 460 | 18.94 | | 26.5 |
| 480 | 19.79 | 32.75 | 27.75 |
| 500 | 20.65 | 34 | 29 |
| 520 | 21.50 | | 30 |
| 540 | 22.35 | 37 | 31.25 |
| 560 | 23.20 | | 32.75 |
| 580 | 24.06 | 40 | 34 |
| 600 | 24.91 | 41.25 | 35.75 |
| 620 | 25.76 | 43 | 37.25 |
| 640 | 26.61 | 44.25 | 39 |
| 660 | 27.47 | 46 | 36.25 |
| 680 | 28.29 | 47.25 | 34.75 |
| 700 | 29.14 | 48.75 | 35 |
| 710 | 29.56 | 49.25 | 35.25 |
| 720 | 29.97 | 50 | 35.25 |
| 730 | 30.39 | 49.25 | 35.25 |
| 740 | 30.81 | 46.50 | 36 |
| 750 | 31.23 | 45 | 37 |
| 760 | 31.65 | 44 | 38 |
| 770 | 32.06 | 43.50 | 39 |
| 780 | 32.48 | 43.25 | 40 |

(continued)

**Table 2.11** (continued)

| T (°C) | FEM (mV) | Dilation ($\cdot 10^{-2}$ mm) | |
|--------|----------|-------------------------------|---------|
|        |          | Heating | Cooling |
| 790 | 32.89 | 43.25 | 41.25 |
| 800 | 33.30 | 43.25 | 42.25 |
| 810 | 33.71 | 44 | 43.25 |
| 820 | 34.12 | 45 | 44.5 |
| 830 | 34.53 | 46 | 45.75 |
| 840 | 34.93 | 47 | 46.75 |

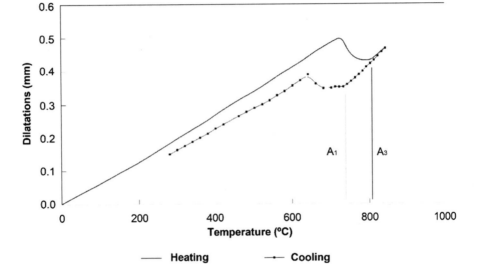

**Fig. 2.35**  Dilatometric curve

where $L_0$ is the initial length of the specimen and $\Delta L$ are the dilations obtained in the test. If we replace in Eq. (2.60), we obtain

$$\alpha = \frac{1}{50.4125} \cdot \left[ \frac{(50.4875 - 50.4125)}{(700 - 600)} \right] = 1.49 \cdot 10^{-5}\,{}^{\circ}\mathrm{C}^{-1} \qquad (2.64)$$

which coincides rather well with the linear expansion coefficient of the $\alpha$-ferrite.

3.   We do Jominy's test of the steel and we represent the corresponding curve.

The values obtained in Jominy's test are collected in Table 2.12.

Figure 2.36 shows the graphical representation of the data collected in Table 2.12. The austenitization temperature was 860 °C, with 30 min of permanence in

**Table 2.12** Values of Jominy's test

| Distance (mm) | HRC |
|---|---|
| 1.5 | 53 |
| 3.5 | 52 |
| 5.5 | 51.5 |
| 7.5 | 51 |
| 9.5 | 50 |
| 11.5 | 49 |
| 13.5 | 46.5 |
| 18.5 | 40.5 |
| 23.5 | 37 |
| 28.5 | 35 |
| 33.5 | 33.5 |
| 38.5 | 33 |
| 43.5 | 33 |
| 48.5 | 32.5 |
| 53.5 | 32 |
| 58.5 | 32 |
| 80.0 | 32 |

the furnace and cooling with water stream for 12 min, where the temperature of the water stream was 22 °C.

4. Now we study the metallography of the steel in normalized state, full (regenerative) annealing, and globular annealing. We also determine the corresponding hardness values (Table 2.13).

NOTE: The ASTM E-140 was used for the conversion of hardness values. These results remind us of the great importance of the heat treatments in the mechanical properties of the steels, consequence of the allotropic and solubility changes of the carbon in solid state

The structures exhibited by the steel after the heat treatments were.

Normalizing: As it is shown in Figs. 2.37 and 2.38, micrographs taken at 100 × and 600 × respectively, the structure exhibited by this axle with air cooling is of bainitic type, which indicates the rather good hardenability of this steel. It is possible to see in Fig. 2.38 some martensitic zones with light grey color.

Full (regenerative) annealing: As it is possible to see in Figs. 2.39 and 2.40, micrographs at 100 × and 600 × respectively, the structure of this steel with slow cooling (in furnace) is ferritic-pearlitic.

Globular annealing: This treatment was carried out at a temperature of 700 °C for 24 h. It is possible to see in Figs. 2.41 and 2.42 micrographs at 100 × and 600 ×

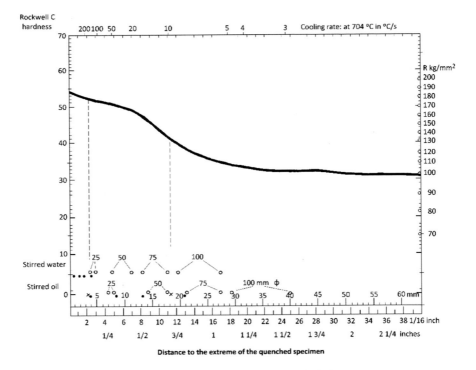

Fig. 2.36  Graphical representation of the data collected in Table 2.12

**Table 2.13** Hardness values of the steel in different conditions

|                               | HRC | HB  |
|-------------------------------|-----|-----|
| Normalizing                   | 33  | 311 |
| Full (regenerative) annealing | 13  | 198 |
| Globular annealing            | 9   | 183 |

**Fig. 2.37** Normalizing. Bainitic structure. 100x

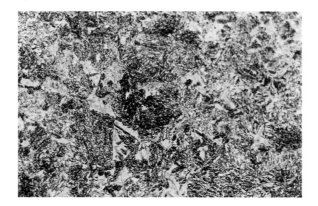

**Fig. 2.38**  Normalizing. Martensitic areas (light grey). 600x

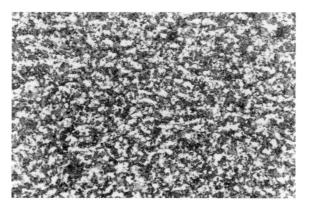

**Fig. 2.39**  Full annealing. Ferritic-pearlitic structure. 100x

**Fig. 2.40**  Full annealing. Ferritic-pearlitic microstructure. HRC≃13. 600x

**Fig. 2.41**  Globular annealing. Globular carbides in a ferritic matrix. 100x

**Fig. 2.42**  Globular annealing. Globular carbides in a ferritic matrix. HRC≃9. 600x

respectively, the globulization that was produced as a consequence of this treatment, being the final structure formed by globular carbides into the ferritic matrix.

5. We obtain the hardness of the steel in quenching state (Fig. 2.15). Change of the hardness as a function of the tempering treatments performed for 1 h at 200, 300, 400, 500, and 600 °C, respectively. Results are collected in Table 2.14 and represented in Fig. 2.43.

**Questions**:

1. We study in this question the influence of the alloy elements (Cr, Ni, Mo) on the critical points of the steel. We select the temperatures for forging, quenching, annealing, and normalizing as a function of the data obtained by dilatometry test. Furthermore, we obtain the approximate value of $M_S$

**Table 2.14** Hardness for different states

|  | HRC | HB (5/E-140 ASTM) |
|---|---|---|
| As-quenched state (25 °C) | 55 | 560 |
| As-tempered at 200 °C | 50 | 482 |
| As-tempered at 300 °C | 49 | 468 |
| As-tempered at 400 °C | 45 | 421 |
| As-tempered at 500 °C | 36 | 336 |
| As-tempered at 600 °C | 31 | 294 |

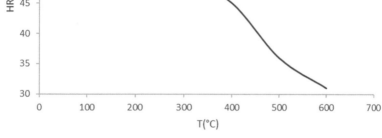

**Fig. 2.43** HRC as a function of the heat treatment temperature

Chromium is an alphagenous element with particular characteristics since for high chromium contents, this element behaves as alphagenous (%Cr > 7) while for low chromium contents, it behaves as gammagenous element (for a carbon content lower than 0.4%). The steel considered in this case study has low chromium content (1.09%), and, therefore, this behaves as gammagenous element, that is to say, it tends to reduce $A_3$ and $A_1$.

Nickel is a gammagenous element, it increases the field of stability of the austenite, and, therefore, it also tends to decrease $A_3$ and $A_1$.

Molybdenum, on the other hand, is an alphagenous element as well as a carbide-forming element that increases the domain of stability of the ferrite, and therefore, it will tend to increase $A_3$ and $A_1$.

It is necessary to fully austenitize the steel to carry out all these treatments. Thus, we will take a temperature of 40–60 °C above the $A_3$. Since $A_3 = 818$ °C, the austenitization temperature will be: $818 + 50 \simeq 860$ °C).

We calculate the $M_S$ by means of Steven's equation (corrected by Irving):

$$Ms(°C) = 561 - 474 \cdot \%C - 33 \cdot \%Mn - 17 \cdot \%Ni - 17 \cdot \%Cr$$
$$- 21 \cdot \%Mo - 11 \cdot \%W - 11 \cdot \%Si \tag{2.65}$$

$$Ms(°C) = 561 - 474 \cdot 0.36 - 33 \cdot 0.74 - 17 \cdot 1.04 - 21 \cdot 0.16 - 11 \cdot 0.30 = 341.6\,°C \tag{2.66}$$

2.    We determine the ideal critical diameter of the steel for structures 50% and 99% martensitic, calculated considering Jominy's test

Starting from the carbon percentage that the steel has (0.36% C), we see, according to the graph that relates hardness and carbon percentage for different percentages of martensitic transformation, that for the 50%, an approximate hardness of 41 HRC is obtained while for the 99%, an approximate hardness of 52 HRC is obtained (Fig. 2.44 rather similar to Fig. 2.15). Using these values in Jominy's curve, we see that these hardness values are obtained at 18.5 and 3.5 mm from the quenched end, respectively. With these distances, we use the graph that relates Jominy's distances—$D_{Ci}$ ideal, and we obtain the following values: $D_{Ci}$ 50% martensite, 96 mm; $D_{Ci}$ 99% martensite, 38 mm (Fig. 2.45).

3.    We evaluate, from the chemical composition and the austenitic grain size of the steel, the bainitic and pearlitic hardenabilities at 50% and 99% respectively. $\phi$50% pearlite, $\phi$99% pearlite, $\phi$50% bainite, and $\phi$99% bainite are the maximum $D_{ci}$ diameters necessaries to avoid the related transformations in the

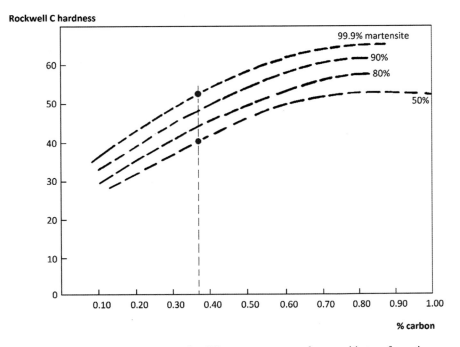

**Fig. 2.44** Hardness-carbon percentage for different percentages of martensitic transformation

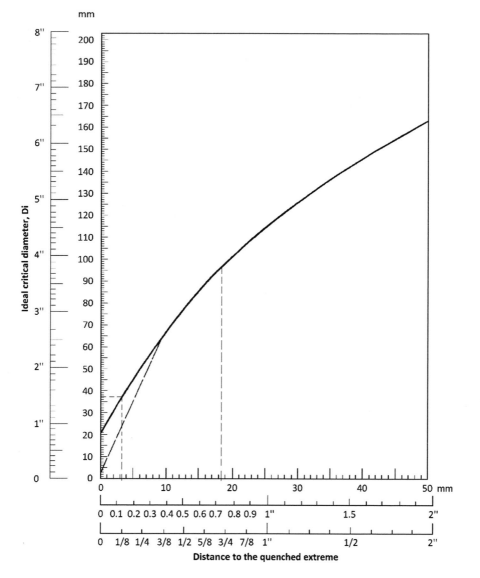

**Fig. 2.45** Curve that relates ideal critical diameters and Jominy's distances

same quantity. We calculate the critical diameters, and we compare them with those obtained in question 2.

The chemical composition of our steel is: 0.36% C, 0.30% Si, 0.74% Mn, 0.025% P, 0.017% S, 1.04% Cr, 0.16% Mo, and 0.063% Al.

We assume that the austenitic grain size of the steel is 7 ASTM. We calculate the hardenabilities using Grossman equation:

$$\phi 50\% \text{ pearlite} = (\%C)^{0.5} \cdot K_{50} \cdot \left[1 + \left(K_{pe} \cdot \%e\right)\right] \text{ inches} \qquad (2.67)$$

$$\phi 99\% \text{ pearlite} = (\%C)^{0.5} \cdot K_{99} \cdot \left[1 + \left(K_{pe} \cdot \%e\right)\right] \text{ inches} \qquad (2.68)$$

In our steel, and considering Tables 2.15, 2.16, and 2.17, we obtain

$$
\begin{aligned}
\phi 50\% \text{ pearlite} = {} & (0.36)^{0.5} \cdot 0.338 \\
& \cdot [(1 + 0.64 \cdot 0.30) \cdot (1 + 4.10 \cdot 0.74) \cdot (1 + 2.83 \cdot 0.025) \\
& \cdot (1 + (-0.62) \cdot 0.017) \cdot (1 + 2.33 \cdot 1.04) \\
& \cdot (1 + 3.14 \cdot 0.16)] = 135 \text{ mm } (5.3 \text{ inches})
\end{aligned}
$$
$$(2.69)$$

$$
\begin{aligned}
\phi 99\% \text{ pearlite} = {} & (0.36)^{0.5} \cdot 0.254 \\
& \cdot [(1 + 0.64 \cdot 0.30) \cdot (1 + 4.10 \cdot 0.74) \cdot (1 + 2.83 \cdot 0.025) \\
& \cdot (1 + (-0.62) \cdot 0.017) \cdot (1 + 2.33 \cdot 1.04) \\
& \cdot (1 + 3.14 \cdot 0.16)] = 101.4 \text{ mm } (4.0 \text{ inches})
\end{aligned}
$$
$$(2.70)$$

Bainitic hardenabilities are given by

$$\phi 50\% \text{ bainite} = 0.494 \cdot (\%C)^{0.5} \cdot K_{50} \cdot [1 + (K_{be} \cdot \%e)] \text{ inches} \qquad (2.71)$$

$$\phi 99\% \text{ pearlite} = 0.272 \cdot (\%C)^{0.5} \cdot K_{50} \cdot [1 + (K_{be} \cdot \%e)] \text{ inches} \qquad (2.72)$$

**Table 2.15** Bainitic and pearlitic hardenabilities (1)

| Hardenability | Pearlitic | | Bainitic | |
|---|---|---|---|---|
| | 50 | 99 | 50 | 99 |
| Diameter in inches | $\sqrt{\%C} \cdot K_2 \cdot K_4$ | $\sqrt{\%C} \cdot K_2 \cdot K_5$ | $0.494 \cdot \sqrt{\%C} \cdot K_3$ | $0.272 \cdot \sqrt{\%C} \cdot K_3$ |

**Table 2.16** Bainitic and pearlitic hardenabilities (1)

| Element | $K_2$(is the product of the following values) | $K_3$(is the product of the following values) |
|---|---|---|
| Mn | $1 + 4.10 \cdot \%Mn$ | $1 + 4.10 \cdot \%Mn$ |
| P | $1 + 2.83 \cdot \%P$ | $1 + 2.83 \cdot \%P$ |
| S | $1 - 0.62 \cdot \%S$ | $1 - 0.62 \cdot \%S$ |
| Si | $1 + 0.64 \cdot \%Si$ | $1 + 0.64 \cdot \%Si$ |
| Cr | $1 + 2.33 \cdot \%Cr$ | $1 + 1.16 \cdot \%Cr$ |
| Ni | $1 + 0.52 \cdot \%Ni$ | $1 + 0.52 \cdot \%Ni$ |
| Mo | $1 + 3.14 \cdot \% \cdot Mo$ | 1 |
| Cu | $1 + 0.27 \cdot \%Cu$ | $1 + 0.27 \cdot \%Cu$ |

**Table 2.17** Bainitic and pearlitic hardenabilities (1)

| Grain size | $K_4$ | $K_5$ |
|------------|-------|-------|
| 1 | 0.546 | 0.410 |
| 2 | 0.504 | 0.378 |
| 3 | 0.465 | 0.350 |
| 4 | 0.429 | 0.323 |
| 5 | 0.397 | 0.298 |
| 6 | 0.366 | 0.275 |
| 7 | 0.338 | 0.254 |
| 8 | 0.312 | 0.234 |
| 9 | 0.288 | 0.217 |
| 10 | 0.266 | 0.200 |

Replacing:

$$\phi 50\% \text{ bainite} = 0.494 \cdot (0.36)^{0.5}$$
$$\cdot [(1 + 0.64 \cdot 0.30) \cdot (1 + 4.10 \cdot 0.74) \cdot (1 + 2.83 \cdot 0.025)$$
$$\cdot (1 + (-0.62) \cdot 0.017) \cdot (1 + 1.16 \cdot 1.04)] \tag{2.73}$$
$$= 84.6 \text{ mm } (3.3 \text{ inches})$$

$$\phi 99\% \text{ bainite} = 0.272 \cdot (0.36)^{0.5}$$
$$\cdot [(1 + 0.64 \cdot 0.30) \cdot (1 + 4.10 \cdot 0.74) \cdot (1 + 2.83 \cdot 0.025)$$
$$\cdot (1 + (-0.62) \cdot 0.017) \cdot (1 + 1.16 \cdot 1.04)] \tag{2.74}$$
$$= 46.6 \text{ mm } (1.8 \text{ inches})$$

If we analyze the obtained results, for 50%, it is possible to appreciate that the pearlitic nose is more delayed with respect to the y-axis (axis of temperatures) than the bainitic nose. Thus, the hardenability at 50% of bainite will be the ideal critical diameter (84.6 mm $\simeq$ 3.3 inches).

Analogously, for the hardenability of 99%, we obtain that the bainitic nose is closer to the temperature axis than the pearlitic nose. Therefore, the ideal critical diameter will be 46.6 mm (1.8 inches).

If we compare these results with those obtained in the real (experimental) part of the case study, we can check that the theoretical values are greater than those experimentally obtained.

4. We calculate the potential hardness and the softening coefficients for tempering treatments performed for 1 h at 200, 300, 400, 500, and 600 °C, respectively

We compare the obtained hardness values with the experimental results, and we determine what treatment was given to the steel.

The potential hardness is given by the following equation (Calvo Rodes 1956):

**Table 2.18** Potential
hardness (1). The value of A
that is replaced in this table is
the addition of the values of A
obtained with the support in
Table 2.19

| %C | Potential hardness |
| --- | --- |
| 0.20 | 78 + A |
| 0.25 | 80 + A |
| 0.30 | 82 + A |
| 0.35 | 83 + A |
| 0.40 | 85 + A |
| 0.45 | 87 + A |
| 0.50 | 89 + A |
| 0.55 | 91 + A |
| 0.60 | 92 + A |
| 0.65 | 93 + A |
| 0.70 | 94 + A |
| 0.75 | 95 + A |

$$H_P = K_C + \sum K_e \cdot \%e \ (HRC) \tag{2.75}$$

From Tables 2.18 and 2.19, and for our chemical composition, we obtain: $K_C =$ 83.4; Si, $K_e = 4$; Mn, $K_e = 4$; Cr, $K_e = 1.5$ for $T_{tempering} < 300\ °C$, $K_e = 5$ for $T_{tempering} > 600\ °C$; Mo, $K_e = 2.5$ for $T_{tempering} < 300\ °C$, $K_e = 20$ for $T_{tempering} > 600\ °C$.

The potential hardness will take the following value:

$$H_P = 83.4 + 4 \cdot 0.3 + 4 \cdot 0.74 + 1.5 \cdot 1.04 + 2.5 \cdot 0.16 = 89.5\ HRC\ (hard) \tag{2.76}$$

$$H_P = 83.4 + 4 \cdot 0.3 + 4 \cdot 0.74 + 5 \cdot 1.04 + 20 \cdot 0.16 = 96\ HRC\ (tough) \tag{2.77}$$

The values for the softening coefficient are calculated from Fig. 2.46 for 1 h of treatment. Values are collected in Table 2.20.

Thus, the hardness H of our specimens with tempering will be equal to:

**Table 2.19** Potential
hardness (2)

| Elements | For hard tempering (T < 300 °C) | For tough tempering (T > 600 °C) |
| --- | --- | --- |
| Mn | 4·%Mn | 4·%Mn |
| Si | 4·%Si | 4·%Si |
| Ni | 1·%Ni | 1·%Ni |
| Co | 0.5·%Cr | 0.5·%Co |
| Cr | 1.5·%Cr | 5·%Cr |
| Mo | 2.5·%Mo | 20·%Mo |
| V | – | 50·%V |

$$H = H_P + M \tag{2.78}$$

We compare the measured values with the calculated values, and we see that the measured values are rather similar to those theoretically calculated (Table 2.21).

The axle has been subjected to a treatment of quenching and tempering with tough tempering. Starting from an average hardness that the axle had in its nucleus (17.5 HRC), we can calculate the tempering temperature for one hour of permanence at this temperature: Potential hardness, $H_P = 96$ HRC; real hardness, $H = 17.5$ HRC; softening coefficient, $M = 78.5$.

For a coefficient $H_d = 78.5$ and 1 h of permanence, the tempering temperature is approximately 740/750 °C, close to the eutectoid temperature of the steel.

5. This question was to evaluate if the steel was prone to decarburizing during the austenitization treatment before the quenching. We have also proposed to evaluate what $CO/CO_2$ protecting atmosphere would we use.

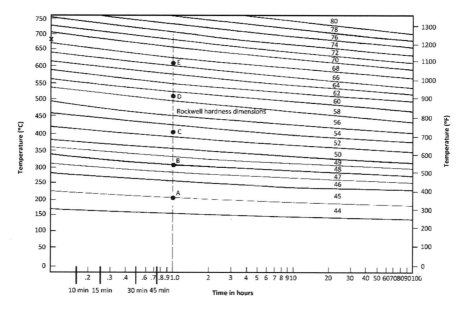

**Fig. 2.46** Plot of curves time–temperature of equivalent tempering treatments

| **Table 2.20** Tempering temperature and softening coefficient | $T_{tempering}$ (°C) | M (HRC) |
|---|---|---|
| | 200 | −45 |
| | 300 | −48 |
| | 400 | −52.8 |
| | 500 | −58.8 |
| | 600 | −66.5 |

**Table 2.21** Calculated and experimental hardness values (c = 15)

| $T_{tempering}$ (°C) | H (HRC) | Experimental hardness (HRC) |
|---|---|---|
| 200 | 89.5–45 = 44.5 | 50 |
| 300 | 96–48 = 48 | 49 |
| 400 | 96–52.8 = 43.2 | 45 |
| 500 | 96–58.8 = 37.2 | 36 |
| 600 | 96–66.5 = 29.5 | 31 |

Generally, in all steels, in more or less degree, decarburizing takes place during their permanence at austenitization temperatures.

It is possible to use a protecting atmosphere to avoid decarburizing where it is necessary to control the $CO/CO_2$ ratio. This is in this manner since the parameter that measures the decarburizing is the difference between the activities of the surface carbon in the steel, and in the atmosphere of the furnace:

$$a_C \text{ steel} < a_C \text{ furnace atmosphere} \rightarrow \text{case hardening} \qquad (2.79)$$

$$a_C \text{ steel} > a_C \text{ furnace atmosphere} \rightarrow \text{decarburizing} \qquad (2.80)$$

6. This question indicated: the axle was broken in service due to bending fatigue. The construction draw plane specified for this part an average hardness of 30 HRC. It is asked whether the steel has been subjected to the suitable heat treatment and how do you improve the fatigue resistance of this steel

Hardness values measured from the periphery to the nucleus of the part give us values very lower than the 30 HRC that the axle should have as average value. This makes us think that the axle was not subjected to the adequate treatment, being the tempering temperature responsible for the hardness loss. The optimal temperature of tempering to obtain such hardness would be around 600 °C (Table 2.21).

A steel will be more resistant to fatigue the greater the $\sigma_y$ and $\sigma_u$. A good manner of improving the resistance to fatigue would be increasing the hardness of the steel, particularly in the periphery, which can be carried out, for instance, by means of a case hardening. Although in this case, considering the measured hardness values and the specifications of the part, a quenching, and tempering treatment, adequately carried out, would be sufficient.

7. We asked in this question about the state in which we would supply the steel to facilitate the machining of the axle

We would supply it in that state that would provide to the steel the lowest hardness possible. Due to the hardness measurements already made, and considering only economic reasons, a regenerative annealing would not represent significative machining problems. A globular annealing, which is the treatment that confers less hardness to the material and, for that reason, better machinability, would result, in this particular case, more expensive than the annealing. The globulizing treatment is frequently used for the machining of tools (alloyed steels, which have great hardenability and, consequently, high hardness by treatment).

8.  The last question was about the specification according to UNE standard of this steel, as well as the specifications by the equivalent AISI, AFNOR, and DIN standards. Properties and applications

The steel used in the manufacture of the axle of the loader corresponds to the specification F-150 UNE-EN ISO 683–1:2019 and UNE-EN ISO 683–2:2019 (original UNE 36,012:1975), construction steel destined for cold deformation. The equivalence to other standards are: AISI, 4137; AFNOR, 35CD4; DIN, 34 CrMo 4; B-5, 708 A 37; EURONORMA, 34 CrMo 4 KD.

The main properties, in quenched and tempered (hardened) states are $\sigma_y \geq$ 560 MPa; $\sigma_u = 780–950$ MPa; E $= 14\%$. This steel is mainly used in the manufacture of crankshafts, axles, tires, etc.

*Solution ends.*

**Exercise 2.11**  Case Study: Molded steels (as-cast state). A sample taken from the casting, corresponding to a mine wagon wheel, in as-molded state, is manufactured with the steel of specification AM-45, of chemical composition 0.26% C and 0.65% Mn.

*Experimental part*:

1.  Determine the microstructure of the steel in the as-molded state. Determine the hardness

It is a hypoeutectoid steel that has as matrix constituent ferrite and as-disperse constituent, pearlite. The main characteristic of this microstructure is that it has Widmanstätten structure, typical of this as-molded steel (great size of austenitic grain). Measured hardness values were 170, 177 and 170 HB, being the average hardness of this specimen 172 HB.

Figures 2.47 and 2.48, micrographs taken at 100 × and 500x, respectively, show the Widmanstätten structure (acicular ferrite), typical from molded steels.

2.  Metallography and hardness of the steel in annealed and normalized states

In annealed state, Figs. 2.49 and 2.50, micrographs at 100× and 500 × respectively, it is possible to see a structure that reminds us that of the as molded state due to its dendritic structure. This structure is formed in annealing treatments carried out at high temperatures and with cooling in furnace by segregation of carbon and manganese, gammagenous elements that would produce the transformation of these zones into pearlite.

Measured hardness values were 158, 158, and 161 HB, being the average hardness of 159 HB.

In normalized state, Figs. 2.51 and 2.52, micrographs at 100× and 500 × respectively, it is possible to appreciate a clear reduction of the grain size and an increase

**Fig. 2.47** As-molded state. Ferritic-pearlitic microstructure, with coarse grain size. BHN≃172. 100x

**Fig. 2.48** As-molded state. Ferritic-pearlitic microstructure, with coarse grain size. 500x

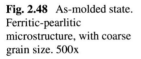
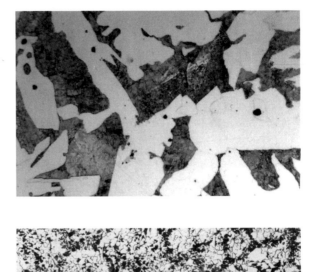

**Fig. 2.49** Annealed state. Ferritic-pearlitic microstructure of ferritic matrix. BHN≃159. 100x

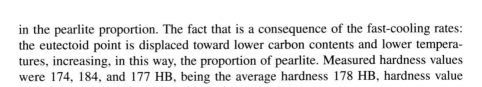

in the pearlite proportion. The fact that is a consequence of the fast-cooling rates: the eutectoid point is displaced toward lower carbon contents and lower temperatures, increasing, in this way, the proportion of pearlite. Measured hardness values were 174, 184, and 177 HB, being the average hardness 178 HB, hardness value

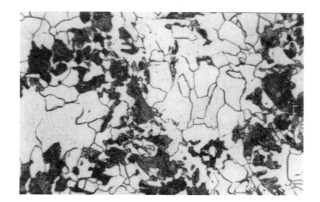

**Fig. 2.50** Annealed state. Ferritic-pearlitic microstructure of ferritic matrix. 500x

**Fig. 2.51** Normalized state.
Ferritic-pearlitic
microstructure. BHN≃178.
100x

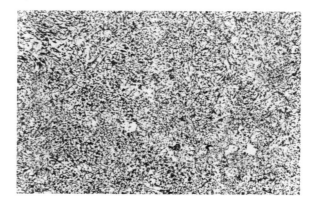

greater than that of the steel in annealed state as was expected due to the increase in the pearlite content and grain size refining. The UTS (Ultimate Tensile Strength) is about HB/3≃59 kg/mm² (580 MPa).

3.    Metallography and hardness of the steel quenched in water

In quenched state, Figs. 2.53 and 2.54, micrographs at 100× and 500x, respectively, a bainitic-martensitic structure resulted from the transformation of the austenite consequence of a fast cooling is observed.

Bainite is appreciated in the ancient austenitic grain boundary, which indicates that its formation takes place by nucleation and growth.

In this case, the hardness of the steel is noticeably greater due to the formation of martensite. The value of hardness that was measured was 48 HRC, which is equivalent to 455 HB (UTS≃150 kg/mm² = 1470 MPa).

*Questions*:

1.    What particular characteristics the structure of the steel in as-molded state have? How does it behave from the mechanics point of view?

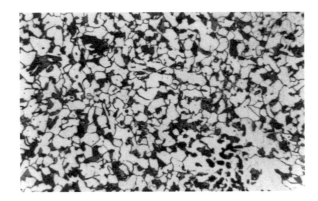

**Fig. 2.52** Normalized state. Ferritic-pearlitic microstructure. 500x

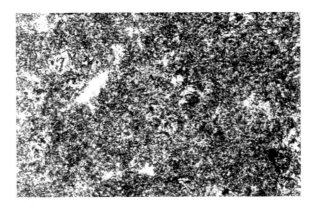

**Fig. 2.53** Quenching microstructure. Complex microstructure formed by bainite-martensite-retained austenite. BHN≃455. 100x

As it was previously indicated, this is a hypoeutectoid steel that has ferrite as matrix constituent and pearlite as disperse constituent. The peculiarity of this sample is that has Widmanstätten structure. It appears when the austenitic grain size is big and the cooling rates are high (air cooling), the formation of ferrite lamellae takes place and progresses toward the inside of the ancient austenitic grain, and the rest transforms into pearlite.

Widmanstätten structure is very undesired from the industrial point of view due to the low mechanical properties conferred to the steel, particularly the low toughness. It is necessary to carry out normalizing heat treatments to make disappearing this morphology.

2.   What is the objective of the heat treatment of an as-molded part? From the annealing and normalizing structures, what do you think that is mechanically more favorable?

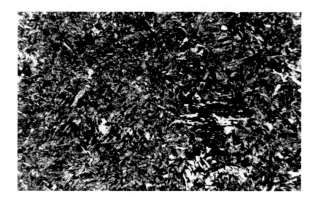

**Fig. 2.54** Quenching microstructure. Structure of acicular bainite-martensite over white background of retained austenite. 500x

As it was already said, the normalizing heat treatment is given to parts in as-molded state with the aim of eliminating the acicular morphology of the ferrite or Widmanstätten structure.

The annealing structure would be similar to that the steel would have if cooled according to the equilibrium diagram. After performing the normalizing, due to the faster cooling, the eutectoid point is displaced toward lower carbon contents and lower temperatures, and greater proportions of pearlite appear. For that reason, the steel in normalized state would have better mechanical properties; greater ultimate tensile strength, and greater toughness (finer microstructure), even when it will have more proportion of pearlite. This one is a pearlite with lower carbon content or "diluted" pearlite, and therefore, the steel has more toughness.

The steel in annealed state will be softer because the matrix is ferritic, and this is the softest constituent of steels, and thus, the ultimate tensile strength is lower.

3. Micrographically, what differences can be found between the annealing, normalizing, and quenching microstructures? What reminds you of the microstructure of this steel in annealed state?

As the cooling rate increases (from the annealing to the quenching), the microstructures become finer. In annealed state, it is possible to appreciate a structure that reminds us that of the as molded state, due to the segregation of the carbon and manganese toward the austenitic grain boundaries, which is favored by a slow cooling from the holding at the austenitization temperature.

In the quenched specimen, the structure is not obtained by nucleation and growth, but the austenite is transformed into martensite by shearing, as it is deduced from the acicular structure instead of equiaxed structure.

4. Is weldable this steel?

The weldability condition is given by the equivalent carbon parameter (Seferian 1962):

$$C_{eq} = \%C + \frac{\%Mn}{6} + \frac{\%Si}{24} + \frac{\%Ni + \%Cu}{15} + \frac{\%Cr + \%Mo}{10} < 0.42 \qquad (2.81)$$

In this case:

$$C_{eq} = 0.26 + \frac{0.65}{6} = 0.37 < 0.42 \qquad (2.82)$$

Therefore, it is weldable. For weldability, it is necessary to use carbon contents at the lowest possible with the purpose of having the greatest $M_S$ temperature possible, and thus the volume increase produced by the martensitic transformation was absorbed by the plastic deformations that avoid the breaking of the material.

*Solution ends.*

## 2.7  Isothermal Treatments

Despite the fact that these treatments are usually called "isothermal treatments," it would be more reasonable calling them "heat treatments with isothermal cooling".

All of them, including isothermal annealing, have in common that the steel parts are heated to the austenitization temperature, and, from that temperature, they are isothermally cooled by submerging them into molten salts, molten lead, or another liquid coolant that allows keeping constant the temperature during the transformation of austenite.

### 2.7.1  *Patenting*

Patenting is an isothermal treatment that is usually given, as final operation, to steel wires of 0.7–0.9% C, which require high mechanical resistance under tension stress because they are used in pre- and post-stressed concrete. The wire—after being austenitized—is submerged into a bath of molten lead (or in salts) at a temperature corresponding to the lower part of the TTT curve of the steel (Fig. 2.55) to achieve these characteristics. This has as objective transforming the austenite into a very fine pearlite, with distances between lamellae of cementite, $S_0$, from 0.1 to 0.2 μm. With this treatment, ultimate tensile strength $\sigma_u$ of approximately 1600 MPa and elongations $A_t$ from 5 to 10% are recorded.

Patenting is sometimes used as an intermediate treatment against work-hardening during wire drawing, to (apart from finally achieving the mentioned high values of $\sigma_u$ or greater) facilitate the operation of wire drawing. In fact, it is convenient for wire drawing to have a structure of fine cementite, also previously achieved by patenting. However, while we are drawing the wire (cold-forming), and the section of the wire is decreased, ferrite is acquiring a work-hardening that progressively

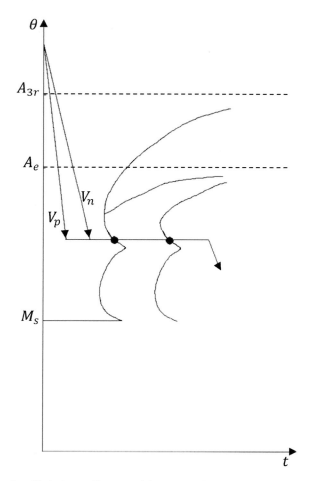

**Fig. 2.55** Patenting. $V_p$ is the cooling rate of the part outside, $V_n$ is the cooling rate in the center of the part

makes it more difficult the process and would end up making the process unfeasible. A new patenting, to eliminate the cold-worked ferrite, allows continuing with the cold reduction of the section (Pero-Sanz et al. 2019).

The ultimate tensile strength that can be obtained in eutectoid binary steels, by means of a suitable combination of fine pearlite and cold-forming due to wire drawing, allows making the fine pearlite the most resistant in tension structural material that is nowadays known, as it can even reach 6000 MPa of ultimate tensile strength.

The wire drawing is the method of industrial forming that allows obtaining the greatest ultimate tensile strengths, $\sigma_u$, in metallic materials (outstanding within them, the pearlite "composite", metal matrix composite material formed by a ceramic material, cementite, and a metallic matrix, ferrite). The peculiarity of the cold wire drawing is that it confers work-hardening—hardening that is simultaneous to the reduction in diameter of the wire—practically without risks of striction. This hardening, in the case of body-centered cubic metals and composites (as in this case), maintains until the maximum deformations nowadays are considered. This allows assuming that the achieving of fibers of great resistance—close to the ideal ultimate tensile strength, $\sigma_u$—(approximately equal to G/5, where G is the shearing modulus) would be only limited by the current technical incapacity of exceeding certain limit reductions. The approximate limits of the ultimate tensile strength, $\sigma_u$, in wire drawing are the following ones (Gil Sevillano 1982): G/50 in cubic metals (body-centered cubic or face-centered cubic), G/20 in metallic glasses, and G/10 in fine pearlite (approximately 6000 MPa) (Pero-Sanz et al. 2019, Fig. 12.24, p. 333).

### 2.7.2  Austempering

"Austempering" (Fig. 2.56) has as objective obtaining a structural part that is fully bainitic. It does not require a subsequent tempering. Bainitic structure has the advantage of being tougher, for the same hardness than that obtained by quenching and low-temperature tempering of steel.

The austempering has as complementary advantage of being a heat treatment without stresses, deformations, and cracks that can be identified in the case of quenching in severe mediums. The transformation of the austenite into bainite in

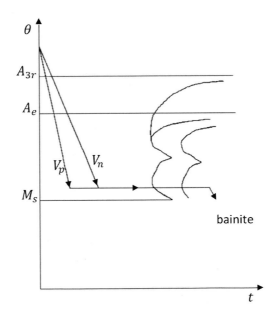

**Fig. 2.56** Austempering

**Fig. 2.57** Martempering

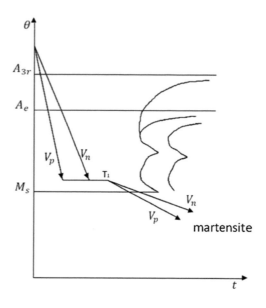

the austempering takes place with volume increase, but at the same time and equal temperature in all points of the part.

As it is obvious, "austempering" cannot be applied to just any steel. It is necessary for the steel to have enough hardenability to ensure that both the periphery and the nucleus would have, in the salts bath, the isothermal temperature before the transformation of austenite into bainite. Otherwise, bainite in the periphery and pearlite in the part's more internal zones could be obtained. In any case, it is not convenient that the steel would have great hardenability to avoid long treatments.

Austempering is usually applied to parts of small diameter, e.g., 10 mm, because in the cooling, it is easier to equalize the temperatures of the part's periphery and nucleus before the transformation of the austenite. For parts of greater diameter, steel with greater hardenability and/or severe cooling is required to avoid pearlitic transformations and allow bainitization of the austenite. Obviously, a steel of very low hardenability, whose TTT curve was that of the pearlitic zone practically tangent to the y-axis, could not be austempered.

Austempering is usually applied to some carbon steels with a carbon content between 0.5 and 1.2% or to low-carbon steels destined for tools. It is also used in some spheroidal ductile irons, e.g., austempered ductile iron (see Chap. 7 in Pero-Sanz et al. 2018a, b).

### 2.7.3 Martempering

Martempering (Fig. 2.57) is the treatment aimed at obtaining martensite in medium hardenability steels without risks of stresses, deformations, and cracks in the part. It is not used in steels of great hardenability, and self-hardenable, because it would be sufficient an air cooling to achieve those objectives.

It is necessary to know with straightforward the $M_S$ temperature to give the martempering treatment to a part, and its hardenability, that is to say, its TTT curve, should allow the aspects mentioned in the following lines. After the austenitization of a part, it should be possible to cold in salts at a temperature $T_1$, only a few degrees above $M_S$, without the transformation of the austenite. At this uniform temperature $T_1$, there should not be another constituent apart from the austenite, both in the periphery and nucleus of the part.

From this temperature $T_1$—without allowing permanence at this temperature as otherwise, the typical transformations of the austempering will begin—the part is air-cooled. As in this cooling from $T_1$ the temperatures between the periphery and the nucleus of the part hardly differ, and the risks of stresses, deformations, and cracks are minimized as all the part is transformed into martensite nearly at the same time.

Martempering is habitually applied to parts of small size. The hardenability of the steel must be similar to that which allowed obtaining martensite both in the periphery and the nucleus of the part by oil cooling from the austenitization temperature. However, if this quenching is applied, there would be risks of cracks, deformations, and stresses in the part (because of the temperature gradient between the periphery and the nucleus).

The above-indicated considerations also explain a habitual industrial practice—sometimes used in steel parts with low hardenability—aimed at achieving analogous results to those obtained with the martempering: double bath quenching. It is a substitute for martempering to achieve similar results, although more economic. After heating the part until its full austenitization, it is rapidly submerged in a coolant with a great heat transfer factor (v. gr. stirred water) to avoid the transformation of the austenite in the pearlitic zone of the TTT curve. It is simply sufficient to shake the part inside of the water. After that—it is assumed that the structures of the periphery and nucleus are still austenitic—the part is submerged in an oil bath (sometimes, it is sufficient with an air cooling). The difference in temperatures between the periphery and nucleus of the part, cooled in oil like this, is small, and the transformation of the austenite into martensite can be simultaneous in all the parts and, for that reason, with little risk of either deformation or cracking. This method of cooling justifies the naming of double bath quenching.

As it is logical if the part has a big size, the first cooling takes place in coolant with great heat transfer factor, and this can be sufficient to avoid the transformation of the peripheric austenite but, on the contrary, the inner austenite might be already transformed into pearlite. In this case, the desired result would not be achieved with the second cooling in oil. And similar situation can happen if the time either between the exit from the austenitization furnace and the first cooling bath or between the immersion in the first bath and the second is long enough.

### 2.7.4 Sub-Zero Treatments

Steels for tools—both carbon steels as alloyed steels—and case-hardened and quenched steels usually have great quantity of residual austenite after the quenching by cooling down to the room temperature. As it was already mentioned, the multiple tempering has as objective the transformation of this residual austenite and, it is the usual treatment to achieve it.

As an alternative to the quenching and multiple tempering, there is a version of quenching without residual austenite that is called sub-zero quenching. In fact, when quenching, there would not be residual austenite if temperatures lower than $M_f$ of the steel were reached during the cooling. A cooling mixture formed by liquid carbon dioxide and acetone is usually employed to achieve these low temperatures. Temperatures lower than $-75\ ^\circ C$, which is lower than the $M_f$ temperature of any steel, are obtained.

However, the sub-zero quenching exhibits several difficulties. If the parts were directly submerged in the coolant ($-75\ ^\circ C$) from the austenitization temperature ($\theta > A_3$), it would be probable the appearance of cracks produced by the great difference in temperatures between the periphery and the nucleus of the part. Therefore, it is usually recommended to make a plain quenching and continuing later with a cooling from the room temperature to the $-75\ ^\circ C$. This way, the temperature difference between the periphery and nucleus is smaller (cooling from 25 °C to $-75\ ^\circ C$). However, this procedure, arrest at room temperature, comprised within $M_S$ and $M_f$, stabilizes the austenite and it might be more complicated to achieve the complete transformation of the austenite into martensite at $-75\ ^\circ C$.

The following procedure is usually employed. After quenching by cooling down to room temperature, a tempering at 160 °C is given before the sub-zero quenching. This way, part of the residual austenite can transform into bainite and, consequently, the harmful effects of a detention at room temperature are attenuated (even more stable, there will be less residual austenite).

## 2.8 Surface Thermochemical Treatments

Together with the massive properties of the alloys, certain surface properties are sometimes required, and occasionally these properties can be thoroughly different from the first ones. We should consider parts where it is demanded, e.g., (1) the requirement of protection against corrosion; (2) or in parts with high peripheric hardness for fatigue and wear resistance, but that they have all of them toughness requirements inside the part. In many cases, the necessity of different properties in the periphery and in the nucleus is solved by means of surface coatings achieved using different techniques such as rapid solidification, diffusion, metallic plating, polymeric coating, passivation treatment, and painting, etc.

It is not the purpose of this section to give information about these different possibilities; it is to complement knowledge previously expressed about heat treatments with details about thermal-chemical treatments that have as a final purpose to enrich the periphery of parts with carbon and/or nitrogen. This seems appropriate if we consider that all countries, when specifying different types of steels—hardened steels, stainless steels, tool steels, etc.—in the national standards, always include steels for case-hardening and for nitriding.

Both case-hardening and nitriding—as the carbonitriding, which at a certain level is a combination of both treatments—peripherally modify the composition of a steel part with the aim of improving its fatigue and wear resistance.

Case-hardening treatment has as objective to peripherally enrich the part with carbon until the carbon content approximately equals to (or greater than) that of the eutectoid. A carburizing environment is required—solid, liquid, or gaseous—to achieve this and this environment should provide active carbon to the periphery for its diffusion toward the part's nucleus (see Chap. 4).

As it is logical, the treatment must be performed at a temperature > $A_{3c}$. In another way, if the temperatures were lower than that of the austenitic state, carbon enrichment would not be achieved. Carbon solubility in alpha iron, in octahedral insertion solid solution, does not exceed 0.02% carbon at temperatures lower than that of the eutectoid. Case-hardened layers, whose thickness varies, according to the requirements, within 0.5 and 1.5 mm, after holding at these high temperatures are obtained. A quenching after the case-hardening allows obtaining a periphery of hard martensite and a tough nucleus, as steels that are adequate for case-hardening have a low-carbon content.

For each material, the variation of the diffusion coefficient, D, as a function of the temperature, T, follows a similar law to that of Arrhenius:

$$D = D_0 \cdot \exp\left(-\frac{Q}{R \cdot T}\right) \qquad (2.83)$$

The activation energy for auto-diffusion in solid state of a given metallic compound is related to its melting point, $T_M$ in K, through an equation of the type:

$$Q = 18 \cdot R \cdot T_M \qquad (2.84)$$

The activation energy, Q, for the carbon diffusion in the austenite is approximately 142 kJ/mol. The approximate value of $D_0$, usually called "frequency factor" is $2.0 \cdot 10^{-5}$ m$^2$/s and the value of R is equal to 8.314 J/mol · K.

The values of D allow calculating the flow J, steady net, of atoms through a geometrical plane:

$$J = -D \cdot \frac{dC}{dx} \quad \text{(first Law of Fick)} \tag{2.85}$$

This way, between parallel planes separated by a distance dx, the value dC is the difference in the concentration of atoms between one plane and another. The first Fick's law is valid if the difference in atomic concentration between both planes does not change with time (stationary regime).

However, the steady state is not habitual. In non-steady states—in which dC changes with time although D does not change—the second Fick's law should be applied:

$$\frac{dC_x}{dt} = D \cdot \frac{d^2 C_x}{dx^2} \tag{2.86}$$

The resolution of this last equation is complex. Applying Fick's second law, e.g., to gaseous carbon diffusion, from a carburizing atmosphere, through the surface of a solid steel of uniform carbon composition $C_0$, the analytical solution would be

$$\frac{C_S - C_x}{C_S - C_0} = \text{erf}\left(\frac{x}{2 \cdot \sqrt{D_C \cdot t}}\right) \tag{2.87}$$

where erf is the mathematical error function (the values of erf(z) as a function of z are known, see Table 2.22), and:

$C_0$, is the initial carbon concentration in the steel,

$C_S$, is the carbon concentration in the steel's solid surface after a time t,

$C_x$, is the carbon concentration in the steel at a distance x from the surface after a time t,

$D_C$, is the diffusion coefficient of the carbon in the steel at the temperature T used to perform case-hardening.

Nitriding has as an objective obtaining a hard peripheric layer without requiring subsequent quenching of the steel. Nitriding treatment is usually applied to quenched

**Table 2.22** Values of erf (z)

| z | erf(z) | z | erf(z) | z | erf(z) | z | erf(z) |
|---|---|---|---|---|---|---|---|
| 0 | 0 | 0.40 | 0.4284 | 0.85 | 0.7707 | 1.60 | 0.9763 |
| 0.025 | 0.0282 | 0.45 | 0.4755 | 0.90 | 0.7970 | 1.70 | 0.9838 |
| 0.05 | 0.0564 | 0.50 | 0.5205 | 0.95 | 0.8209 | 1.80 | 0.9891 |
| 0.10 | 0.1125 | 0.55 | 0.5633 | 1.00 | 0.8427 | 1.90 | 0.9928 |
| 0.15 | 0.1680 | 0.60 | 0.6039 | 1.10 | 0.8802 | 2.00 | 0.9953 |
| 0.20 | 0.2227 | 0.65 | 0.6420 | 1.20 | 0.9103 | 2.20 | 0.9981 |
| 0.25 | 0.2763 | 0.70 | 0.6778 | 1.30 | 0.9340 | 2.40 | 0.9993 |
| 0.30 | 0.3286 | 0.75 | 0.7112 | 1.40 | 0.9523 | 2.60 | 0.9998 |
| 0.35 | 0.3794 | 0.80 | 0.7421 | 1.50 | 0.9661 | 2.80 | 0.9999 |

and tempered parts. Since the nitriding temperature is low, mechanical properties corresponding to fairly high strength are retained through nitriding treatment applied after the steel is appropriately heat-treated. It consists of supplying nitrogen, in an atomic state (active N), usually obtained by thermal dissociation of $NH_3$ (see Chap. 4). It is possible to obtain using a nitriding treatment a very hard fine peripheric layer with thickness <0.5 mm. Its hardness is conferred by the distortion that the submicroscopic nitrides, formed by the reaction of nitrogen with some elements (Al, Cr, V, or Mo), produce in the ferrite. It is true precipitation-hardening or age-hardening. The temperature of this thermochemical treatment is always <590 °C to avoid the formation of nitro-austenite. This nitro-austenite, in the air-cooling that follows the nitriding, produces a eutectoid, braunite, which is unfavorable because, apart from other issues, is brittle.

Another habitual thermochemical treatment to achieve hard surfaces is carbonitriding. This treatment consists of simultaneously providing carbon and nitrogen (actives) in the parts outside. The treatment is followed by oil-quenching to obtain hardness by martensitic transformation of this peripheric layer. Quenching can be performed by simply air-cooling because of the great hardenability conferred by the nitrogen, apart from the carbon, solubilized in the austenite; however, with this low severe cooling, a great proportion of soft austenite is retained. The layers obtained through carbonitriding—from 0.08 to 0.8 mm—are thinner than those obtained through case-hardening (carburizing) or nitriding.

The temperatures required for carbonitriding do not usually exceed 875 °C because it is not necessary to reach temperature $A_3$ of the base steel to achieve a fully austenitic periphery. Even at 650 °C, for instance, carbonitriding could be performed. In fact, the supply of nitrogen notably decreases the temperature $A_3$ of the peripheric layer enriched by solid solution of this element. It is possible to achieve by long holding (hours) at this temperature of 650 °C a small austenitic peripheral layer in the steel. This austenite can receive carbon in solid solution. The inner rest of the steel, obviously, does not austenitize at 650 °C. For that reason, cooling from this temperature would only give martensite in the carbonitrided layer at 650 °C, thus retaining an allotropically non-transformed inner steel part during the whole treatment. This kind of carbonitriding "at low temperatures," i.e., at 600–700 °C, is called "light carbonitriding without quenching of the nucleus."

**Exercise 2.12** There are criteria for the use of steels for case-hardening, as a function of the resistance required and the suitability of the case-hardened layer.

– Considering that steel 1 is adequate for the manufacture of a certain part peripherically case hardened. Would be also for the same use the steel 2?
– Can show any inconvenience the steel 2?

NOTE: We will assume to justify the answer that both steels have the same grain size and that Jominy's curves are hypothetically those in Fig. 2.58.

*Solution starts*

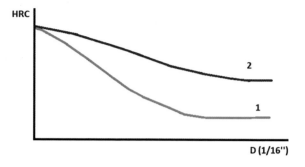

**Fig. 2.58** Jominy's curves of the steels of Exercise 2.12

Steels have similar carbon content. Steel 1, valid, has a hardenability lower than the steel 2. Steel 1 can be, for example, Cr–Mo steel, and steel 2, alloyed also with nickel. See Fig. 2.59 and Table 2.23. Nickel decreases the $T_{case\text{-}hardening}$, increases the hardenability of the case-hardened layer, improves the mechanical resistance and toughness of the nucleus, and having a certain tendency to give retained austenite in the quenching, requires treatments after the case-hardening, it is not sufficient with a quenching from the case-hardening temperature followed by a hard tempering (low at $T < 200°C$). It is proposed for the steel 1, before case-hardening, direct quenching from $A_3 + 40–60 °C$, and hard tempering. For the steel 2, before the case hardening, treatment of incomplete austenitization $A_e < T < A_3$ to quench the case hardened layer and hard tempering. See Fig. 2.60.

*Solution ends.*

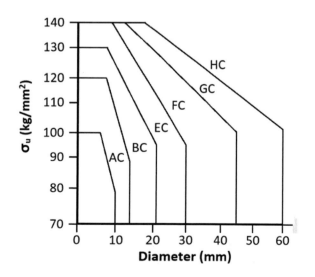

**Fig. 2.59** Graphic for the election of carburizing steels

**Exercise 2.13** A certain steel for case-hardening has a very fine (>7 ASTM) austenitic grain size and, has 4% Ni (0.11–0.17% C, 0.25–0.55% Mn, 0.15–0.40%Si, 0.80–1.20% Cr, 3.75–4.25% Ni) in its chemical composition. The following is requested:

– criteria for the election of this steel. Advantages and disadvantages. Suitable treatments after the case hardening.

*Solution starts*

As in the previous exercise, steel 2, the steel will have a high hardenability, high mechanical resistance, and greater propension than the steel 1 to the formation of retained austenite, which can be removed with the intercritical treatment $A_e < T < A_3$ after the case hardening (to harden the periphery of the part). Situation b) in Fig. 2.60. The predominant criterion is not to find retained austenite in the case hardened layer.

NOTE: For the steel 0.14% C-0.4% Mn-0.30% Si-1% Si-1% Cr-4% Ni

$$
\begin{aligned}
M_S({}^\circ C) &= 561 - 474 \cdot 0.14 - 33 \cdot 0.44 - 11 \cdot 0.30 - 17 \cdot 1 - 17 \cdot 4 \\
&\simeq 393\,{}^\circ C \rightarrow V_a = 100 \cdot \exp[-0.011 \cdot (393 - 20)] \simeq 1.7\%
\end{aligned} \tag{2.88}
$$

which is reasonable.

According to Smithells (1984):

$$
\begin{aligned}
\log V_e(^\circ C/s) &= 4.3 - 3.27 \cdot \%C - \frac{\%Mn + \%Cr + \%ni + \%Mo}{1.6} \\
&= 4.3 - 3.27 \cdot 0.14 - \frac{0.4 + 1 + 4}{1.6} = 4.3 - 0.46 - 3.38 \\
&= 0.465 \rightarrow V_e \simeq 2.9\,^\circ C/s
\end{aligned} \tag{2.89}
$$

equivalent to a normalizing.

The critical diameter is:

**Table 2.23** Proposal of INTA (Spanish Aerospatiale Institute) for case-hardening steels

| Types | %C | %Mn | %Si | %Cr | %Ni | %Mo |
|-------|------|------|------|------|------|------|
| AC | 0.14/0.19 | 0.40/0.70 | 0.25/0.35 | – | – | 0.40/0.50 |
| BC | 0.18/0.23 | 0.40/0.70 | 0.25/0.35 | – | – | 0.40/0.50 |
| EC | 0.18/0.23 | 0.50/0.80 | 0.25/0.35 | 0.40/0.60 | – | 0.40/0.50 |
| FC | 0.18/0.23 | 0.50/0.80 | 0.25/0.35 | 0.40/0.60 | 0.40/0.50 | 0.40/0.50 |
| GC | 0.18/0.23 | 0.50/0.80 | 0.25/0.35 | 0.40/0.60 | 1.00/1.30 | 0.40/0.50 |
| HC | 0.18/0.23 | 0.50/0.80 | 0.25/0.35 | 0.40/0.60 | 1.40/1.70 | 0.40/0.50 |

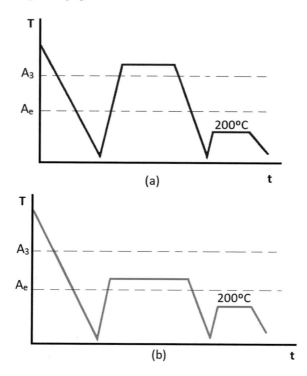

**Fig. 2.60** Heat treatments after case-hardening, in fine-grained steels **a** Medium-resistance steels (low alloy) **b** High-resistance steels (alloyed)

$$D_{cr} \text{ (oil quenching)} \simeq \left(\frac{10000}{2.9}\right)^{1/1.8} \simeq 92 \text{ mm} \qquad (2.90)$$

*Solution ends.*

## 2.9  Hyperquenching and Aging

It was already mentioned in previous pages that steels with less than 0.02% C have ferritic structure at room temperature. However, they have also tertiary—vermicular—cementite, precipitated from carbon dissolved in the ferrite, by reaction with iron atoms, during the cooling, at a temperature lower than the ferrite saturation temperature in carbon, $T_s$ (solvus temperature), see Fig. 1.20 .

This $T_s$ temperature, critical for each steel—greater the carbon percentage and always lower than $A_e$ (727 °C)—is defined by the precipitation line (solvus) of the cementite in the Fe–C equilibrium diagram (for example, the carbon solubilized in the ferrite at 25 °C is only 50 parts per million). The precipitation of tertiary cementite, at temperatures lower than $T_s$, is induced by a mechanism of nucleation and growth: the kinetics of precipitation is expressed, for each low-carbon steel, by means of the corresponding C curve.

If a steel of %C < 0.02, with cementite precipitated in the ferrite matrix, is heated from the room temperature to a temperature greater than $T_S$, cementite decomposes into carbon and iron, and the carbon atoms enter again in solid solution in the ferrite. Later, if a fast cooling of the steel from this temperature is applied, a rate that did not intersect the C curve, once redissolved the carbides, a fully ferritic structure will be obtained at room temperature (with ferrite supersaturated in carbon).

This heat treatment—heating above $T_S$ to redissolve the carbides and fast cooling from this temperature—is called hyperquenching. It can be applied to any alloy— ferrous or not—of monophasic matrix, where the precipitates were susceptible of being redissolved above a certain temperature $T_S$, to later obtain—if the cooling from $T_S$ is adequate—a solid solution supersaturated at room temperature. We call it hyperquenching to distinguish it from the quenching (a name that is exclusively reserved for the heat treatment that provides martensite in steels).

Regarding steels, hyperquenching is not only applied in those with <0.02% C but in many other cases. For example, it is applied in ferritic or austenitic stainless steels with the aim of redissolving the carbides above the critical temperature $T_S$ and obtaining, by the suitable cooling, a steel free of carbides at room temperature. This way, better behavior for a certain time against localized corrosion is achieved.

A supersaturated solution is not thermodynamically stable and, for that reason, a prolonged permanence at room temperature will produce again the precipitates due to a solubility loss of this supersaturated solid solution. However, in this precip- itation at room temperature, the size of the precipitates—because of a nucleation and growth mechanism—will be very small as the diameter of the nucleus depends on the precipitation temperature (smaller size the greater the temperature difference between the $T_S$ and the precipitation temperature).

The possibility of obtaining a supersaturated solid solution by means of hyper- quenching and being possible to govern the size of the precipitates by isothermal holding at a temperature lower than $T_S$ is frequently used in industrial practice to improve the mechanical characteristics of both non-ferrous and ferrous alloys. The heating with isothermal holding after the hyperquenching is usually called either artificial aging or tempering (Pero-Sanz et al. 2017, Sect. 7.1.2., pp. 221–232).

# References

Apraiz JA (1985) Tratamientos térmicos de los aceros, 8th edn, Ed. Dossat, Madrid, Spain
Bain EC, Paxton HW (1961) Alloying elements in steels. American Society for Metals, Metals Park, Ohio, USA
Calvo Rodes R (1956) El acero. Su elección y selección, Ed. Instituto Nacional de Tecnica Aeronáutica, Madrid, Spain

Davenport ES, Bain EC (1930) Transformation of austenite at constant sub-critical temperatures. Trans AIME 901:117–144

Fernández-González D, Ruiz-Bustinza I, González-Gasca C, Piñuela-Noval J, Mo-chón-Castaños J, Sancho-Gorostiaga J, Verdeja LF (2018) Concentrated solar energy applications in materials science and metallurgy. Sol Energy 170(8):520–540

Gil-Sevillano J (1982) Trefilado de aceros al carbono. Ensidesa-Formación, Avilés, Asturias

Grossmann MA, Bain EC (1972) Principios de tratamiento térmico, Ed. Blume, Madrid, Spain

Hollomon JH, Jaffe LD (1945) Time-temperature relations in tempering steel. Trans Amer Inst Min Metall Eng 162:223–249

Koistinen DP, Marburger RE (1959) A general equation prescribing the extent of the austenite-martensite transformation in pure iron-carbon alloys and plain carbon steels. Acta Metall 7(1):59–60

Leslie WC (1981) The physical metallurgy of steels. Hempisphere Publishing Corporation

Pero-Sanz JA (2004) Aceros: Metalurgia física, selección y diseño, 1st edn, Ed. CIE Dossat 2000, Madrid, Spain

Pero-Sanz JA, Quintana MJ, Verdeja LF (2017) Solidification and solid-state transformations of metals and alloys, 1st edn. Elsevier, Boston, USA

Pero-Sanz JA, Fernández-González D, Verdeja LF (2018a) Physical metallurgy of cast irons. Springer International Publishing, Cham, Switzerland

Pero-Sanz JA, Fernández-González D, Verdeja LF (2018b) Materiales para ingeniería: fundiciones férreas. Ed. Pedeca Publishing Press, Madrid, Spain

Pero-Sanz JA, Fernández-González D, Verdeja LF (2019) Structural materials: properties and selection. Springer International Publishing, Cham, Switzerland

Schrader A, Rose A (1966) De ferri metallographia: metallographic atlas of iron, steels and cast irons, vol 2, European Coal and Steel Community. High Authority, Presses académiques européennes, Brussels, Belgium

Seferian D (1962) The metallurgy of welding, Ed. Wiley, London, UK

Smithells CJ (1984) Metal reference book, 6th edn, Ed Butterworth, London, UK

Verdeja JI, Fernández-González D, Verdeja LF (2020) Operations and basic processes in ironmaking. Springer International Publishing, Cham, Switzerland

Verdeja LF, Fernández-González D, Verdeja JI (2021) Operations and basic processes in steel-making. Springer International Publishing, Cham, Switzerland

# Chapter 3
# Thermomechanical Treatments of Steels

## 3.1 Recrystallization

Any metallic material sufficiently cold worked, and subjected later to a suitable heating, gradually recovers the structure—see Fig. 3.1, an ultra-low carbon steel—and properties that it has before the deformation. The material softens, reduces its resistance in the tensile test, increases the elongation, modifies its texture, etc. (Romano et al. 2000a, b; Martínez et al. 2001). The full recovery of the properties is accompanied by the previous existence, or the appearance, of small regular crystalline grains inside of the cold worked matrix. Later, these grains develop until finally obtaining a non-work hardened state of regular grains that receive the name of "recrystallized" state (Fig. 3.2). The heat treatment that has made possible this structure is called recrystallization annealing.

The cold deformation confers work hardening to the metallic materials. The work hardened state involves an increase of energy with respect to that of the non-cold deformed polycrystalline aggregate. Metallic crystals are energetically more stable when they do not contain dislocations. Thus, the work hardened structures tend to evolve toward others with a smaller number of dislocations. However, the work hardened crystals do not spontaneously cede their excess of energy to transform into regular crystals. For that, it is necessary an activation energy conferred by external heating, and a process of atomic diffusion at the heating temperature.

Recrystallization does not involve a phase change but easily a new regrouping of the atoms, by diffusion, to form crystalline groups with a number of dislocations very lower than that available in the work hardened state. In 1 cm$^3$ of work hardened material, the total addition of lengths of existing dislocations even reaches the 40,000 km, while, for the recrystallized state, the estimations are of 1 km (that is to say, $10^5$ cm dislocations/cm$^3$).

Diffusion intervenes in the recrystallization and, for that reason, temperature and time are complementary. On the other hand, as opposed to other solid-state transformations, the recrystallization is irreversible. In the solidification of a pure metal,

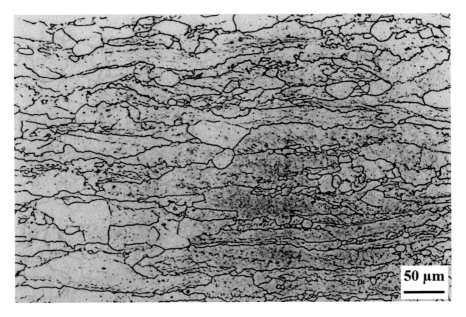

**Fig. 3.1**  Extra-mild steel: work hardened structure, partially recovered

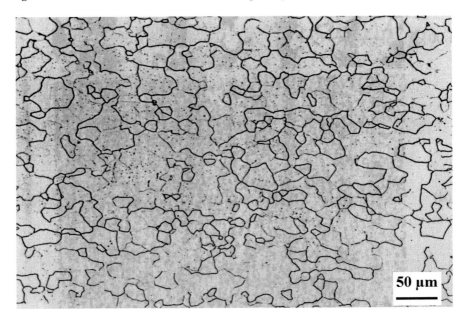

**Fig. 3.2**  Fully recrystallized structure of the steel of Fig. 3.1

for example, the solid melts after a certain time at a definite temperature, while below this temperature, the liquid will solidify again. Something similar happens in the allotropic transformations: the pure iron transforms from the alpha to the gamma variety when the temperature of 912 °C is exceeded, but it transforms again into alpha below this temperature. The same happens with the precipitation: the precipitates of an alloy redissolve, to form solid solution, when the temperature is greater than the solvus one, but reprecipitate when the temperature descends below the critical one. On the contrary, once achieved the recrystallization of a structure, a cooling below the temperature of recrystallization does not come back to the work hardened state.

The activation energy required to diffuse/migrate the atoms and form those small nuclei to recrystallize is smaller the more deformed (in cold) the metal is before the heating. And thus, the temperature $T_1$ that—for the same time of holding—would be required to start the recrystallization is inversely proportional to the previous work hardening of the metal. Therefore, it is necessary to recrystallize at a certain temperature $T_1$, that the metal would have at least a minimum work hardening, which is called critical work hardening. Moreover, the greater the work hardening of the alloy, and therefore greater the number of nuclei of recrystallization that will be formed at this temperature, smaller is the average size of the recrystallized grains at the end of the primary recrystallization at the temperature $T_1$. As it is logical, the recrystallized grain at this temperature will be very big in size with small previous work hardening, for example, with the critical work hardening at the temperature $T_1$ (Pero-Sanz et al. 2017).

The average size of the recrystallized grain also depends on the effect of the impurities. That is to say, it depends on the atoms in solid solution and on the precipitates. The atoms in solid solution usually delay the advance of the boundary surfaces of the nuclei, due to the elastic interaction that the solute atoms produce, curving them. The precipitates or secondary phases also obstacle the advance of those boundary surfaces. As example of important industrial relevance, it is possible to indicate in mild (low carbon) steels recrystallized in batch furnaces, the inhibition produced by the NAl on the grain sub-boundaries {1 0 0}; this favors the growth of sub-grains with textures {1 1 1} very favorable for the deep-drawing operation (Verdeja et al. 2021).

It is possible to say with respect to the formation of the recrystallized nuclei that the impurities or secondary precipitated phases in the work hardened matrix accelerate the formation of nuclei if the size of the secondary phases is big (of around several microns). This is due to the local concentration of stresses that the cold deformation produces on the work hardened matrix that surrounds the particle. The impurities, on the contrary, delay the nucleation if these secondary phases are very small because they reduce the mobility of sub-boundaries of grains in polygonalizable alloys and impede that the sub-grains could be effective nuclei (see Sect. 3.4.2).

Once recrystallized the structure, if the permanence is extended at the recrystallization temperature—or at other greater temperatures—the resulting effect is a growth of the average grain size. This growth, generally continuous, can be discontinuous in some cases.

When this grain growth is discontinuous, due to a strong inhibition of the grain growth—either due to dispersed phases of great fineness that stop (as pin obstacles), "pinning", the advance of the grain boundaries, or by a strong texture of primary recrystallization or by thinness of the sheet, etc.—the following situation can be verified: after a certain time, few grains experience a sudden growth until dimensions of several centimeters, while the rest of the grains disappear due to "cannibalistic" growth of the others. When this happens, it is possible to say that a secondary recrystallization has taken place (it is possible to appreciate a texture of secondary recrystallization different from the texture of primary recrystallization), although the big size secondary grains do not come from a new nucleation: they are simple grains of primary recrystallization that disproportionally grow at the expenses of their neighboring grains.

In some cases, the nature of the atmosphere where the recrystallization is performed can particularly contribute to the secondary recrystallization, governed by the surface energy, which is usually called tertiary recrystallization.

This is the case that usually takes place during the annealing of Fe-3% Si sheets, with 0.1 mm in thickness, destined to electric transformers, when it is aimed to obtain the Goss texture {1 1 0} <1 0 0>. This texture is advantageous because <1 0 0> is a direction of easy magnetization and, therefore, a smaller stack of sheets is required to manufacture the nucleus of the electric transformer than in the case of having a different texture (or the polycrystalline aggregate would not have texture). If the atmosphere of the annealing furnace contains traces of oxygen, it is observed that, after long holding at 1000–1200 °C, a texture {0 0 1} <1 0 0> appears, which is still more favorable for magnetic sheet, with grain twofold oriented, called "cube texture". This happens even if the dispersion of inclusions of MnS was adequate to obtain, by secondary recrystallization, the Goss texture. If the atmosphere is replaced in sheets with cube texture by other of very dry hydrogen, or by high vacuum, the Goss texture will appear again—by a new secondary recrystallization. A new change of atmosphere, at these temperatures, would provide again the "cube" texture.

The starting material is an ingot of Fe-3% Si, with inclusions of MnS—grain growth inhibitor—hot rolled and with a long annealing under protecting atmosphere. Several cold reductions follow this process—normally 2 or 3—of the 50% to the 70% with intermediate annealing between 800 and 900 °C. The texture of final recrystallization contains a component of average intensity of Goss texture and other weak—in approximately 2% in volume—of cube texture {1 0 0} <1 0 0>. The material with this texture of primary recrystallization is subjected to an annealing of 24 h at 1100 °C under high vacuum, or in dry hydrogen atmosphere, generating the texture of Goss type as the MnS redissolves. If the cube texture wants to be developed, from the 2% existing in the texture of final primary recrystallization, it is necessary to play with the surface energies as driving force (which is only possible in sheets of very fine thickness) and taking advantage of the favorable influence that the oxygen—and sulphur—impurities in the surrounding atmosphere confer to the growth of {1 0 0} grains: as the surface energy is reduced. As the plane of minimum surface energy for the Fe-Si is the (1 1 0) in a pure atmosphere (under vacuum), this will grow in quantity. However, the (1 0 0) plane is that of minimum energy in presence of $O_2$ or $H_2S$.

The driving force for secondary grains in bidimensional growth can be expressed by means of the following equation:

$$\Delta G = 2 \cdot \frac{\gamma_{GB}}{r} + 2 \cdot \frac{\Delta\gamma_S}{t} \tag{3.1}$$

where $\gamma_{GB}$ is the grain boundary energy, r is the radius of the primary grain, $\Delta\gamma_S$ is the difference between energies of surface of the secondary grain in growth and the primary, and t is the thickness of the sheet. It is obvious that the second term increases its contribution when the thickness is small and the values of $\Delta\gamma_S$ are negative (this parameter is strongly influenced by the presence of impurities in the surrounding atmosphere).

It is possible to appreciate for heating temperatures smaller than that of the primary recrystallization, certain variation of properties of the metal or alloy, although there is negligible modification of the work hardened structure visible when using the optical microscope. Texture is not modified, too. This stage of partial recovery of properties, without modification of the microstructure, is called recovery, and the corresponding heat treatment—without reaching the recrystallization—is the recovery annealing. During the recovery, it is possible to appreciate, within other manifestations, the disappearance of internal stresses; recovery of electrical properties (considering that the cold deformation increases the electrical resistivity); partial recovery of mechanical properties (for example, loss of hardness and increase of the elongation); and texture modifications as measured by X-ray diffraction techniques.

Recrystallization annealing is usually carried out, in the practice, by continuous heating (Pero-Sanz et al. 1999; Martínez et al. 2001; Asensio et al. 2001). Even if the annealing furnace where the sample to be recrystallized is at a constant temperature, the work hardened sample usually progressively acquires this temperature. The loss of work hardening during the heating in some materials—if they have low stacking fault packing defects—can be so great that the material recovers but hardly recrystallizes.

Both the recrystallization annealing and the simple recovery annealing are usually called annealing against the work hardening.

## 3.2 Hot Deformation

The information indicated in the previous section referred to the heat treatments called annealing against the work hardening (recovery annealing and/or static recrystallization annealing). That is to say, holding in furnace, at suitable temperature, of work hardened structures by previous cold deformation (sheets, profiles, rods or wires) with the aim of softening them.

But when a metallic material, of recrystallized structure, is hot deformed at adequate working temperature (in the case of the steels at $T > A_3$), hardening by deformation and softening of the deformed material are simultaneously produced.

That is to say, two mechanisms of creation and elimination of crystalline defects compete simultaneously, as antagonist: the conferred work hardening during this deformation and the dynamic recovery of the structure (or its dynamic recrystallization) at the forming temperature. Generally, equilibrium in the competition between work hardening and dynamic recovery/recrystallization is reached: an equilibrium regime is achieved in such a way that—from a certain deformation—the material plastically deforms without acquiring work hardening when a constant stress is applied, at the forming temperature.

The flow stress depends on the forming temperature and on the hot deformation rate, but it does not depend on the reduction degree for each metallic material, always and when a certain hot deformation of the material is exceeded. The relation is of the following type:

$$A \cdot (\sinh \alpha \cdot \sigma)^m = Z \qquad (3.2)$$

where A and m are not constant, $\sigma$ is the applied stress and Z—equal to the product of $d\varepsilon/dt$ and $\exp(Q/R \cdot T)$ —is the Zener-Hollomon parameter.

It is habitually called hot forming—hot worked (wrought steel)—of a metal or alloy, to the process that produces great real deformations $\varepsilon$ ($\varepsilon$ between 0.5 and 5) at deformation rates, $d\varepsilon/dt$, between $10^{-2}$ and $10^3$ s$^{-1}$, which requires forming at high temperatures. Thus, the hot rolling involves deformation rates between 1 and $10^2$ s$^{-1}$, and real deformations, $\varepsilon$, between 0.2 and 0.5. The hot extrusion, in a single pass, involves rates between 10 and $5 \cdot 10^2$ s$^{-1}$, which allows deformations $\varepsilon$ between 2 and 5. The hot forge rates between $10^{-2}$ and 5 s$^{-1}$ and deformations, at once, between 0.2 and 0.5. The hammer forge 10 s$^{-1}$ < $d\varepsilon/dt$ < $10^3$ s$^{-1}$ and 0.2 < $\Sigma$ of $\varepsilon$ < 2.

It is convenient that the temperature $T_H$ of the hot deformation, to largely obtain these ductility improvements, was $T_H/T_M > 0.75$ (where $T_M$ is the melting point of the material in Kelvin degrees). Above $0.5 \cdot T_M$, the self-diffusion can take place, and it is necessary to make possible, at least, the diffusion of vacancies toward—or from—the dislocations and, therefore, achieving a dynamic recovery/recrystallization by climbing of the dislocations produced during the deformation.

The climbing of the dislocations importantly intervenes, together with other movements of dislocations, during any type of hot deformation. This participation is almost exclusive, with vacancies diffusion, when the deformation is carried out at high temperature, habitually lower than those used for the hot deformation, but always above $0.5 \cdot T_M$, by means of the application of small stresses into the glide plane that produces slow deformation rates (as also happens in the creep and in the superplastic forming).

When great stresses are used at high temperatures to achieve very fast deformations, the dynamic recovery/recrystallization is usually noticeable. This recovery is due to not only the climbing but also due to the annihilation of dislocations. The softening produced by climbing is added in this recovery together with the complementary softening resulted from, curiously, the great stress used to achieve this fast deformation (which does not only produce dislocations but also facilitates their movement at high temperature—by sliding and cross slip, apart from climbing—to produce sub-grains). In any case, dynamic recrystallization or recovery of the material occurs (in this last case with a sub-grain size proportional to $Z^{-1}$) as a result of the thermomechanical process (Verdeja et al. 2021).

## 3.3  Improvements by Hot Forming of the Solidification Structures

The as-cast structure has physical heterogeneities—columnar structure, macro-shrinkage, micro-shrinkage, blowholes, internal cracks, etc.—and chemical heterogeneities of the typical dendritic segregation.

The hot forming, whose primary aim is to provide the final shape of the part, complementarily allows to correct some of the physical and chemical heterogeneities of the as-cast structure.

Therefore, for example, the flattening of the dendritic grains at high temperature approaches the nucleus of the crystals to their periphery and, with this, the homogenization of the dendritic structure is favored because the distances that the atoms should cover during the diffusion are smaller. Obviously, the major segregation of an ingot cannot be removed by hot forge, and neither is removed by homogenization annealing (except hydrogen, carbon, oxygen, nitrogen, and boron).

On the other hand, the work hardening conferred by the deformation, if the temperature at which it is formed is adequate, induces the recrystallization of the structure and contributes to sound the part. For example, the internal discontinuities, if their free surfaces are perfectly clean, almost disappear during the forming. The nucleus of recrystallization, located at one and other side of the separation surfaces, during their growth can cross this border and the discontinuity is welded. On the contrary, when the free surface is oxidized by contact with the air (v. gr. regarding the external macro-shrinkage in ingots), the cavity cannot be welded by recrystallization during the hot mechanical forming and must be scrapped.

If the deformation rate and the temperature are adequately combined, the physical heterogeneities are corrected and it is possible to obtain, also, a grain refinement with respect to that of the as-cast structure. On the other hand, as the ferrous alloys undergo transformations, allotropic, and by solubility change, these effects complement the recovery and/or recrystallization during the hot deformation.

Moreover, the forming of ingots or semi-finished products obtained by solidification guides the intermetallic precipitates, if they are available, and the possible inclusions as well. If the inclusions are plastic, they deform in the forming direction. If the inclusions are brittle, they might be crushed, apart from being aligned in that direction. The few deformable inclusions—alumina, aluminates, some types of sulphides, etc.—noticeably decrease the hot ductility due to the formation of holes during the deformation—damage—in the inclusion-matrix interface.

Biphasic structures—v. gr. gamma + alpha—also affect the ductility. Therefore, two phases, whose individual ductility was high, behave with ductility significantly lower than when they take part, jointly, of a biphasic material. This is due to, probably, the weakness of the interphases. Logically, the chemical segregations also noticeably affect the hot ductility. It is sufficient to remember the possibility of intergranular melting (burning), and the interest of heat treatments of homogenization before the forge in the case of as-cast structures.

**Exercise 3.1** Homogenization heat treatment.

*Solution starts:*

It is often of interest to be able to calculate the time taken for an inhomogeneous alloy to reach complete (if possible) homogeneity, as, for example, in the elimination of segregation in castings and products (long products, slabs) produced by continuous casting.

The simplest composition variation that can be solved mathematically, applying Fick's equations, is if $c_B$ (segregated element) varies sinusoidally with distance as shown in Fig. 3.3. In this case, atoms diffuse down the concentration gradients, and regions with negative curvature, such as between $x = 0$ and $x = L$, decrease in concentration, while regions between $x = L$ and $x = 2 \cdot L$, increase in concentration. The curvature is zero at $x = L/2$, $x = 3 \cdot L/2$ and $x = 5 \cdot L/2$, so the concentrations at these points remain unchanged with time (Porter and Easterling 1981). Consequently, the concentration profile after certain time reduces to that indicated by the dashed line in Fig. 3.3.

At time $t = 0$, the concentration profile is given by

$$c = \bar{c} + \beta_0 \cdot \sin \frac{\pi \cdot x}{1} \tag{3.3}$$

**Fig. 3.3** The effect of diffusion on a sinusoidal variation of composition

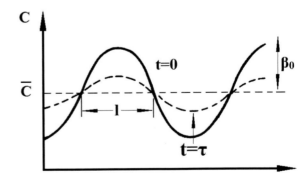

as $\bar{c}$ is the mean composition and $\beta_0$ is the amplitude of the initial concentration profile. Assuming $D_B$ is independent of concentration, the solution of the second Fick's equation is

$$\frac{\partial C_B}{\partial t} = D_B \cdot \frac{\partial^2 C_B}{\partial x^2} \tag{3.4}$$

that satisfies the initial condition, which is

$$c = \bar{c} + \beta_0 \cdot \sin\left(\frac{\pi \cdot x}{l}\right) \cdot \exp\left(-\frac{t}{\tau}\right) \tag{3.5}$$

where $\tau$ is a constant called the relation time and is given by

$$\tau = \frac{l^2}{\pi^2 \cdot D_B} \tag{3.6}$$

Thus, the amplitude of the concentration profile after a time $t(\beta)$ is given by c at $l/2$, that is to say:

$$\beta = \beta_0 \cdot \exp\left(-\frac{t}{\tau}\right) \tag{3.7}$$

In other words, the amplitude of the concentration profile decreases exponentially with time and after a sufficiently long time approaches so that $c = \bar{c}$ everywhere. The rate at which this occurs is determined by the relation time $\tau$. After a time $t = \tau$ and $\beta = \beta_0/e$, that is, the amplitude has decreased to $1/2.72$ of this value to $t = 0$. The solution distribution at this stage would therefore appear as shown by the dashed line in Fig. 3.3. After a time $t = 2 \cdot \tau$, the amplitude is reduced by a total of $1/e^2$, i.e., by about one order of magnitude. From the equation:

$$\tau = \frac{l^2}{\pi^2 \cdot D_B} \tag{3.8}$$

it can be seen that the rate of homogenization increases rapidly as the wavelength of the fluctuation decreases. The initial concentration profile will not usually be sinusoidal, but in general any concentration profile can be considered as the sum of an infinite series of sine waves of varying wavelength and amplitude and each wave decays at a rate determined by its own $\tau$. Thus, the short wavelength terms die away very rapidly, and homogenizations will ultimately be determined by $\tau$ for the longest wavelength component.

Application:

A low-alloy ingot contains 2.0% Ni and 0.4% C. It is slowly solidified so that the dendrite arm spacing is 500 μm. How long a time will be required to reduce the coring of carbon by 90%? And, to reduce the coring of nickel by the same amount?

Data: $D_C^\gamma = 2.23 \cdot 10^{-6}\,cm^2/s$ (at 1200 °C), $D_{Ni}^\gamma = 7.03 \cdot 10^{-11}\,cm^2/s$ (at 1200 °C).

– For the carbon:

$$\tau = \frac{l^2}{\pi^2 \cdot D_B} = \frac{(250 \cdot 10^{-4}\,cm)^2}{\pi^2 \cdot 2.23 \cdot 10^{-6}\,cm^2/s} = 28.4\,s \tag{3.9}$$

$$\frac{\beta}{\beta_0} = 0.1 = \exp\left(-\frac{t}{\tau}\right) \rightarrow t = -28.4 \cdot \ln 0.1 = 65.4\,s \tag{3.10}$$

– For the nickel:

$$\tau = \frac{l^2}{\pi^2 \cdot D_B} = \frac{(250 \cdot 10^{-4}\,cm)^2}{\pi^2 \cdot 7.03 \cdot 10^{-11}\,cm^2/s} = 9 \cdot 10^5\,s \tag{3.11}$$

$$\frac{\beta}{\beta_0} = 0.1 = \exp\left(-\frac{t}{\tau}\right) \rightarrow t = -9 \cdot 10^5 \cdot \ln 0.1 = 2.07 \cdot 10^6\,s(24\,days\,) \tag{3.12}$$

Note: The removal of the segregation, inherited from the solidification, is possible and very fast in the elements of small atomic radius (hydrogen, carbon, oxygen, nitrogen) that form interstitial solid solution by homogenization. However, it is industrially not possible for the elements in substitutional solid solution (manganese, chromium, nickel, molybdenum). It is necessary to use the hot forge (rolling, extrusion) that will "partially" achieve it as maximum, which approaches the periphery and nucleus of the solidification dendrites (see section dedicated to banded structure).

*Solution ends.*

### 3.3.1 Forge Fibering and Crystalline Texture. Anisotropy of Properties

Each type of forming usually confers a characteristic forge fibering (mechanical)—by aligning of secondary phases: precipitates, inclusions, constituents, bands—and a specific crystalline texture or crystallographic preferential orientation of the grains. As a consequence of one and other, anisotropy of properties is produced between the lengthwise, LD, and the cross/transverse, DT, directions in the plane of the sheet.

Therefore, for example, the texture {1 0 0} parallel to the rolling plane induces in the sheets a mechanical anisotropy in "V": the yield strength according to the LD direction is smaller than in the DT direction and is minimum at 45° to the rolling direction. On the other hand, the texture {1 1 3} parallel to the rolling plane gives an anisotropy of toughness of opposite sign with shape of "A" by grains cleavage along the crystallographic directions <1 0 0>. With this texture, the ductile–brittle transition temperature (or impact transition temperature) determined in the rolling direction is smaller than that of the cross direction, and both smaller than the DBTT in intermediate directions such as 45°. According to Inagaki et al. (1977), the anisotropy of yield strength decreases and the average value of the yield strength increases when the rolling texture transforms from the {1 0 0}-{1 1 3}-{1 1 2} to the {1 1 0}-{3 3 2}-{1 1 1}; in other words, when there is not preferential orientation of grains with crystallographic planes {1 0 0}, or neighboring, parallel to the rolling plane.

With frequency, on the contrary, it is possible to mention that in long products, as a consequence of the forge fibering conferred by the inclusions, the lengthwise (RD) characteristics and, concretely, the toughness are greater than in the cross direction (DT).

### 3.3.1.1  Anisotropy and Embrittlement by Hydrogen

The embrittlement by hydrogen (hydrogen-induced cracking, HIC) is well known in high-carbon steels, which are especially prone to the absorption of this element during the elaboration and solidification. After forging the steel, monoatomic hydrogen—available in interstitial solid solution—diffuses, tends to molecularly recombine, produces localized pressures, and—especially if there are stresses in the material—produces very fine cracks with the subsequent decrease of toughness. Historically, this influence of the hydrogen on the embrittlement was firstly identified by the fracture surface in some broken rails of the ancient trains of North America.

Obviating the absorption of hydrogen during the elaboration, avoiding the stresses in the material, and—in the corresponding case—using isothermal heating of long duration with the aim of achieving the hydrogen diffusion out of the steel surface and transfer it to ambient are remedies against this type of embrittlement.

HIC is also usually presented in low-carbon steels with significant number of inclusions. It was also checked, because of very localized increase of hardenability, in quench and temper steels that the monoatomic hydrogen diffuses toward the inclusions and enriches them in this element.

For example, HIC is present in weldable steels with <0.20% C and noticeably contributes to this the anisotropy of forge and textures. It is observed, for example, in high-resistance tubes—manufactured in microalloyed steels—through which gas or petroleum with dissolved $H_2S$ circulates. The nascent monoatomic hydrogen preferentially diffuses toward the segregated zones (MnS inclusions, pearlite bands, carbides) and tends to form fissures (cracks) parallel to the rolling plane.

It seems to be sufficient to eliminate these risks of HIC in low-carbon steels to use degassed steel, of very low sulphur content, treated with calcium, homogeneous

| Element | Host element | $D_0$ (m²/s) | Q (kJ/mol) |
|---------|-------------|--------------|-----------|
| Hydrogen | Fe-γ | $6.3 \cdot 10^{-7}$ | 43.1 |
| Hydrogen | Fe-α | $1.0 \cdot 10^{-7}$ | 13.4 |

**Table 3.1** Data to calculate the hydrogen diffusion in the iron

in composition and without pearlite bands (%C < 0.02%, see Sect. 3.4.1.2). We have seen, however, that the HIC defect also appears in the ferrite, matrix constituent of the steel. Everything makes us to assume that the crystallographic orientation of the ferrite grains—its texture—is a factor to be taken into account. It has been checked that, sheets with texture {1 1 0}, {1 1 3}, {1 1 2}—as those obtained in very hard rolling processes—favors the HIC even in steels of very clean metallurgy. The interstitial diffusion of the hydrogen between planes {1 0 0} could be one of the reasons.

The embrittlement by hydrogen is usually also presented in other cases. Thus, for example, in the heat-affected zone (HAZ) that appears during the welding, if the steel contains hydrogen, or there is absorption of this element in the molten bead, and there are stresses (v. gr. by formation of very tetragonal martensite). This is the interest of using adequate electrode coatings and, in certain cases, the convenience of preheating before welding.

It seems convenient to mention that it is not adequate to confuse the embrittlement mentioned in this section with the incidence that the hydrogen in interstitial solid solution has in the aging of very low-carbon steels due to the Cottrell effect (Table 3.1).

**Exercise 3.2** Compare the results of the hydrogen diffusion at 450 °C and 1000 °C when the grain size is 1000 μm for the Fe-α ($d_\alpha$) and Fe-γ ($d_\gamma$), respectively. Compare the results also for a flat rough product of the continuous casting having a thickness equal to 20 cm. What consequences of industrial relevance can be deduced?

Data: $a_\gamma = 3.639$ Å, atomic weight of the iron: 55.85 g·mol⁻¹.

*Solution starts:*

(a) The distance of diffusion <x>, corresponding to the distance between the center of the grain and the periphery, is approximately half of the grain size. Analogously, the macrosegregated zones in rough passes are found in the center of the rough product and, therefore, the diffusion distance will be half the thickness.

– In the case of the diffusion of hydrogen in the Fe-α at 450 °C (723 K):

The diffusion coefficient at 723 K is calculated as follows:

$$D = D_0 \cdot \exp\left(-\frac{Q}{R \cdot T}\right) = 1.0 \cdot 10^{-7} (m^2/s) \cdot \exp\left(-\frac{13400\frac{J}{mol}}{8.314\frac{J}{mol \cdot K} \cdot 723K}\right)$$

$$= 9.293 \cdot 10^{-9} m^2/s \tag{3.13}$$

And the time required for the diffusion is

In the grain (1000 µm in diameter):

$$t = \frac{\langle x \rangle^2}{D} = \frac{\left(500\mu m \cdot \frac{1\,m}{10^6\mu m}\right)^2}{9.293 \cdot 10^{-9} m^2/s} = 26.9\,s \qquad (3.14)$$

In the rough product (20 cm in thickness):

$$t = \frac{<x>^2}{D} = \frac{\left(10\,cm \cdot \frac{1\,m}{10^2\,cm}\right)^2}{9.293 \cdot 10^{-9} m^2/s} = 1.076 \cdot 10^6\,s(12.5\ days) \qquad (3.15)$$

- In the case of the hydrogen diffusion in the Fe-γ at 1000 °C (1273 K):

$$D = D_0 \cdot \exp\left(-\frac{Q}{R \cdot T}\right) = 6.3 \cdot 10^{-7} (m^2/s) \cdot \exp\left(-\frac{43100\frac{J}{mol}}{8.314\frac{J}{mol \cdot K} \cdot 1273K}\right)$$
$$= 1.073 \cdot 10^{-8} m^2/s \qquad (3.16)$$

And the time required for the diffusion is

In the grain (1000 µm in diameter):

$$t = \frac{<x>^2}{D} = \frac{\left(500\mu m \cdot \frac{1\,m}{10^6\mu m}\right)^2}{1.073 \cdot 10^{-8} m^2/s} = 23\,s \qquad (3.17)$$

In the rough product (20 cm in thickness):

$$t = \frac{<x>^2}{D} = \frac{\left(10\,cm \cdot \frac{1\,m}{10^2\,cm}\right)^2}{1.073 \cdot 10^{-8} m^2/s} = 9.31 \cdot 10^5\,s(10.8\ days) \qquad (3.18)$$

Conclusions:

- The elimination of the hydrogen microsegregation is feasible in very short periods.
- It is possible the attenuation of the hydrogen microsegregation in parts of small thickness (<x> <10 cm). For example, old pit furnaces still operating in integrated steel factories in thick plate lines are used with small thickness rough products that are kept in contact with the atmosphere (↓ $a_H$ (atm)) for 4–5 days. It is applicable to steels sensitive to Stress Corrosion Cracking (SCC) to which is required H < 1.5 ml/100 g or acid gas.
- The homogenization in ferritic phase is cheaper than the homogenization in austenitic phase due to the temperatures involved in the process.
- The existence of inclusions involves traps for the diffusion of the hydrogen where the concentrations could be locally high and, this way, two atoms of hydrogen

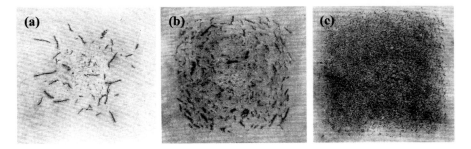

**Fig. 3.4** 140 mm bloom of bearing steel. Rolled and cooled at three different cooling rates. Hair-line cracks have appeared in the center of the bloom. **a** After rolling, air cooled to 200 °C followed by furnace cooling; **b** After rolling, air cooled to room temperature; **c** After rolling, air cooled to 200 °C followed by water quench. These micrographs show sections through hair-line cracks, whose direction is only random in 585/1. The cooling rate affects the number, size, and distribution of the cracks

react and give $H_2$ (g) and the joint between the inclusion and the metallic matrix is broken.

*Solution ends.*

**Exercise 3.3** Hydrogen-induced cracking (HIC).

*Solution starts:*

Figure 3.4 has been taken from the 140 mm bloom of bearing steel rolled and cooled at three different cooling rates. Hair-line cracks appeared in the center of the bloom. The first macrostructure is obtained after rolling, air cooled at 200 °C, followed by furnace cooling. The second one after rolling, air cooled at room temperature. The third one after rolling at 200 °C followed by a water quench. The hair-line cracks are formed below 200 °C. The cooling rates affect the number, size, and distribution of cracks (Schrader and Rose 1966).

The appearance of "flaking" is associated with a certain number of factors. The principal cause is too high hydrogen content. Below a certain value, which may be obtained by vacuum casting, the susceptibility to flaking disappears. The hydrogen content decreases by diffusion toward the surface of the worked product; the flakes are no longer found in thin sections not in heavy sections, if the cooling rate is very slow during the passage, through the critical zone from 400 to 200 °C (this is, in fact, one remedy for this type of defect). As a result of this outward diffusion, flake formation does not occur in the periphery of forged or rolled products; the cracks are therefore unoxidized and can be welded up by subsequent working with judiciously distributed passes. The number and disposition of the flakes and the volume of metal involved depend upon the cooling rate below the temperature at which they form: 200 °C.

Certain alloy elements, such as nickel, molybdenum, chromium, and manganese, increase the susceptibility to flake formation. The hardenability of the segregated zones may even give residual austenite in which the hydrogen can concentrate. Thus, although there is no close relationship between flakes and segregations—this type of crack is often found to cross large dendrites in blooms—ghost and axial V segregates favor the appearance.

The combination of conditions necessary for flaking is thus found in large forging ingots of alloy steel, but flakes are also found in thin products, for example, in 40 mm plate, even in low-alloy steel.

Internal stresses have a directional effect on the flakes which are oriented in the direction of working of the part. In blooms, for example, the cracks are perpendicular to the cambered faces, and in plates they are parallel to the plane of the rolling. In this latter case, the banded structure will not play an essential role except where there is considerable segregation.

Application:

A stress relieving treatment at 450 °C is given to avoid the precipitation of hydrogen flakes in a slab of 20 cm in thickness and the average grain size is 1000 $\mu$m. Calculate the time required to avoid the minor segregation (coring) and the macrosegregation of the hydrogen (macrosegregation at the scale of the slab) and comment if they are reasonable.

Data: for the diffusion of the hydrogen in the $Fe_\alpha$, $D_0^H = 10^{-3}$ cm$^2$/s; $Q = 13.4$ kJ/mol.

Note: We use the same model than that followed to eliminate the segregation of carbon in alloy elements during the heat treatment of homogenization annealing.

$$D_H^\alpha = D_0^H \cdot \exp\left(-\frac{Q}{R \cdot T}\right) = 10^{-3} \, cm^2/s \cdot \exp\left(-\frac{13400\frac{J}{mol}}{8.314\frac{J}{mol \cdot K} \cdot 723K}\right)$$

$$= 1.1 \cdot 10^{-4} \, cm^2/s \tag{3.30}$$

– Coring:

$$\tau = \frac{l^2}{\pi^2 \cdot D} = \frac{(0.1 \, cm/2)^2}{\pi^2 \cdot 1.1 \cdot 10^{-4} \, cm^2/s} = 2.30 \, s \tag{3.31}$$

To eliminate it in a 99%:

$$\frac{\beta}{\beta_0} = 0.01 = \exp\left(-\frac{t}{\tau}\right) \to t = -2.3 \cdot \ln 0.01 = 11 \, s \tag{3.32}$$

There is no problem in the duration of the process.

– Macrosegregation (l = 20/2 cm):

$$\tau = \frac{l^2}{\pi^2 \cdot D} = \frac{(20\,\text{cm}/2)^2}{\pi^2 \cdot 1.1 \cdot 10^{-4}\,\text{cm}^2/\text{s}} = 92110\,\text{s} \tag{3.33}$$

To eliminate the macrosegregation in a 99%:

$$\frac{\beta}{\beta_0} = 0.01 = \exp\left(-\frac{t}{\tau}\right) \rightarrow t = -92110 \cdot \ln 0.01 = 4.9\ \text{days (long)} \tag{3.34}$$

$$\frac{\beta}{\beta_0} = 0.1 = \exp\left(-\frac{t}{\tau}\right) \rightarrow t = -92110 \cdot \ln 0.01 = 2.45\ \text{days (long)} \tag{3.35}$$

It is necessary to apply long treatments of stress relieving in the case of slabs of 20 cm in thickness. It is possible to reduce the hydrogen content from 10 to 1 ppm ($\beta/\beta_0 = 0.1$) or even 0.1 ppm ($\beta/\beta_0 = 0.01$) with these treatments.

If the rough slab had 10 cm in thickness (half thickness, $l = 5$ cm), the times required to eliminate the segregation would be divided by 4. That is to say:

$$\tau = \frac{l^2}{\pi^2 \cdot D} = \frac{(10\,\text{cm}/2)^2}{\pi^2 \cdot 1.1 \cdot 10^{-4}\,\text{cm}^2/\text{s}} = 23028\,\text{s} \tag{3.36}$$

$$\frac{\beta}{\beta_0} = 0.01 = \exp\left(-\frac{t}{\tau}\right) \rightarrow t = -23028 \cdot \ln 0.01 = 1.23\ \text{days} \tag{3.37}$$

$$\frac{\beta}{\beta_0} = 0.1 = \exp\left(-\frac{t}{\tau}\right) \rightarrow t = -23028 \cdot \ln 0.01 = 14.7\ \text{h} \tag{3.38}$$

It is possible to reduce the trapped hydrogen from 10 ppm to 1–0.5 ppm with these annealing treatments aimed at alleviating stresses, and this avoids the hydrogen-induced cracking of the manufactured product. It is convenient to not forget that these are long treatments that are deleterious for the economy and productivity of the process.

*Solution ends.*

## 3.4   Thermomechanical Treatments of the Austenite Before Their Allotropic Transformation

The austenitic structure of the steel can be recrystallized, or recovered, at the end of the forming in gamma state (above the $A_{3C}$ temperature) and later, during the cooling, this gamma structure will allotropically transform. We will comment two consequences of this transformation posterior to the forming in this section: the banded structure and the controlled rolling.

Complementarily, considering that it is also possible to deform a steel in gamma state at temperatures lower than $A_1$, we will comment in Sect. 3.4.3 a thermomechanical treatment called ausforming LT.

### 3.4.1 Banded Structure

It seems convenient to refer here to a particular type of fibering, called banded structure—generally alternate bands of ferrite and pearlite—that usually appears in low-alloy steels once cooled at air after their forming in gamma state (Schrader and Rose 1966). Figure 3.5b, c illustrates this banded structure that is legacy of the as-cast state.

The steel with alternated bands of ferrite and pearlite can be considered as a composite material formed by pearlite fibers (integrated at the same time by a ceramic material—cementite—and a metallic material—ferrite) in a metallic matrix of ferrite. This banded structure accentuates the anisotropy of properties between the direction of forming and their normal. This can be advantageous for some directional behaviors but, in general, the banded structure is not favorable.

For example, in operations of lengthwise machining, as the milling, the surface finishing is worse if the structure is banded that when not: because the bands of ferrite have less hardness than those of pearlite and, for that reason, the chip removal is different depending on the constituent. In the turning, on the contrary, as a consequence of being a non-lengthwise operation, the degree of surface finishing hardly depends on the banded structure.

The banded structure is also disadvantageous if vibrations want to be reduced. Therefore, for example, the gear assemblies showing structure in bands are noisier than those that do not have bands.

#### 3.4.1.1  Origin of the Banded Structure

During the solidification process of the steel, the dendritic surroundings enrich in elements in solid solution, both alphagenous and gammagenous (we should remember that the partition coefficient K of the elements in the iron is <1, see Table 3.2). Also, the manganese sulphides usually appear—although not in solid solution, obvious, but as precipitates—in the periphery of the dendrites. However, these sulphides appear in the steel (they are inclusions habitually deformable) once formed aligned in the forming direction.

Carbon easily diffuses during the forge in gamma state of the as-cast steel, and therefore the carbon content of the formed austenite is almost equal in the zones that solidified first and enriched in other elements. But this does not happen like this with the other alloy elements that, as a consequence of being in substitutional solid solution in the Fe, diffuse more difficultly than the carbon (whose solid solution is of insertion). And the responsible of the banded structure is precisely the alphagenous

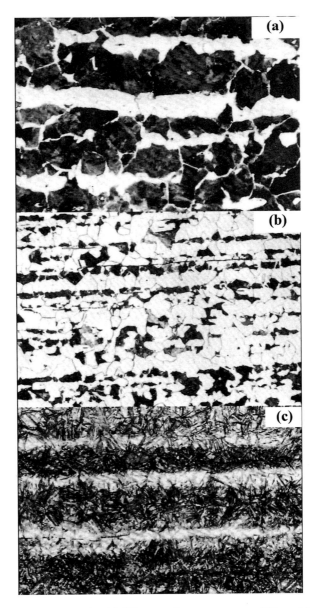

**Fig. 3.5** Banded structure in steels with different carbon percentages: **a** 180 mm bloom of medium-carbon Mn-Mo steel, longitudinal section, as-rolled condition; white bands consisting of ferrite grains alternate with of pearlite; the large austenite grain size can be seen in places (0.35%C, 0.35%Si, 1.50%Mn, 0.22%Mo) (50x); **b** Carbon steel rolled to 30 mm diameter bar, section in the direction of the fiber structure; 880 °C 45 min/furnace; clearly visible banded structure: alternating white ferrite and dark pearlite bands, with some inclusion (0.195%C, 0.20%Si, 0.54%Mn, 0.032%P, 0.022%S) (100x); **c** Banded structure after stepped cooling, medium-carbon Cr–Mo steel, longitudinal section of a 5-mm-thick plate; 900 °C 45 min/350 °C 30 s/water; very marked structure due to the presence of two differently etching constituents: dark needles of bainite and light bands of martensite (0.33%C, 0.22%Si, 0.67%Mn, 0.027%P, 0.025%S, 1%Cr, 0.28%Mo) (100x)

**Table 3.2** Partition coefficient, K, of some elements in the iron

|            | Delta iron | Gamma iron |
|------------|------------|------------|
| Oxygen     | 0.02       | 0.02       |
| Sulphur    | 0.02       | 0.02       |
| Phosphorus | 0.13       | 0.06       |
| Carbon     | 0.13       | 0.36       |
| Nitrogen   | 0.28       | 0.54       |
| Copper     | 0.56       | 0.88       |
| Silicon    | 0.66       | 0.95       |
| Molybdenum | 0.80       | 0.60       |
| Nickel     | 0.80       | 0.95       |
| Manganese  | 0.84       | 0.95       |
| Cobalt     | 0.90       | 0.95       |
| Tungsten   | 0.95       | 0.50       |
| Chromium   | 0.95       | 0.85       |

or gammagenous character of these elements: as a consequence of being different the values of $A_3$ (and also of $A_1$) in the segregated zones and in the non-segregated zones.

The genesis of the banded structure is produced as follows. We assume, for example, that the elements in solution were only alphagenous v. gr. small phosphorus and silicon contents (as it happens in carbon steels of Fig. 3.5a). When the temperature descends once hot formed the steel, the first zones that transform from gamma to alpha are precisely the most enriched in those elements as their temperature $A_3$ is higher. As a consequence of this, the carbon atoms expelled of these ferrite grains pass later to the contiguous austenite, enriching it in carbon and, therefore, decrease the $A_3$. Thus, these zones transform later than the first ones during the continuous cooling and give origin to the bands of pearlite. The sulphides would be included in the ferrite bands.

If the solute elements were all gammagenous, instead of alphagenous, the banded structure also appears. It would result—by analogous reasoning—that the segregated zones would transform in the last time to give pearlite. These pearlite bands, at the same time, would be contoured by the bands of ferritic grains previously formed, and the sulphides would appear located in the pearlite bands instead of being located in the ferrite bands (this happens when the alloying elements are alphagenous).

The banded structure sometimes does not appear. This happens if the chemical composition of the steel is such that, by an adequate dosage of antagonist elements alphagenous and gammagenous, it is possible to transform at the same time from gamma to alpha both the segregated zones as the non-segregated ones (as a consequence of achieving the same value of $A_3$ in both zones). But in this case, although bands of ferrite and pearlite would not appear, the chemical heterogeneities survive. The bands are only masked: the segregated zones—rich in elements in solid solution—have more hardenability than those with negative segregation. Therefore, if the structure is again austenitized, and later continuously and rapidly cooled, it is possible to appreciate these hardenability differences: alternated bands of martensite and bainite, or bainite and pearlite, or martensite and pearlite, etc. appear—according to the TTT curves corresponding to the segregated and non-segregated zones, see Fig. 3.5c.

### 3.4.1.2  Factors to Remove or Mask the Banded Structure

All those factors that contribute to achieve a uniform chemical composition, at macroscopic scale before the forming, are favorable to avoid the banded structure. Therefore, a remedy to eliminate the possibility of banded structures is the homogenization annealing at very high temperature (that is more efficient if it is performed after a forge previous to the definitely hot forging).

It is less important the chemical heterogeneity, responsible of the bands, if the solidification takes place with a crystallization in dendrites of small size. Steels with fine grain of solidification do not usually exhibit the banded structure, because it is homogenized with greatest easiness during the hot forming. And, in any case, that chemical heterogeneity can be fastly removed during a homogenization annealing. The grain of solidification of small size can be obtained by means of an increase of the nucleation rate of the solid, by faster cooling of the melt, or—for the same rate—using ingots of smaller size, or by means of grain refinement clustering agents, etc. (Pero-Sanz et al. 2017).

The banded structure is not observed in extra-mild steels with carbon contents lower than 0.02%C because the formation of pearlite is not possible. It is especially representative in steels whose carbon content is between 0.17% and 0.53% C. These steels whose solidification takes place with peritectic reaction (see Fig. 1.16) are more prone to the banded structure, that is, the high- or very low-carbon content steels.

In fact, steels with carbon content comprised between 0.17 and 0.53% C, whose structure of primary solidification should be only gamma (see Fig. 1.16), have a discontinuity in the chemical composition as a consequence of the peritectic reaction (liquid+delta $\rightarrow$ gamma), which takes place after the end of the solidification. Thus, as example, the proportions of delta solid (of 0.09% C) and gamma solid (of 0.17% C) in a steel of 0.13% C, at the end of the peritectic solidification, are almost equal. And after, the previous gamma regions still have almost twice carbon than the others once the allotropic delta $\rightarrow$ gamma transformation is carried out. This way, the local carbon discontinuity is added to the heterogeneity in composition of the substitutional solutes.

On the other hand, whenever there is peritectic reaction, both in ferrous and non-ferrous alloys, the equilibrium is not complete (see Fig. 3.6). Even steels of carbon content between 0.09 and 0.02% C, whose equilibrium solidification structure is only delta, can also exhibit peritectic reaction by non-equilibrium solidification.

When the chemical heterogeneity responsible of the bands cannot be eliminated, it is possible to mask its structure with the aim of facilitating the regularity in machining

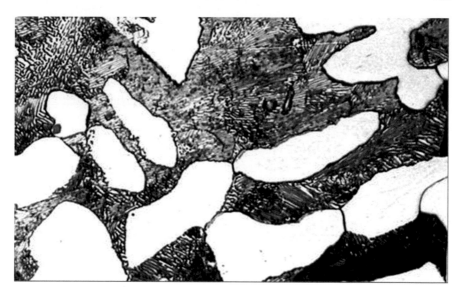

**Fig. 3.6** Incomplete peritectic reaction. Molded steel, 0.38% C Above: Nucleus, ferrite dendrites surrounded by pearlite, 100x; Below: Nucleus, 600x

operations; although the bands potentially subsist, they will appear in posterior heat treatments.

Thus, for example, the banded structure will not appear when the cooling of the steel, instead of being carried out at air from the forming temperature, is isothermally performed in salts at temperatures—below $A_e$—belonging to the lower pearlitic zone of the TTT curve. This way, it is possible to simultaneously transform the segregated zones and the non-segregated zones, to only give a pearlitic structure (more favorable for machining operations). Proceeding in this way, it is possible to mask the bands, but not eliminating them, since the heterogeneity regarding the chemical composition will still exist. If this pearlitic structure is austenitized again and an air cooling from this temperature was carried out—for example, a normalizing treatment after the machining—the banded structure will appear again.

The bands can be also masked by means of a full austenitization annealing, if the holding at temperatures greater than $A_3$ for growth of the gamma grain is extended. The objective of this treatment is to achieve that the diameter of the austenitic grain could exceed the distance existing between the segregated and non-segregated zones, contiguous and parallel. Thus, if the width of the chemical heterogeneity is localized inside of the austenitic grain, bands will not appear during the gamma/alpha transformation that is produced during the continuous cooling. Anyway, this growth annealing—from which the low toughness inherited from the coarse austenitic grain is derived—only has justification to improve the machining easiness, and it is, for that reason, an intermediate treatment, not a final treatment for parts in service.

### 3.4.2  Controlled Rolling

It is understood by controlled rolling the hot forming technique that achieves a noticeable grain refinement of the resulting ferrite, by means of three kinetics:

– that of the carbides, nitrides, and carbonitrides of microalloying elements (Ti, Nb) precipitation available in very small quantity in the chemical composition of the considered steel (<0.1%).
– that of the recrystallization of the hot deformed austenite.
– that of the austenite $\rightarrow$ ferrite allotropic transformation (precipitation of vanadium if available).

We schematically illustrate in Fig. 3.7 the changes of structure from gamma to alpha, in three different processes of rolling in gamma phase for the same steel. It is possible to check that with a suitable rolling, it is possible to impede the recrystallization of the austenitic structure and achieving at the end of the rolling, at temperature greater than $A_3$, a structure of work hardened austenite. The structure like this obtained by this controlled rolling, as they later allotropically transform during the cooling, provides regular grains of alpha phase with a size noticeably smaller than that obtained from grains of austenite that would have recrystallized at the end of the rolling.

The structure refinement like this achieved provides important improvements of the yield strength and the toughness in the case of weldable ferritic-pearlitic steels. In these steels, the delay of the nucleation is easily achieved, and with this, the recrystallization of the austenite during the rolling, if the steel contains nanometric

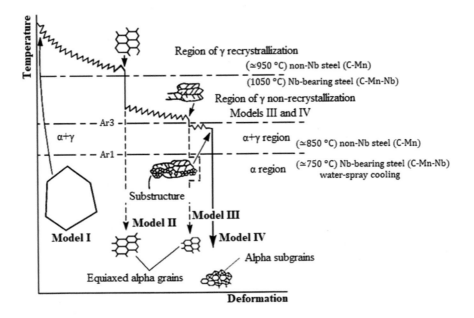

**Fig. 3.7** Scheme of three rolling processes: resulting structures

particles (of around 0.002–0.003 μm, that is to say, 2–3 nm or 20–30 Å). In fact, the fine particles lead to a uniform increase in the density of dislocations and drag their movement, and therefore the nucleation is delayed. Although the delay in the recrystallization can be attributed to more than one mechanism, the particles can fix both the grain sub-boundaries and the grain boundaries, and with this delay both the nucleation and the migration of grain boundaries. The obstacle that these nano-metric precipitates imply can even impede the recrystallization at this temperature (increasing, therefore, the temperature to make possible the recrystallization). The temperature of non-recrystallization can be estimated as follows, where each element represents their percentage in the composition of the steel:

$$T_M(\text{ in }°C) = 887 + 464 \cdot \%C + 6445 \cdot \%Nb - 644 \cdot (\%Nb)^{1/2} + 732 \cdot \%V$$
$$- 230 \cdot (\%V)^{1/2} + 890 \cdot \%Ti + 363 \cdot \%Al - 357 \cdot \%Si \qquad (3.39)$$

The particles of precipitated secondary phases, especially if they are finely dispersed, influence in the recrystallization, both static and dynamic, of any metal or solid solution. They affect to the mechanisms that intervene during the recovery—which at the same time affect the nucleation—and there is experimental evidence that they also act on the movement of the grain boundaries. With this, it is difficult to point out if the presence of precipitates has more influence in the nucleation or in the movement of the grain boundaries during the recrystallization process: since any experimental measurement of the nucleation rate has associated part of the grain growth and, therefore, it is not possible to separate both effects. Regarding the static nucleation, it is possible to say that the second phases precipitated in the work hard-ened matrix accelerate the formation of nuclei if they are particles with a size greater than 100 nm (we do not include here the non-metallic inclusions or impurities of big size—of around 1 mm— which are always unfavorable). This is due to the local concentration of stresses that the cold deformation produces on the work hardened matrix that surrounds the particles for sizes between 10 and 20 μm. Particles of intermediate size, between 10 and 20 nm—between 0.01 and 0.02 μm— are exces-sively small to favor the nucleation and, on the other hand, they are excessively big to stop the movement of the dislocations or sub-grains. Definitely, they do not have influence on the recrystallization rate, although, on the contrary, they can inhibit the grain growth stage and are usually employed v. gr. to limit the grain size in steels (see Sect. 2.3.1).

The recrystallization of the austenite is impeded if the steel contains a certain amount of niobium precipitated as nitride or carbonitride. When rolling these steels in gamma state, the austenite deforms and acquires work hardening, but it does not recrystallize during the finishing stage of the hot rolling process because the precipitates impede it. When transforming the work hardened austenite to give ferrite, during the cooling, very fine grains (more than 31,000 grains of ferrite per square millimeter, 12 ASTM grain size) are obtained.

Other forming processes of the austenite at high temperature, different from the controlled rolling, carried out with the same aim of obtaining work hardened austenite before the allotropic transformation that would produce ferritic-pearlitic structures are called HT ausforming (high-temperature ausforming).

**Exercise 3.4** It is assumed that the mechanical properties of the final product, univocal function of its microstructure, are characterized in the current case by a single parameter: the grain size. This is the reason of the importance of calculating the evolution of grain size along the rolling process with the aim of obtaining the suitable mechanical properties (yield strength, ultimate tensile strength, ductility, ductile–brittle transition temperature).

We proceed to hot controlled rolling according to the model II in Fig. 3.7 of the deep-drawing quality (DDQ) steel (%C < 0.6; %Mn < 0.03; %Si < 0.02; %S < 0.02; %P < 0.02; 0.03 < %Al < 0.06). It does not contain any other microalloying agents (Ti, Nb, V) except aluminum which plays an important role in the homogenization heat treatment (around 6 h at 1200 °C) to avoid the austenitic grain growth and try that both nitrogen (around 60 ppm) and aluminum were in solid solution in the austenite, avoiding its precipitation both during the controlled rolling (roughing in reversing mill and finishing in tandem mill, semicontinuous mill) and during the coiling, before the water shower, that is carried out at around 600 °C. The sheet leaves the process with a thickness of 4.5 mm and is cold rolled at 70% until a thickness of 2.6 mm. Finally, it is subjected to a discontinuous recrystallization annealing (batch annealing) at 650 °C with slow cooling rate until reaching an average value of the Lankford coefficient of anisotropy of $\bar{r} \simeq 1.6$.

We use the empirical relations determined by C. M. Sellars and J. A. Whiteman for the calculation in each pass of the Zener-Hollomon (Z) parameter, the time of recrystallization ($t_{x=0.95}$), and the time of grain growth (Sellars and Whiteman 1979; Dutta and Sellars 1986, 1987):

$$Z = \dot{\varepsilon} \cdot \exp\left(\frac{37440}{T}\right) [s^{-1}] \tag{3.40}$$

$$t_{x=0.95} = K \cdot Z^{-0.375} \cdot \varepsilon^{-4} \cdot D_0^2 \cdot \exp\left(\frac{Q_{rex}}{R \cdot T}\right) \tag{3.41}$$

where $Q_{rex}$ represents a constant of the material, which, in this case, is equal to 57,400 J/mol (activation energy for the recrystallization), and K depends on the fraction of recrystallized material, X, in this case, X = 95% of mild steel, and the value of K is

$$K = 3.54 \cdot 10^{-21} [s^{0.625} \cdot \mu m^{-2}] \tag{3.42}$$

The time of growth is calculated as follows:

$$t_{growth} = t_{between\ passes} - t_{x=0.95} \tag{3.43}$$

Additionally, we take as temperature between passes the average temperature between them.

The mean grain size at the end of the recrystallization process depends only on two factors: the energy stored during the deformation and the density of possible sites for the nucleation. This way, the size of the recrystallized grain is

$$D_{rex} = K' \cdot \left[ \frac{1}{n \cdot \alpha} \cdot \ln\left(\frac{Z}{A}\right) \right]^{-0.67} \cdot \varepsilon^{-1} \cdot D_0^{0.5} \tag{3.44}$$

where

$$K' = 25\left[ (m^2/MN)^{0.67} \cdot \mu m^{1/2} \right] \tag{3.45}$$

$$n \cdot \alpha = 6.7 \cdot 10^{-2}\left[ m^2/MN \right] \tag{3.46}$$

$$A = 8.5 \cdot 10^9\left[ s^{-1} \right] \tag{3.47}$$

for the mild steel.

Regarding the grain growth, with short times of permanence at high temperature, the variation with the time of the grain diameter is adjusted by the following equation:

$$D^{10} - D_{rex}^{10} = M \cdot t_{growth} \tag{3.48}$$

where M is given by an equation similar to that of the thermally activated processes:

$$M = K'' \cdot \exp\left(-\frac{Q}{R \cdot T}\right) \tag{3.49}$$

and in the case of the mild steel:

$$M = 3.87 \cdot 10^{32} \cdot \exp\left(-\frac{400000J/mol}{R \cdot T}\right) \tag{3.50}$$

We also use the following equations:

- Linear speed of the rolls:

$$v = 2 \cdot \pi \cdot R \cdot \frac{rpm}{60} \tag{3.51}$$

- Equivalent deformation:

$$\bar{\varepsilon} = 1.155 \cdot \ln\left(\frac{h_0}{h_f}\right) \tag{3.52}$$

- Mean equivalent deformation rate:

$$\bar{\dot{P}} = \frac{1.155 \cdot v}{\sqrt{R \cdot \Delta h}} \cdot \ln\left(\frac{h_0}{h_f}\right) \qquad (3.53)$$

The roughing is performed in 7 passes (four high reversing mill) and the finishing in 7, semicontinuous hot rolling mill, whose geometry and treatment process is

- Dimensions of the rough product (slab): 230 mm × 1030 mm × 8998 mm.
- Intermediate plate (slabbing mill): 40 mm × 41,797 mm (plate obtained at the end of the roughing).
- Weight: 16,733 kg.

Homogenization treatment: 396 min (time of permanence in the furnace), temperature at the exit of the furnace: 1201 °C.

Roughing mill: diameter of the working cylinders, 1119.2 mm; diameter of the back (supporting) cylinders, 1304.5 mm.

Values of $t_{pause}$ (s) are known for the roughing.

In the case of the finishing, the distance between cages being about 5500 mm and the $t_{pause}$ (s) is calculated as follows:

$$t_{pause}\ (s) = \frac{5500 mm}{v_{linear}\ (m/min) \cdot \frac{1000 mm}{1 m} \cdot \frac{1 min}{60\,s}} \qquad (3.54)$$

*Start of the exercise:*

We collect in Tables 3.3, 3.4, 3.5, 3.6, 3.7, 3.8 and 3.9 the results for each pass using the equations indicated in the enunciation of the exercise.

We collect in Table 3.10 a summary of the roughing process.

**Table 3.3** Results for the pass 1

| Pass 1 | |
|---|---|
| $h_0$ (mm) | 230 |
| $h_f$ (mm) | 197 |
| $\bar{\varepsilon}$ | 0.179 |
| $\bar{\dot{\varepsilon}}\ (s^{-1})$ | 3.185 |
| T (K) | 1460 |
| Z | $4.4 \cdot 10^{11}$ |
| $t_{x=0.95}$ (s) | 4.43 |
| $t_{growth}$ (s) | 7.37 |
| $D_0$ ($\mu$m) | 500 |
| $D_{rex}$ ($\mu$m) | 203.96 |
| $D_{final}$ ($\mu$m) | 203.96 |
| M | $1.89 \cdot 10^{18}$ |

**Table 3.4** Results for the pass 2

| Pass 2 | |
|---|---|
| $h_0$ (mm) | 197 |
| $h_f$ (mm) | 164 |
| $\bar{\varepsilon}$ | 0.212 |
| $\bar{\dot{\varepsilon}}$ (s$^{-1}$) | 4.053 |
| T (K) | 1453 |
| Z | $6.3 \cdot 10^{11}$ |
| $t_{x=0.95}$ (s) | 0.40 |
| $t_{growth}$ (s) | 8.40 |
| $D_0$ ($\mu$m) | 203.96 |
| $D_{rex}$ ($\mu$m) | 103.7 |
| $D_{final}$ ($\mu$m) | 104.6 |
| M | $1.61 \cdot 10^{18}$ |

**Table 3.5** Results for the pass 3

| Pass 3 | |
|---|---|
| $h_0$ (mm) | 164 |
| $h_f$ (mm) | 131 |
| $\bar{\varepsilon}$ | 0.259 |
| $\bar{\dot{\varepsilon}}$ (s$^{-1}$) | 5.325 |
| T (K) | 1447 |
| Z | $9.2 \cdot 10^{11}$ |
| $t_{x=0.95}$ (s) | 0.05 |
| $t_{growth}$ (s) | 9.55 |
| $D_0$ ($\mu$m) | 104.6 |
| $D_{rex}$ ($\mu$m) | 57.3 |
| $D_{final}$ ($\mu$m) | 82.0 |
| M | $1.41 \cdot 10^{18}$ |

After the roughing process in 7 passes (Table 3.10), there is a time delay of 44 s between the roughing and the finishing processes. In this period, the grain grows up to 82.49 $\mu$m (from 72.62 $\mu$m at the end of the roughing process), while the temperature decreases from 1118°C to 1070 °C. We present in Tables 3.11, 3.12, 3.13, 3.14, 3.15 and 3.16 the results corresponding to the finishing stage.

We appreciate in Table 3.17, a stagnation of the austenitic grain size in around 30.6 $\mu$m due to the noticeable growth of the specific surface area of the grain that stimulates the growth between passes. After the allotropic transformation $\gamma \rightarrow \alpha$ (at around 890 °C), each grain of austenite generates between 2 and 3 grains of

**Table 3.6** Results for the pass 4

| Pass 4 | |
| --- | --- |
| $h_0$ (mm) | 131 |
| $h_f$ (mm) | 98 |
| $\bar{\varepsilon}$ | 0.335 |
| $\bar{\dot{\varepsilon}}$ ($s^{-1}$) | 7.443 |
| T (K) | 1438 |
| Z | $1.5 \cdot 10^{12}$ |
| $t_{x=0.95}$ (s) | 0.01 |
| $t_{growth}$ (s) | 7.99 |
| $D_0$ ($\mu$m) | 82.0 |
| $D_{rex}$ ($\mu$m) | 36.7 |
| $D_{final}$ ($\mu$m) | 78.6 |
| M | $1.14 \cdot 10^{18}$ |

**Table 3.7** Results for the pass 5

| Pass 5 | |
| --- | --- |
| $h_0$ (mm) | 98 |
| $h_f$ (mm) | 71 |
| $\bar{\varepsilon}$ | 0.372 |
| $\bar{\dot{\varepsilon}}$ ($s^{-1}$) | 9.989 |
| T (K) | 1429 |
| Z | $2.4 \cdot 10^{12}$ |
| $t_{x=0.95}$ (s) | 0.01 |
| $t_{growth}$ (s) | 11.19 |
| $D_0$ ($\mu$m) | 78.7 |
| $D_{rex}$ ($\mu$m) | 30.6 |
| $D_{final}$ ($\mu$m) | 79.7 |
| M | $9.24 \cdot 10^{18}$ |

ferrite (Tanaka 1981; Tamura et al. 1988). That is to say, the grain size in the ferritic-pearlitic steel, at room temperature, will be comprised within 10 (9.5 ASTM) and 15 (8.5 ASTM) $\mu$m. It is not possible to obtain very fine grains, except in the case of the controlled rolling (Verdeja et al. 2021).

The final microstructure of the material is around 15 $\mu$m (ferritic grain size), being very difficult to obtain finer grains if the steel is not microalloyed with niobium. It is possible to estimate the non-recrystallization temperature of the austenite (Barbosa et al. 1987 and Belzunce et al. 1995) as follows:

$$T_{nr}(°C) = 887 + 464 \cdot \%C + \left[6445 \cdot \%Nb - 644 \cdot (\%Nb)^{1/2}\right]$$

**Table 3.8** Results for the pass 6

| Pass 6 | |
|---|---|
| $h_0$ (mm) | 71 |
| $h_f$ (mm) | 52 |
| $\bar{\varepsilon}$ | 0.360 |
| $\bar{\varepsilon}$ (s$^{-1}$) | 12.365 |
| T (K) | 1411 |
| Z | $4.1 \cdot 10^{12}$ |
| $t_{x=0.95}$ (s) | 0.01 |
| $t_{growth}$ (s) | 6.79 |
| $D_0$ ($\mu$m) | 79.7 |
| $D_{rex}$ ($\mu$m) | 29.9 |
| $D_{final}$ ($\mu$m) | 72.6 |
| M | $6.02 \cdot 10^{17}$ |

**Table 3.9** Results for the pass 7

| Pass 7 | |
|---|---|
| $h_0$ (mm) | 52 |
| $h_f$ (mm) | 40 |
| $\bar{\varepsilon}$ | – |
| $\bar{\varepsilon}$ (s$^{-1}$) | – |
| T (K) | 1392 |
| Z | – |
| $t_{x=0.95}$ (s) | – |
| $t_{growth}$ (s) | – |
| $D_0$ ($\mu$m) | – |
| $D_{rex}$ ($\mu$m) | – |
| $D_{final}$ ($\mu$m) | – |
| M | – |

$$+ \left[ 732 \cdot 0V - 230 \cdot (\%V)^{1/2} \right] + 890 \cdot \%Ti + 363 \cdot \%Al - 357 \cdot \%Si \tag{3.55}$$

The composition of our steel is %C, 0.03; %Mn, 0.15; %Si, 0.01; %Al, 0.045. The non-recrystallization temperature of the austenite is

$$T_{nr}(°C) = 887 + 464 \cdot 0.03 + 363 \cdot 0.045 - 357 \cdot 0.15 = 863.7\,°C \tag{3.56}$$

The austenite recrystallizes above 863.7 °C in our steel, and as the temperature of entrance to the last stand is greater (930 °C), thus all the austenite is recrystallized, and all the process of controlled rolling takes place in recrystallized austenite phase.

**Table 3.10**  Rolling process (roughing) of a carbon-manganese steel (DDQ steel)

| Pass | $h_0$(mm) | $h_f$(mm) | $\bar{\varepsilon}$ | $\bar{\dot{\varepsilon}}(s^{-1})$ | T (K) | $t_{x=0.95}$(s) | $t_{pause}$(s) | $t_{growth}$(s) | $D_0(\mu m)$ | $D_{rex}(\mu m)$ | $D_{final}(\mu m)$ |
|---|---|---|---|---|---|---|---|---|---|---|---|
| 1 | 230 | 197 | 0.179 | 3.185 | 1460 | 4.43 | 11.8 | 7.37 | 500.00 | 203.96 | 203.96 |
| 2 | 197 | 164 | 0.212 | 4.053 | 1453 | 0.40 | 8.8 | 8.40 | 203.96 | 103.70 | 104.63 |
| 3 | 164 | 131 | 0.259 | 5.325 | 1447 | 0.05 | 9.6 | 9.55 | 104.64 | 57.28 | 81.91 |
| 4 | 131 | 98 | 0.335 | 7.443 | 1438 | 0.01 | 8.0 | 7.99 | 82.04 | 36.69 | 78.58 |
| 5 | 98 | 71 | 0.372 | 9.989 | 1429 | 0.01 | 11.2 | 11.19 | 78.71 | 30.58 | 79.58 |
| 6 | 71 | 52 | 0.360 | 12.365 | 1411 | 0.01 | 6.8 | 6.79 | 79.70 | 29.93 | 72.51 |
| 7 | 52 | 40 | – | – | – | – | – | – | 72.62 | – | – |

**Table 3.11** Results for the pass 1 (finishing)

| Pass 1 | |
|---|---|
| $h_0$ (mm) | 25 |
| $h_f$ (mm) | 15.82 |
| $\bar{\varepsilon}$ | 0.528 |
| $v_{linear}$ (m/min) | 50.4 |
| $\bar{\dot{\varepsilon}}$ $(s^{-1})$ | 6.896 |
| T (K) | 1198 |
| Z | $2.6 \cdot 10^{14}$ |
| $t_{x=0.95}(s)$ | 0.78 |
| $t_{growth}(s)$ | 5.764 |
| $t_{pause}(s)$ | 6.54 |
| $D_0$ ($\mu$m) | 82.49 |
| $D_{rex}$ ($\mu$m) | 14.71 |
| $D_{final}$ ($\mu$m) | 38.89 |

**Table 3.12** Results for the pass 2 (finishing)

| Pass 2 | |
|---|---|
| $h_0$ (mm) | 15.82 |
| $h_f$ (mm) | 10.46 |
| $\bar{\varepsilon}$ | 0.478 |
| $v_{linear}$ (m/min) | 79.6 |
| $\bar{\dot{\varepsilon}}$ $(s^{-1})$ | 15.363 |
| T (K) | 1201 |
| Z | $5.3 \cdot 10^{14}$ |
| $t_{x=0.95}$ (s) | 0.18 |
| $t_{growth}$ (s) | 3.969 |
| $t_{pause}$ (s) | 4.146 |
| $D_0$ ($\mu$m) | 38.89 |
| $D_{rex}$ ($\mu$m) | 10.67 |
| $D_{final}$ ($\mu$m) | 37.85 |

Finally, the temperature $Ar_3$ (temperature of the transformation of the austenite into ferrite during the cooling) of the steel is

$$Ar_3(°C) = 910 - 310 \cdot \%C - 80 \cdot \%Mn + 0.35 \cdot (t - 8) \qquad (3.57)$$

where t is the thickness plate (mm):

$$Ar_3(°C) = 910 - 310 \cdot 0.03 - 80 \cdot 0.15 + 0.35 \cdot (4.49 - 8) = 887.5 °C \quad (3.58)$$

**Table 3.13** Results for the pass 3 (finishing)

| Pass 3 | |
|---|---|
| $h_0$ (mm) | 10.46 |
| $h_f$ (mm) | 7.89 |
| $\bar{\varepsilon}$ | 0.326 |
| $v_{linear}$ (m/min) | 120.4 |
| $\bar{\dot{\varepsilon}}$ (s$^{-1}$) | 22.745 |
| T (K) | 1198 |
| Z | $8.5 \cdot 10^{14}$ |
| $t_{x=0.95}$ (s) | 0.73 |
| $t_{growth}$ (s) | 2.009 |
| $t_{pause}$ (s) | 2.741 |
| $D_0$ ($\mu$m) | 37.85 |
| $D_{rex}$ ($\mu$m) | 15.02 |
| $D_{final}$ ($\mu$m) | 35.00 |

**Table 3.14** Results for the pass 4 (finishing)

| Pass 4 | |
|---|---|
| $h_0$ (mm) | 7.89 |
| $h_f$ (mm) | 6.34 |
| $\bar{\varepsilon}$ | 0.253 |
| $v_{linear}$ (m/min) | 159.7 |
| $\bar{\dot{\varepsilon}}$ (s$^{-1}$) | 29.608 |
| T (K) | 1201 |
| Z | $1.0 \cdot 10^{15}$ |
| $t_{x=0.95}$ (s) | 1.43 |
| $t_{growth}$ (s) | 0.636 |
| $t_{pause}$ (s) | 2.066 |
| $D_0$ ($\mu$m) | 35.00 |
| $D_{rex}$ ($\mu$m) | 18.43 |
| $D_{final}$ ($\mu$m) | 31.53 |

Both the $Ar_3$ and the $T_{nr}$ are lower than the final finishing temperature of the steel and, for that reason, the process takes place, as it was mentioned, in austenitic phase. There is always recrystallization of the austenite in each pass. We will have a ferritic-pearlitic structure at room temperature with grain size of around 20 $\mu$m.

Now we are going to calculate the yield strength and the ductile–brittle transition temperature. We use the following equations (Quintana 2013; Fernández et al. 2013; Pickering 1978; Petite et al. 1998):

**Table 3.15** Results for the pass 5 (finishing)

| Pass 5 | |
|---|---|
| $h_0$ (mm) | 6.34 |
| $h_f$ (mm) | 5.12 |
| $\bar{\varepsilon}$ | 0.247 |
| $v_{linear}$ (m/min) | 198.8 |
| $\bar{\dot{\varepsilon}}$ (s$^{-1}$) | 40.718 |
| T (K) | 1200 |
| Z | $1.4 \cdot 10^{15}$ |
| $t_{x=0.95}$ (s) | 1.16 |
| $t_{growth}$ (s) | 0.496 |
| $t_{pause}$ (s) | 1.660 |
| $D_0$ ($\mu$m) | 31.53 |
| $D_{rex}$ ($\mu$m) | 17.55 |
| $D_{final}$ ($\mu$m) | 30.65 |

**Table 3.16** Results for the pass 6 (finishing)

| Pass 6 | |
|---|---|
| $h_0$ (mm) | 5.12 |
| $h_f$ (mm) | 4.49 |
| $\bar{\varepsilon}$ | 0.152 |
| $v_{linear}$ (m/min) | 245.9 |
| $\bar{\dot{\varepsilon}}$ (s$^{-1}$) | 42.692 |
| T (K) | 1209 |
| Z | $1.2 \cdot 10^{15}$ |
| $t_{x=0.95}$ (s) | 5.80 |
| $t_{growth}$ (s) | −4.455 |
| $t_{pause}$ (s) | 1.342 |
| $D_0$ ($\mu$m) | 30.65 |
| $D_{rex}$ ($\mu$m) | – |
| $D_{final}$ ($\mu$m) | 30.65 |

$$\sigma_y (MPa) = 15.4 \cdot \left( 3.5 + 2.1 \cdot \%Mn + 5.4 \cdot \%Si + 23 \cdot \%N_{free}^{1/2} + 1.13 \cdot d^{-1/2} \right)$$
$$(3.59)$$

and:

$$DBTT\ (^\circ C,\ 27J) = -19 + 44 \cdot \%Si + 700 \cdot \%N_{free}^{1/2} - 11.5 \cdot d^{-1/2} + 2.2 \cdot f_P$$
$$(3.60)$$

**Table 3.17** Rolling process (finishing) of a carbon-manganese steel (DDQ-Class R3)

| Pass | $h_0$(mm) | $h_f$(mm) | $\varepsilon$ | $\bar{\dot{\varepsilon}}(s^{-1})$ | T (K) | $t_{x=0.95}(s)$ | $t_{pause}(s)$ | $t_{growth}(s)$ | $D_0(\mu m)$ | $D_{rex}(\mu m)$ | $D_{final}(\mu m)$ |
|---|---|---|---|---|---|---|---|---|---|---|---|
| 1 | 25 | 15.82 | 0.528 | 6.896 | 1198 | 0.78 | 6.548 | 5.764 | 82.49 | 14.71 | 38.89 |
| 2 | 15.82 | 10.46 | 0.478 | 15.363 | 1201 | 0.18 | 4.146 | 3.969 | 38.89 | 10.67 | 37.85 |
| 3 | 10.46 | 7.89 | 0.326 | 22.745 | 1198 | 0.73 | 2.741 | 2.009 | 37.85 | 15.02 | 35.00 |
| 4 | 7.89 | 6.34 | 0.253 | 29.608 | 1201 | 1.43 | 2.066 | 0.636 | 35.00 | 18.43 | 31.53 |
| 5 | 6.34 | 5.12 | 0.247 | 40.718 | 1200 | 1.16 | 1.660 | 0.496 | 31.53 | 17.55 | 30.65 |
| 6 | 5.12 | 4.49 | 0.152 | 42.692 | 1209 | 5.80 | 1.342 | -4.455 | 30.65 | – | 30.65 |
| 7 | 4.49 | – | – | – | – | – | – | – | – | – | – |

where d is the grain diameter expressed in millimeters, %Si and %$N_{free}$ are the compositions in mass percent, and $f_P$ is the pearlite fraction. The percentage of pearlite in our steel is approximately 2.5%. This way (the ferrite grain size is 20 $\mu$m):

$$\sigma_y(MPa) = 53.9 + 15.4 \cdot \left[ 2.1 \cdot 0.15 + 1.13 \cdot \left( 15\mu m \cdot \frac{1mm}{10^3 \mu m} \right)^{-1/2} \right]$$
$$= 200.84 MPa \tag{3.61}$$

$$DBTT \ (^\circ C, 27J) = -19 - 11.5 \cdot \left( 15\mu m \cdot \frac{1mm}{10^3 \mu m} \right)^{-1/2} + 2.2 \cdot 2.5 = -107.4\,^\circ C \tag{3.62}$$

*End of the exercise.*

### 3.4.3 LT Ausforming

Other thermomechanical treatments have as objective to deform the austenite at temperatures lower than $A_1$ and transform later the work hardened austenite into martensite by cooling. This is the case of the thermomechanical treatment of ausforming called LT ausforming (low-temperature ausforming) to distinguish it from the forming of the austenite at temperatures greater than $A_3$.

An isothermal cooling of the austenite is carried out until the metastable zone, above the $M_S$ temperature of the steel, and the austenite is deformed in this zone— by rolling, extrusion, hammer forge, etc.—before the beginning of any transformation, as it is schematized in Fig. 3.8. Later, the deformed austenite is cooled for its transformation into martensite.

That forming of the austenite at low temperature is only possible if the steel has great hardenability, or its TTT curve has a discontinuity between the C curves of pearlitic and bainitic transformations. It is applied with satisfactory results, for instance, in dies.

This isothermal forming of the austenite produces, in this range of temperatures, sub-grains of austenite, and, with this, the characteristics of the martensite obtained when cooling this structure of non-recrystallized austenite are improved. Martensite like this obtained is finer, less tetragonal, and with greater aptitude to achieve a fine precipitation (nanometric) of the cementite and carbides during the tempering of the martensite. It is possible to obtain with the LT ausforming, comparatively to the classic quenching and tempering practices, noticeable improvements of $\sigma_y$ and $\sigma_u$ without significant deterioration of the ductility.

It is possible to achieve in low hardenability steels similar results to those obtained in the LT ausforming by means of the explosive forming (considering that

**Fig. 3.8**  LT ausforming

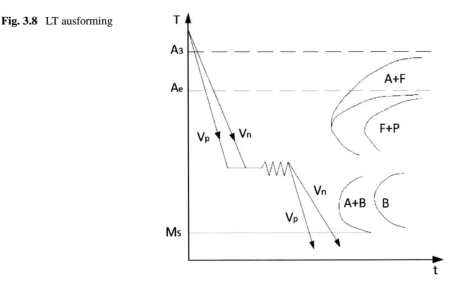

the time required for this forming is smaller than that required for the isothermal transformation of the austenite).

**Exercise 3.5**  Ausforming process.

*Solution starts*:

We said that an austenite is unstable at temperature below $A_{r3}$ so steels are selected with sufficient alloy content to allow time for the cycle of treatment represented in Fig. 3.8 to be carried out (ausforming). The TTT curves must contain, at this temperature range (450–550 °C), a large bay or clear separation of the bainite and pearlite ranges. Figure 3.9 shows some aspects of the austenite worked at 500 °C to different degrees on the TTT medium-carbon low-alloyed steel: 0.36% C-0.36% Si-0.86% Mn-2.96% Cr-1.34% Ni-0.34% Mo-0.041%P-0.019% S in weight percent.

**Fig. 3.9**  Deformation cycle in the metastable austenite range (ausforming) of the alloy of Fig. 3.10a, b

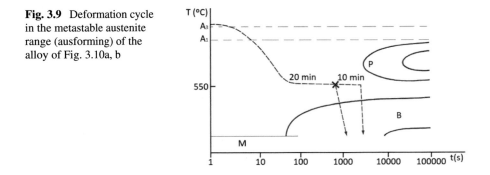

We ask for calculating $A_{3c}$, $A_e$, and $M_S$ according to formulae of Andrews and Steven (corrected by Irving), respectively, and $B_S$ (bainite start according to Pickering):

$$A_e = 727 - 10.7 \cdot \%Mn - 16.9 \cdot \%Ni + 29.1 \cdot \%Si + 16.9 \cdot \%Cr = 756\,°C \tag{3.63}$$

$$A_{3c} = 912 - 203 \cdot \sqrt{\%C} - 30 \cdot \%Mn - 15.2 \cdot \%Ni + 44.7 \cdot \%Si$$
$$+ 31.5 \cdot \%Mo = 770\,°C \tag{3.64}$$

$$M_S = 561 - 474 \cdot \%C - 33 \cdot \%Mn - 17 \cdot \%Ni - 17 \cdot \%Cr - 21 \cdot \%Mo$$
$$- 11 \cdot \%Si = 278\,°C \tag{3.65}$$

$$B_S = 83.0 - 270 \cdot \%C - 90 \cdot \%Mn - 37 \cdot \%Ni - 70 \cdot \%Cr - 83 \cdot \%Mo = 370\,°C \tag{3.66}$$

Calculations of $A_{3c}$, $A_e$, $M_S$, and $B_s$ coincide rather well with the TTT curve.

Figure 3.10 shows the microstructure of the steel heat treated at 900 °C (1 h), followed by isothermal holding at 450 °C and water quenching. The austenite has not been deformed and presents the typical equiaxed microstructure with straightforward grain boundaries (Schrader and Rose 1966).

On the other hand, the next figure shows the microstructure of the austenite after being deformed 60% at 450 °C. There are highly elongated grains containing some very closely spaced slip lines. The dark horizontal bands (bainite) indicated the banded structure of the initial steel.

*Solution ends.*

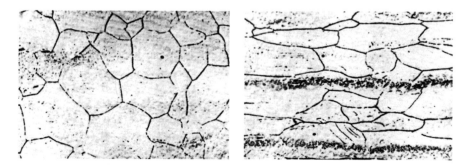

**Fig. 3.10** 16 mm square section sheet. Medium carbon alloy steel normalized. Longitudinal section through the fiber structure after dynamic compression. Deformation of austenite at 450 °C. 900 °C,1 h/450 °C, 20 min/deformation/water. (a) no deformation, the prior austenite grain boundaries are indicated by a black network, the dark etching acicular matrix is of bainite and martensite (b) about 60% deformation, elongated grains containing very closely spaced slip lines, the dark horizontal bands (bainite) indicate the banded structure of the initial material (500x)

## 3.5    Thermomechanical Treatments of the Austenite During Its Allotropic Transformation

### 3.5.1   Isoforming

When the steel, previously austenitized, is formed at temperatures close to $A_e$, in the upper zone of the TTT curve, corresponding to the pearlitic transformations—see Fig. 3.11—the thermomechanical treatment is called isoforming. It can be applied to low-alloy steels when the characteristics of the process would allow to make an isothermal forming.

The deformation, like this produced, generates great density of dislocations, and also facilitates, by polygonization of the ferrite, the formation of sub-grains of ferrite of around 0.3 μm. And it favors the precipitation of cementite inside the ferrite sub-grains as small carbides of non-lamellar nature due to the presence of great number of crystalline defects inside these sub-grains.

As a consequence of the above-mentioned points, the mechanical characteristics of the steel formed in this manner are greater than those that would have with lamellar pearlitic structure. The toughness, elongation, and Wöhler fatigue limit are significantly improved, while $\sigma_y$ and $\sigma_u$ are moderately improved.

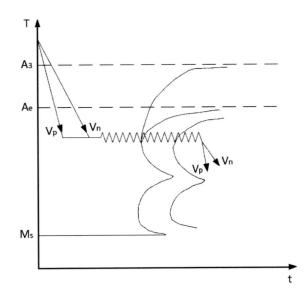

**Fig. 3.11**  Isoforming

It is possible to apply thermomechanical treatment in some steels similar to the pearlitic isoforming, but in the zone of bainitic transformation, when its TTT curve is adequate.

### 3.5.2 TRIP Effect

Austenite can be transformed into martensite by mechanical forming. We call $M_d$ in steels, by analogy with the $M_S$, to that temperature below which the formation of 1% of martensite by mechanical deformation of the austenite is possible.

The $M_d$ temperature (in °C) is always greater than $M_S$ (in °C). This seems logical because if it is possible to obtain martensite by means of the energy that confers the temperature difference, then it would be possible to obtain martensite with smaller temperature difference if mechanical energy is also provided.

There are empirical equations that allow calculating, as a function of the chemical composition of the steel, the $M_{d30}$ temperature below which it is possible to obtain more than 50% "mechanical" martensite for a real deformation, $\varepsilon$, equal to 0.30.

Originally, this thermomechanical treatment at temperature intermediate between $M_d$ and $M_S$ was applied to some fully austenitic steels with aim of achieving, simultaneously, great mechanical resistance—as a consequence of the work hardening and the formation of martensite—and an improved ductility. For example, a steel with composition 0.3% C, 8.5% Ni, 2% Mn, 9% Cr, 2% Si, 4% Mo, after the suitable thermomechanical treatment, reaches in the tensile test, according to Pickering: 1450 MPa of yield strength, $\sigma_y$; 1500 MPa of ultimate tensile strength, $\sigma_u$; and elongation to fracture of 50%.

In this, as in other austenitic steels, the transformation produced during the deformation increases the rate of work hardening and the amount of uniform deformation—increases the coefficient n—previous to the beginning of the striction or plastic instability. This is the reason of the name of steels of plasticity improved by deformation or—more frequently—TRIP steels (transformation-induced plasticity).

The thermomechanical treatment of the austenite at temperatures in the range $M_d$–$M_S$ is also nowadays used in low-alloy steels taking advantage of the possibilities offered by the continuous annealing, which allows achieving 20% of residual austenite suitable for mechanical transformation.

**Exercise 3.6** TRIP steel.

*Solution starts:*

The austenite in some of the less highly alloyed austenitic stainless steel may be transformed into martensite. This can occur in the solution-treated conditions in which the $M_S$ temperature is above room temperature or it may occur during refrigeration in more stable alloy in which $M_S$ is below room temperature, and thus martensite may be transformed at any temperature below the $M_d$.

Apart from cobalt, almost all alloy elements depress the temperature. The following equation has been specified in austenitic stainless steel:

$$M_S(°C) = 502 - 810 \cdot \%C - 1230 \cdot \%N - 13 \cdot \%Mn$$
$$- 30 \cdot \%Ni - 12 \cdot \%Cr - 54 \cdot \%Cu - 46 \cdot \%Mo \qquad (3.67)$$

in weight percent alloy addition.

A typical equation for the $M_{d30}$ temperature at which 50% martensite is produced under the action of a true strain of 0.3 is

$$M_{d30}(°C) = 497 - 462 \cdot (\%C + \%N) - 9.2 \cdot \%Si$$
$$- 8.1 \cdot \%Mn - 13.7 \cdot \%Cr - 20 \cdot \%Ni - 18.5 \cdot \%Mo \qquad (3.68)$$

in weight percent alloy addition also:

We propose (0.1% C, 8.5% Ni, 2% Mn, 9% Cr, 2% Si, 4% Mo):

(a) Calculate $M_S$ and $M_d$ for the steel composition listed above:

$$M_S(°C) = 502 - 810 \cdot 0.1 - 1230 \cdot 0.01 - 13 \cdot 2 - 30 \cdot 8.5$$
$$- 12 \cdot 9 - 46 \cdot 4 = -164.3\,°C \qquad (3.69)$$

well below room temperature.

$$M_{d30}(°C) = 497 - 462 \cdot 0.1 - 9.2 \cdot 2 - 8.1 \cdot 2 - 13.7 \cdot 9$$
$$- 20 \cdot 8.5 - 18.5 \cdot 4 = 48.9\,°C \qquad (3.70)$$

above room temperature.

So, the cited steel can undergo the TRIP effect by strain transformation of austenite which also improves ductility.

However, because of the cost of the alloys, the difficulty of large plastic deformation at low temperature (room temperature), and problems with joining, these steels will only be used in special applications of flat-rolled products or wire.

(b) It has been known that if a metastable austenitic stainless steel such as familiar type 301 (UNS number S30100) with average composition in weight percent, 0.1% C-2% Mn-1% Si-18% Cr-8% Ni, low N, is plastically deformed at room temperature, some of the austenite in the most severe strain portions of the specimen will transform into martensite, and such transformations, by locally increasing the rate of work hardening, inhibit necking and increase the uniform elongation. We ask to calculate

$M_S$ and $M_d$ temperatures and demonstrate that is partially subjected to the TRIP effect.

According to the formula giving Ms and $M_d$ and the chemical composition of the 301 steel, we obtain

$$M_S(°C) = 502 - 810 \cdot 0.1 - 13 \cdot 2 - 30 \cdot 8 - 12 \cdot 18 = 61\,°C \qquad (3.71)$$

above room temperature

$$M_{d30}(°C) = 497 - 462 \cdot 0.1 - 9.2 \cdot 1 - 8.1 \cdot 2 - 13.7 \cdot 18 - 20 \cdot 8 = 18.8\,°C \qquad (3.72)$$

at room temperature.

So, the 301 austenitic stainless steel experiences TRIP effect when deformed at room temperature, for example, in tensile test.

Figure 3.12 represents the effect of the nickel content in steels with 17% Cr (Pickering 1965). It shows that as more martensite is formed during continuing straining, and because the martensite is strong, the flow stress rapidly increases as

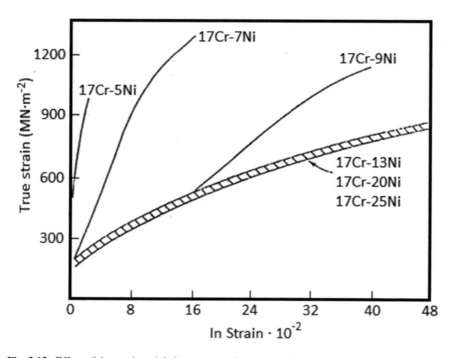

**Fig. 3.12** Effect of decreasing nickel content on the stress–strain curves of stainless steels

the martensite itself starts to participate in the deformation. This produces an increase in work hardening rate to a level much above that for stable steel (no TRIP) as the same strain.

*End of the exercise.*

## 3.6   Thermomechanical Treatments After the Transformation of the Austenite

### 3.6.1   Pearlite Forming

Similar results to those of the isoforming can be obtained in carbon steels, or few alloyed—increase of $\sigma_y$, of $\sigma_u$, and simultaneously of the ductility and the toughness—, by means of the thermomechanical treatment of the forming of the pearlite already transformed (pearlite forming) illustrated in the Fig. 3.13. This forming must be carried out at high temperatures, although always lower than $A_e$ to avoid the austenitization of the structure. The improvement of the mechanical characteristics by means of this thermomechanical treatment is due to the globulization of the pearlitic cementite—favored by the forming—and to the recovery of the ferrite.

Starting from very fine pearlitic structures, it is very advantageous for the ductility during the forming, both for the pearlite forming and cold forming. Thus, for instance, in the production of high-resistance steel wires (0.7% C), for the cold wire drawing, the best characteristics are achieved when the steel has a previous structure of very fine pearlite. This pearlite is obtained by means of the isothermal treatment of patenting previously mentioned: by isothermal cooling in molten lead to achieve its transformation in the lower zone of the pearlitic curve (see Sect. 2.7.1).

**Fig. 3.13** Pearlite forming

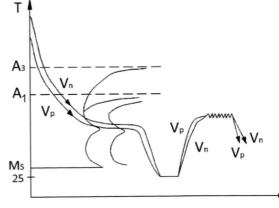

The ductility during the forming is always greater if the starting structure is of globulite (globular pearlite). This structure—advantageous for a subsequent pearlite forming—is also practically indispensable to allow cold forming some steels of high-carbon content (for instance, balls for bearings).

Globulite can be obtained—as it was already mentioned—by means of a previous heat treatment: by incomplete austenitization of the steel, at temperatures in the range $A_e$–$A_3$ followed by air cooling (if the steel has low hardenability) or cooling in salts in the pearlitic region (if the steel is very hardenable). And in low- and medium-alloy steels by subcritical annealing (by means of prolongated heating of the pearlite at temperature close to $A_e$).

### 3.6.2 Martensite Forming

It is possible to improve by thermomechanical treatment of the martensite the mechanical characteristics of several quenched steels. The advantages of this thermomechanical treatment are derived from the increase of dislocations in the martensite, by cold deformation and of greater number of possible places for the precipitates of carbides during the tempering. Increase of $\sigma_y$ and fatigue limit of around 25%. This treatment is applied to some steels for matrices and martensitic stainless steels.

This thermomechanical treatment is performed, in some cases, following one of the two following varieties. By means of quenching, cold deformation of the martensite—v. gr. by stretching in wire drawplate—and a posterior tempering at 200–400 °C of the deformed structure (Fig. 3.14a). Or by means of hot deformation at 650 °C of the non-tempered martensite (Fig. 3.14b). One and other thermomechanical processes are usually designated with the same name: marforming.

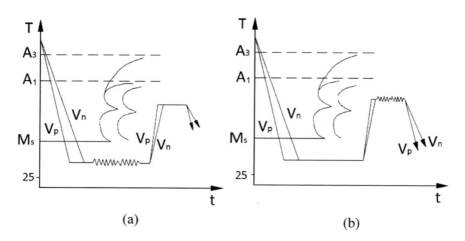

(a)                                          (b)

**Fig. 3.14** Thermal–mechanical treatment of the martensite

Analogously, maraging steels can improve their characteristics by cold forming of the Fe–Ni martensite (body-centered cubic and supersaturated), before the artificial aging of this martensite. This thermomechanical treatment is also called marforming.

# References

Asensio J, Romano G, Martínez VJ, Verdeja JI, Pero-Sanz JA (2001) Ferritic steels: Optimization of hot-rolled textures through cold rolling and annealing. Mater Charact 47(2):119–127

Barbosa R, Boratto F, Yue S, Jonas JJ (1987) The influence of chemical composition on the recrystallisation behaviour of microalloyed steels. In: Proceedings symposium on processing, microstructure and properties of HSLA steel, Pittsburgh, 1987, pp 51–56

Belzunce FJ, Rodríguez C, Fernández B (1995) Rolling procedure and thickness effects on strength and fracture behavior of microalloyed steels. Mater Sci Technol 11:252–257

Dutta B, Sellars CM (1986) Strengthening of austenite by niobium during hot rolling of microalloyed steel. Mater Sci Technol 2:146–153

Dutta B, Sellars CM (1987) Effect of composition and process variables on Nb (C, N) precipitation in niobium microalloyed austenite. Mater Sci Technol 3:197–205

Fernández S, Quintana MJ, García JO, Verdeja LF, González R, Verdeja JI (2013) Superplastic HSLA steels: microstructure and failure. J Fail Anal 13:368–376

Inagaki H, Kurihara K, Kozasu I (1977) Influence of crystallographic texture on strength and toughness of control-rolled high tensile strength steel. ISIJ Int 17:75–81

Martínez VJ, Verdeja JI, Pero-Sanz JA (2001) Interstitial free steel: Influence of α-phase hot-rolling and cold-rolling reduction to obtain extra-deep drawing quality. Mater Charact 46(1):45–53

Pero-Sanz JA, Ruiz-Delgado M, Martínez V, Verdeja JI (1999) Annealing textures for drawability: influence of the degree of cold rolling reduction for low-carbon and extra low-carbon ferritic steels. Mater Charact 43(5):303–309

Pero-Sanz JA, Quintana MJ, Verdeja LF (2017) Solidification and solid-state trans-formations of metals and alloys, 1st edn. Elsevier, Boston, USA

Petite MM, Monsalve A, Gutiérrez I, Zaitegui J, Larburu JI (1998) Modelización de la evolución microestructural durante el recocido continuo de chapas de acero bajo en carbono. Predicción de las propiedades mecánicas. Revista De Metalurgia De Madrid 34:333–337

Pickering FB (1965) High strength steels and their uses today. In: Metallurgical achievements (edited by Alexander WO), selection of papers presented at the Birmingham metallurgical society's diamond jubilee session. Elsevier, Boston, USA

Pickering FB (1978) Physical metallurgy and design of steels. Applied Science Publishers, London, United Kingdom

Porter DA, Easterling KE (1981) Phase transformation of metals and alloys. Van Nostrand Reinhold, New York, USA

Quintana MJ (2013) Room temperature and high temperature behavior of low carbon steels manufactured by controlled rolling process, PhD thesis, September 2013, Escuela Técnica Superior de Ingenieros de Minas de Oviedo, University of Oviedo, Oviedo, Asturias, Spain

Romano G, Verdeja JI, Pero-Sanz JA (2000a) Aceros Para Carrocerías. Revista De Minas 20:50–56

Romano G, Verdeja JI, Pero-Sanz JA (2000b) Aceros Para Hojalata. Revista De Minas 20:15–21

Schrader A, Rose A (1966) De ferri metallographia: metallographic atlas of iron, steels and cast irons. Volume 2, European Coal and Steel Community. High Authority, Presses académiques européennes, Brussels, Belgium

Sellars CM, Whiteman JA (1979) Recrystallization and grain growth in hot rolling. Metal Sci, 187–194

Tamura I, Ouchi C, Tanaka T, Sekine H (1988) Thermomechanical processing of high strength low alloy steels. Butterworths, London, United Kingdom

Tanaka T (1981) Controlled rolling of steel plate and strip. Int Metals Rev 4:185–212

Verdeja LF, Fernández-González D, Verdeja JI (2021) Operations and basic processes in steel-making. Springer International Publishing, Cham, Switzerland

# Chapter 4
# Controlled Atmospheres in Furnaces for Heat Treatments

## 4.1 Introduction

Heat treatments of alloys usually take place in an environment where it is not always possible to maintain inalterable the chemical composition of the surface. This is because it is almost unavoidable that chemical reactions between the steel parts and the atmosphere of the furnace could take place.

At habitual temperatures of heat treatment, oxygen can form oxides in the periphery of the parts. In the case of steels, apart from oxides—inappropriately known in steelworks as "calamine"—peripheral decarburizations or precipitation of soot can occur. Controlled atmospheres are usually employed to protect the metallic surface from these drawbacks.

However, it is sometimes interesting to have an environment that could modify the composition of the surface during the heat treatment. This is the case, for instance, of the thermochemical treatment of steels whose aim is to enrich the periphery of the part in carbon, as the case hardening, or in other elements, as the nitriding or the carbonitriding.

The control of the gaseous atmospheres for the protection in heat treatments of steels answers to the principles that are analogous to those that allow determining the compositions of the suitable atmospheres for the case hardening of steels or to obtain metallic powders in powder metallurgy by reduction of oxides. We are going to consider in the following sections, first the formation and reduction of oxides with general character, and, later, we are going to study the topic of the decarburizing and case hardening of steels.

## 4.2   Formation and Dissociation of Metallic Oxides: Ellingham's Diagram

The formation of a metallic oxide $M_xO_y$, at the temperature T, by reaction of a solid metal M with the oxygen (if the reaction reaches the reversibility) corresponds to a heterogeneous equilibrium, which adopts the following equation per mol of oxygen:

$$O_2(g) + \frac{2 \cdot x}{y} \cdot M(s) \leftrightarrow \frac{2}{y} M_xO_y \tag{4.1}$$

We call *heterogeneous* to an equilibrium when, as in this case, apart from gases, solids and/or liquids are also involved in the equilibrium reaction. This is the case, for instance, of the reaction:

$$\alpha \cdot A + \beta \cdot B + \mu \cdot M \leftrightarrow \lambda \cdot L + \gamma \cdot G + \delta \cdot D \tag{4.2}$$

where A, B, G, and D are gases, M solid and L liquid. The number of moles of each one of these solids are $\alpha$, $\beta$, $\gamma$, $\delta$, $\mu$, and $\lambda$, respectively.

The variation of free energy $\Delta G_T$ that accompanies this reaction, at constant temperature and pressure, is equal to the difference between the free energy of the products of reaction L, G, D and the free energy of the reagents A, B, M. If $\Delta G_T < 0$, the reaction will progress to the right and for $\Delta G_T > 0$, to the left. Only when $\Delta G_T = 0$, the reaction is in equilibrium.

The value of $\Delta G_T$ is connected to the variation of standard free energy $\Delta G_T{}^0$ and to the thermodynamic activities of the available phases by the following equation:

$$\Delta G_T = \Delta G_T^0 + R \cdot T \cdot \ln\left(\frac{a_L^\lambda \cdot a_G^\gamma \cdot a_D^\delta}{a_A^\alpha \cdot a_B^\beta \cdot a_M^\mu}\right) \tag{4.3}$$

where $a_A$, $a_B$, $a_M$, $a_L$, $a_G$, $a_D$ are the activities of the phases involved in the reaction, R is the constant of the gases, and $\Delta G_T$ is the standard free energy, that is to say, the free energy $\Delta G_T$ when all the activities are equal to one. The quotient:

$$k_{eq} = \frac{a_L^\lambda \cdot a_G^\gamma \cdot a_D^\delta}{a_A^\alpha \cdot a_B^\beta \cdot a_M^\mu} \tag{4.4}$$

is called equilibrium constant $k_{eq}$ of the reaction at the temperature T. Thus:

$$\Delta G_T = \Delta G_T^0 + R \cdot T \cdot \ln k_{eq} \tag{4.5}$$

and, for that reason, when $\Delta G_T$ is equal to zero, in the equilibrium:

$$\ln k_{eq} = -\frac{\Delta G_T^0}{R \cdot T} \tag{4.6}$$

For a working pressure of the system equal to 1 atmosphere (or for a value close to one, as happens in industrial atmospheres), the activities of the solids $a_M$ and liquids $a_L$ can be considered equal to one; and, regarding gases, their activities can be replaced by their partial pressures $P_G, P_D, P_A, P_B$, since they can be considered as "ideal gases" as the industrial atmosphere approximate to the ideal gas. Therefore, it is possible to replace the value $\left( a_L^\lambda \cdot a_G^\gamma \cdot a_D^\delta / a_A^\alpha \cdot a_B^\beta \cdot a_M^\mu \right)$ in this heterogeneous equilibrium by the value $\left( P_G^\gamma \cdot P_D^\delta / P_A^\alpha \cdot P_B^\beta \right)$. This way, Eq. (4.3) can be also written as follows, Eq. (4.7), as a function of the volume concentrations $n/n_r$ of the gases and of the total pressure of the mixture, $P_T$:

$$\Delta G_T = \Delta G_T^0 + R \cdot T \cdot \ln \frac{\left(\frac{n_G}{n_T}\right)^\gamma \cdot \left(\frac{n_D}{n_T}\right)^\delta}{\left(\frac{n_A}{n_T}\right)^\alpha \cdot \left(\frac{n_B}{n_T}\right)^\beta} \cdot P_T^{\gamma+\delta-\alpha-\beta} \tag{4.7}$$

It is adequate to remind that in a mixture where the moles of each available gas were respectively $n_1, n_2, n_3$, etc., (where the total number of moles of gas is $n_T = n_1 + n_2 + n_3 + \cdots$), the partial pressure of each gas—$P_1, P_2, P_3$, etc.—can be calculated as a function of the total pressure of the mixture $P_T$. As:

$$P_1 \cdot V = n_1 \cdot R \cdot T \tag{4.8}$$

and:

$$P_T \cdot V = n_T \cdot R \cdot T \tag{4.9}$$

we obtain:

$$P_1 = P_T \cdot \left(\frac{n_1}{n_T}\right) \tag{4.10}$$

$$P_2 = P_T \cdot \left(\frac{n_2}{n_T}\right) \tag{4.11}$$

$$P_3 = P_T \cdot \left(\frac{n_3}{n_T}\right) \tag{4.12}$$

etc. It is also verified that $P_T$ is equal to the addition of partial pressures since:

$$V \cdot (P_1 + P_2 + P_3 + \cdots) = R \cdot T \cdot (n_1 + n_2 + n_3 + \cdots) \tag{4.13}$$

In Eq. (4.7), $\Delta G_T$ does not depend on the total pressure, $P_T$, when this pressure is 1 atmosphere, or it is verified that $\gamma + \delta - \alpha - \beta$ is equal to zero.

The value $\left[ (n_G/n_T)^\gamma \cdot (n_D/n_T)^\delta / (n_A/n_T)^\alpha \cdot (n_B/n_T)^\beta \right] \cdot P_T^{\gamma+\delta-\alpha-\beta}$ is called equilibrium constant $k_P$ at the temperature T:

$$k_{eq} = k_P \cdot P_T^{\gamma+\delta-\alpha-\beta} \qquad (4.14)$$

If we apply the above mentioned, the free energy of formation of a solid metallic oxide $M_xO_y$ by reaction of the solid metal M with the oxygen at a certain temperature T:

$$O_2(gas) + \frac{2 \cdot x}{y}M(solid) \leftrightarrow \frac{2}{y}M_xO_y(solid) \qquad (4.15)$$

will be given—where $P_{O_2}$ is the oxygen partial pressure—by:

$$\Delta G_T = \Delta G_T^0 + R \cdot T \cdot \ln \frac{1}{P_{O_2}} = \Delta G_T^0 - 8.314 \cdot T \cdot \ln P_{O_2} \qquad (4.16)$$

Therefore, the oxygen partial pressure in the equilibrium (to avoid that reaction (4.16) progresses neither to the right nor to the left) will be that where $\Delta G_T = 0$. If the oxygen partial pressure in the atmosphere is greater than that of the equilibrium, it results that $\Delta G_T < 0$ and the metal oxidizes. On the contrary, for oxygen partial pressure smaller than that of equilibrium, and thus $\Delta G_T > 0$, the oxide reduces to metal. From this, it is deduced that for the same external pressure of oxygen, it is more probable that a metal could oxidize at lower than at higher temperature.

The Ellingham's diagram for oxides (Fig. 4.1) collects the values of the reference free energy, $\Delta G_T^0$, of formation of oxides of different elements, per mol of oxygen, at different temperatures. A straight line corresponds in the diagram to each formed oxide—see, for example, the formation of the $Cu_2O$—of equation:

$$\Delta G_T^0 = -a + b \cdot T(K) \qquad (4.17)$$

Equation that is similar to the thermodynamic equation:

$$\Delta G_T^0 = \Delta H_T^0 - T \cdot \Delta S_T^0 \qquad (4.18)$$

and, thus, the origin ordinate of each straight line represents $\Delta H_T^0$ (that is to say, the formation of an oxide is an exothermal reaction); and the slope of the straight line is $-\Delta S_T^0$. See also in the diagram that many straight lines are parallel between them, which indicates that $\Delta S_T^0$ is nearly equal for many of the oxides; and that the straight lines of oxides formation with volume decrease (for example, $2H_2 + O_2 \rightarrow 2H_2O$) appear in the diagram with positive slope (as the disorder decreases and, consequently, the $\Delta S_T$ is $< 0$).

On the other hand, it is appreciated in some of the elements whose slope is positive (cases of the manganese and nitrogen) that the straight line changes the slope from a certain temperature; this indicates a change of state in the metal at this temperature; and, for that reason, the entropy of formation of the oxide is different from the previous one (it is necessary to add the variation of entropy due to the change of state to the entropy of the previous state). The metal has more than one line in the

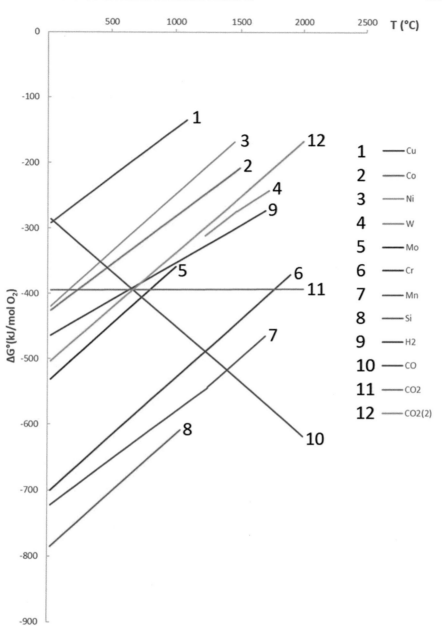

**Fig. 4.1** Variation of the free energy of formation of oxides as a function of the temperature. M: melting point; E: boiling point (from data in Table 4.1)

Ellingham's diagram when an element has different degrees of oxidation, as it is the case of the iron—that can form $Fe_2O_3$, FeO or $Fe_3O_4$—. The formation of oxides without volume change (this is the case of $C + O_2 \rightarrow CO_2$) appears in the diagram as horizontal lines as they are reactions whose $\Delta S_T^0$ is equal to zero. The lines corresponding to the formation of oxides with volume increase ($\Delta S_T^0 > 0$) logically have negative slope: this is the case of the reaction corresponding to $2C + O_2 \rightarrow 2CO$.

Regarding $\Delta H_T^0$, it was previously indicated that corresponds to the origin ordinate of each line. However, it is convenient to advise that the origin ordinate is not exactly the difference of enthalpies of products minus reagents at any temperature. Although this difference, analogously to difference of entropies of products and reagents, can be considered constant always and when physical transformations are not observed (allotropic change, melting—M—, or sublimation—B—).

For the oxygen pressure existing in the atmosphere, metallic oxides whose value of $\Delta G_T^0$ is very small (of around $-120$ kJ at 1200 °C, as happens for instance with the $Cu_2O$) will be spontaneously reduced when the temperature is increased: they are called easily dissociable oxides. For other oxides, its $\Delta G_T^0$—always negative—is so big (greater than $-760$ kJ at 1200 °C, as happens with the aluminum, magnesium, zirconium, thorium, calcium, etc.) that the atmospheric oxygen pressure, $P_{O_2}$, is always oxidizing and the elements oxidize at any temperature: they are called refractory oxides. As the value of $\Delta G_T^0$ of formation of the oxide determines the greater or smaller oxide formation from the metal, those whose $\Delta G_T^0$ at 1200 °C is between $-440$ and $-60$ kJ are usually considered oxides that are difficultly reduced. It is considered that easily reducible oxides are those whose $\Delta G_T^0$ at 1200 °C is around $-440$ kJ.

As complement of Fig. 4.1, we also include in Table 4.1 some of the standard free energies of oxides formation.

It is convenient, in the case of the iron (carbon steels), to indicate and provide details in a separated manner about the different oxidized species in equilibrium with the oxygen (Fe–O binary system): two components (Fe and O) and five compounds (Fe, $O_2$, $Fe_2O_3$, $Fe_3O_4$ and FeO). Even when we can use the representation of the standard free energies of formation of the different iron oxides with the temperature to explain-develop the thermodynamic possibilities of the reduction with gas and carbon, it is also possible to use the Fig. 4.2 to study the alternatives that can have the steels to oxidize or to be protected in its contact with the atmospheres used in the heat treatments.

From Fig. 4.2, it is possible to confirm that the nature of the oxides that contact with the steel at temperatures below 660 K is not the same than that developed at 1273 K. Equally, in strongly oxidizing environments, as it is usually the case of the full austenitization heat treatments of the carbon steels, even when the oxide layer that is in contact with the metal is FeO, over this last one other of $Fe_3O_4$ can be formed and, finally, in contact with the atmosphere other of $Fe_2O_3$.

**Exercise 4.1** Invariant point in the Fe–O system.

*Solution starts*

**Table 4.1**  Standard free energies (in kJ/mol) for the formation of some oxides, per mol of oxygen

| Temperature range | Reaction | Number of reactions in Fig. 4.1 | Free energy | $T_M$ (°C) |
|---|---|---|---|---|
| 25–1084 °C | $2Cu(s) + \frac{1}{2}O_2(g)$ $\rightarrow Cu_2O(s)$ | 1 | $\Delta G_T^0 =$ $-168 + 0.074 \cdot T(K)$ | 1083 |
| 25–1495 °C | $Co(s) + \frac{1}{2}O_2(g)$ $\rightarrow CoO(s)$ | 2 | $\Delta G_T^0 =$ $-235 + 0.074 \cdot T(K)$ | 1495 |
| 25–1452 °C | $Ni(s) + \frac{1}{2}O_2(g) \rightarrow$ $NiO(s)$ | 3 | $\Delta G_T^0 =$ $-236 + 0.088 \cdot T(K)$ | 1453 |
| 25–1371 °C | $Fe(s) + \frac{1}{2}O_2(g)$ $\rightarrow FeO(s)$ | – | $\Delta G_T^0 =$ $-265 + 0.065 \cdot T(K)$ | 1536.5 |
| 25–560 °C | $3Fe(s) + 2O_2(g)$ $\rightarrow Fe_3O_4(s)$ | – | $\Delta G_T^0 =$ $-1105 + 0.328 \cdot T(K)$ | 1536.5 |
| 1227–1470 °C | $W(s) + \frac{3}{2}O_2(g)$ $\rightarrow WO_3(s)$ | 4 | $\Delta G_T^0 =$ $-815 + 0.232 \cdot T(K)$ | 3410 |
| 1470–1727 °C | $W(s) + \frac{3}{2}O_2(g)$ $\rightarrow WO_3(s)$ | 4 | $\Delta G_T^0 =$ $-736 + 0.186 \cdot T(K)$ | 3410 |
| 25–1000 °C | $Mo(s) + O_2(g)$ $\rightarrow MoO_2(s)$ | 5 | $\Delta G_T^0 =$ $-584 + 0.177 \cdot T(K)$ | 2610 |
| 25–1898 °C | $2Cr(s) + \frac{3}{2}O_2(g)$ $\rightarrow Cr_2O_3(s)$ | 6 | $\Delta G_T^0 =$ $-1129 + 0.264 \cdot T(K)$ | 1875 |
| 25–1244 °C | $Mn(s) + \frac{1}{2}O_2(g)$ $\rightarrow MnO(s)$ | 7 | $\Delta G_T^0 =$ $-383 + 0.073 \cdot T(K)$ | 1245 |
| 1244–1700 °C | $Mn(l) + \frac{1}{2}O_2(g)$ $\rightarrow MnO(s)$ | 7 | $\Delta G_T^0 =$ $-402 + 0.086 \cdot T(K)$ | 1245 |
| 400–1410 °C | $Si(s) + O_2(g) \rightarrow$ $SiO_2(s)$ | 8 | $\Delta G_T^0 =$ $-903 + 0.175 \cdot T(K)$ | 1404 |
| 25–1700 °C | $H_2(s) + \frac{1}{2}O_2(g)$ $\rightarrow H_2O(g)$ | 9 | $\Delta G_T^0 =$ $-249 + 0.057 \cdot T(K)$ | |

(continued)

**Table 4.1**  (continued)

| Temperature range | Reaction | Number of reactions in Fig. 4.1 | Free energy | $T_M$ (°C) |
|---|---|---|---|---|
| 25–2000 °C | $C(s) + \frac{1}{2}O_2(g) \rightarrow CO(g)$ | 10 | $\Delta G_T^0 = -118 - 0.084 \cdot T(K)$ | |
| 25–2000 °C | $C(s) + O_2(g) \rightarrow CO_2(g)$ | 11 | $\Delta G_T^0 = -395 + 0.001 \cdot T(K)$ | |
| 25–2000 °C | $2CO(g) + O_2(g) \rightarrow 2CO_2(g)$ | 12 | $\Delta G_T^0 = -554 + 0.17 \cdot T(K)$ | |

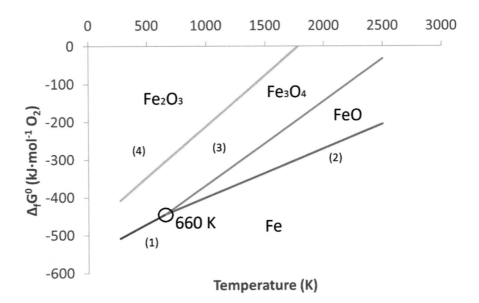

**Fig. 4.2**  Diagram $\Delta_f G^0/T$ for the iron compounds of the Fe–O system. (1) $\Delta_f G^0$ of the $Fe_3O_4$; (2) $\Delta_f G^0$ of the FeO; (3) $\Delta_f G^0$ of the $Fe_3O_4$ from the FeO; and (4) $\Delta_f G^0$ of the $Fe_2O_3$ from the $Fe_3O_4$

We have analyzed from the thermodynamics point of view in Verdeja et al. (2020) the succession of phases that are produced during the reduction of pellets. It was studied using the Ellingham's diagram as that shown in Fig. 4.2, where the stability zones for each one of the compounds of the Fe–O system are defined. Two situations might be considered:

– Reduction at low temperature: T < 387 °C (660 K).
– Reduction at high temperature: T > 387 °C (660 K).

Similar considerations can be extracted if the Fe–O phase diagram shown in Fig. 4.3 is used (elaborated with data of Massalski et al. 1986 and Verdeja et al. 2020), although in this thermodynamic calculation, the temperature, at which the

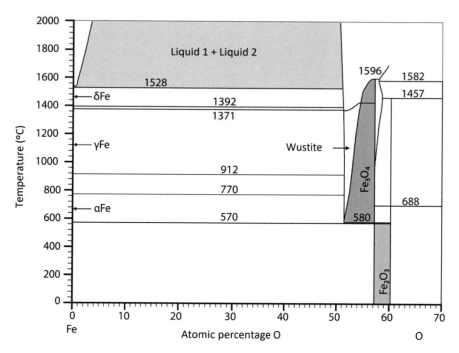

**Fig. 4.3** Fe–O binary diagram

invariant equilibrium between the phases Fe–FeO–Fe$_3$O$_4$–O$_2$ (gas) takes place is 570 °C, that is to say, 183 °C above the estimated temperature calculated using the thermodynamic data in Appendix 1 (Appendices).

Now we calculate the invariant point in the case of considering stoichiometric wüstite (iron (II) oxide). The following reactions and the corresponding standard free energies are collected from the Appendix 1 (Appendices):

$$2Fe(s) + O_2(g) \rightarrow 2FeO(s) \tag{4.19}$$

$$\Delta G^0(eq.\,4.19) = -530.0 + 0.130 \cdot T \left(kJ \cdot mol^{-1}O_2\right) \tag{4.20}$$

$$3/2Fe(s) + O_2(g) \rightarrow 1/2Fe_3O_4(s) \tag{4.21}$$

$$\Delta G^0(eq.\,4.21) = -552.5 + 0.164 \cdot T \left(kJ \cdot mol^{-1}O_2\right) \tag{4.22}$$

$$6FeO(s) + O_2(g) \rightarrow 2Fe_3O_4(s) \tag{4.23}$$

$$\Delta G^0(eq.\,4.23) = -592.0 + 0.224 \cdot T\left(kJ \cdot mol^{-1}O_2\right) \tag{4.24}$$

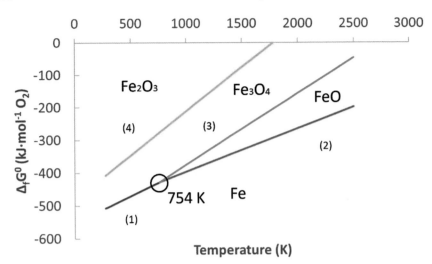

**Fig. 4.4** Diagram $\Delta_f G^0/T$ for the iron compounds of the Fe–O system. (1) $\Delta_f G^0$ of the $Fe_3O_4$; (2) $\Delta_f G^0$ of the FeO; (3) $\Delta_f G^0$ of the $Fe_3O_4$ from the FeO; and (4) $\Delta_f G^0$ of the $Fe_2O_3$ from the $Fe_3O_4$. Potential Neutral of Oxygen (PNO), Potential of Oxygen of the System (POS) and oxygen equilibrium pressure coincide at the triple point, as it will be later indicated

We must equalize the standard energies of the corresponding Ellingham's straight lines to find the temperature of the invariant point of Fig. 4.2. For example, if we equalize the straight lines 1 and 2 (Eqs. (4.19) and (4.21)) of Fig. 4.2:

$$-530.0 + 0.130 \cdot T = -552.5 + 0.164 \cdot T \tag{4.25}$$

we obtain as result the temperature of 661.7 K. For the equilibriums described by the Eq. (4.21), straight line 1, and the Eq. (4.23), straight line 3:

$$-552.5 + 0.164 \cdot T = -592.0 + 0.224 \cdot T \tag{4.26}$$

which indicates a temperature of 658.3 K. Finally, for the equilibriums between the straight line 2, reaction (4.19), and the straight line 3, reaction (4.23):

$$-530.0 + 0.130 \cdot T = -592.0 + 0.224 \cdot T \tag{4.27}$$

In this case, the temperature is 659.6 K, which is interpreted as the temperature of the invariant point of the Fe–O system where equilibrium between the four phases can take place (three solid and one gaseous, oxygen).

Similar calculations can be repeated in the case of considering non-stoichiometric wüstite. The following reactions and the corresponding standard free energies are collected from the Appendix 1 (Appendices):

$$2\text{Fe}(s) + \text{O}_2(g) \rightarrow 2\text{FeO}(s)\,(\text{non-stoichiometric}) \qquad (4.28)$$

$$\Delta G^0(\text{eq. 4.28}) = -528.0 + 0.132 \cdot T\,(\text{kJ} \cdot \text{mol}^{-1}\text{O}_2) \qquad (4.29)$$

$$3/2\text{Fe}(s) + \text{O}_2(g) \rightarrow 1/2\text{Fe}_3\text{O}_4(s) \qquad (4.30)$$

$$\Delta G^0(\text{eq. 4.30}) = -552.5 + 0.164 \cdot T\,(\text{kJ} \cdot \text{mol}^{-1}\text{O}_2) \qquad (4.31)$$

$$6\text{FeO}(s) + \text{O}_2(g) \rightarrow 2\text{Fe}_3\text{O}_4(s) \qquad (4.32)$$

$$\Delta G^0(\text{eq. 4.32}) = -594.0 + 0.219 \cdot T\,(\text{kJ} \cdot \text{mol}^{-1}\text{O}_2) \qquad (4.33)$$

We must equalize the standard energies of the corresponding Ellingham's straight lines to find the temperature of the invariant point of Fig. 4.2. For example, if we equalize the straight lines 1 and 2 (Eqs. (4.28) and (4.30)) of Fig. 4.2:

$$-528.0 + 0.132 \cdot T = -552.5 + 0.164 \cdot T \qquad (4.34)$$

we obtain as result the temperature of 765.6 K. For the equilibriums described by the Eq. (4.30), straight line 1, and the Eq. (4.32), straight line 3:

$$-552.5 + 0.164 \cdot T = -594.0 + 0.219 \cdot T \qquad (4.35)$$

which indicates a temperature of 754.5 K. Finally, for the equilibriums between the straight line 2, reaction (4.28), and the straight line 3, reaction (4.32):

$$-528.0 + 0.132 \cdot T = -594.0 + 0.219 \cdot T \qquad (4.36)$$

The temperature is 758.6 K in this case, which is interpreted as the temperature of the invariant point of the Fe–O system where equilibrium between the four phases can take place (three solid and one gaseous, oxygen).

As it was previously demonstrated, the temperature of 660 K is the invariant temperature of the system of two components Fe–O. This invariant point is reached as a consequence of that the variance-degrees of freedom of the system are zero as a result of the application of the phase rule:

$$F + L = 2 + 2;\ L = 0 \text{ when } F = 4 \qquad (4.37)$$

That is to say, the temperature of 660 K is defined when there are four phases in equilibrium in the binary system of two components Fe–O. Now, for the diagram of Fig. 4.2, it is necessary to calculate the standard free energy per mol of $O_2$ associated to the invariant point. The value of the standard free energy for the formation of the FeO and $Fe_3O_4$ both from the Fe as from the FeO is calculated. The average value of the three standard free energies per mol of oxygen is of $-444.21 \; kJ \cdot mol^{-1}$ $O_2$. Consequently, the coordinates of the invariant point of the Fe–O system in an Ellingham's diagram would be: (660 K, $-444.21 \; kJ \cdot mol^{-1} \; O_2$). However, if instead stoichiometric iron (II) oxide, wüstite is used, the invariant point displaces to: (754 K, $-428.87 \; kJ \cdot mol^{-1} \; O_2$). Therefore, small changes in the characteristics of the oxide involve significant changes in the invariant point.

*Solution ends.*

## 4.3   Reduction of Oxides with Gases

The reduction of metallic oxides—of non-volatile metals at the temperature at which the reaction is verified—is usually carried out in industrial processes by means of reductant gases. Two are the reasons that advise the selection of a reductant agent of gaseous nature always and when this is possible. On the one hand, it facilitates the contact of the reductant agent and the oxide; reaction rate increases with this. On the other hand, it is interesting—by economy—that the reductant agent, apart from gaseous at room temperature, to be not expensive. CO (that in the reaction will pass to $CO_2$) or/and $H_2$ (that in the reaction will pass to $H_2O$) are generally used as reductant reagents. The reduction of oxides can be also achieved with solid agents— by metallothermic reduction, v. gr. by means of aluminum (that in the reaction will transform into $Al_2O_3$)—but its study was already considered in other books (Pero-Sanz et al. 2017, 2018).

In what is next, we primarily analyze the behavior of metals and their oxides in presence of binary atmospheres comprised by mixtures of two gases (CO and $CO_2$, Sect. 4.3.1, or $H_2$ and $H_2O$, see Sect. 4.3.2) and, posteriorly, we will study the answer in presence of several gases.

### 4.3.1   Reduction of an Oxide $M_xO_y$ in the Presence of a CO–$CO_2$ Gaseous Mixture

We consider the reaction:

$$\frac{2 \cdot x}{y} M(s) + O_2(g) \rightarrow \frac{2}{y} M_xO_y \qquad (4.38)$$

at temperature T, being $\left(\Delta G_T^0\right)_1$ the corresponding standard free energy (known by means of the Ellingham's diagram for this reaction and this temperature T); it is, for that reason, known the oxygen partial pressure, $P_{O_2}$, that should have the atmosphere to avoid the oxidation of the metal (and, also to avoid the reduction of the oxide of this metal):

$$\ln P_{O_2} = \frac{\left(\Delta G_T^0\right)_1}{-8.3144 \cdot T} \tag{4.39}$$

On the other hand, we also take into account the following reaction:

$$2CO(s) + O_2(g) \rightarrow 2CO_2(g) \tag{4.40}$$

at temperature T, whose standard free energy:

$$\left(\Delta G_T^0\right)_{eq.\ 4.40} = -554 + 0.17 \cdot T(K) \tag{4.41}$$

allows calculating the oxygen partial pressure $P_{O_2}'$ in this equilibrium between gases by means of the following equation:

$$\ln P_{O_2} = \left[\frac{\left(P_{CO_2}\right)^2}{\left(P_{CO}\right)^2 \cdot \left(P_{O_2}'\right)}\right] = \frac{\left(\Delta G_T^0\right)_1}{-8.3144 \cdot T} \tag{4.42}$$

If the metal is in the presence of a binary atmosphere of CO and $CO_2$ where $P_{O_2}' > P_{O_2}$, the metal will oxidize. If, on the contrary, $P_{O_2}' < P_{O_2}$, the atmosphere will be reductant. And if the atmosphere in contact with the metal reaches the same potential of oxygen (PNO, Potential Neutral of Oxygen, where the pressure of $O_2$ in equilibrium in the reaction (4.40) is equal to $P_{O_2}$ of the Eq. (4.39)), the equilibrium between two condensed phases, M and $M_xO_y$, and a gas formed by $O_2$, CO and $CO_2$ is reached. Therefore, replacing in the Eq. (4.42) the value of $P_{O_2}'$ by the value of $P_{O_2}$ taken from the Eq. (4.39), we have the function that receives the name of neutral potential of oxygen of this metal in presence of the binary mixture of CO and $CO_2$:

$$\ln \frac{\left(P_{CO_2}\right)^2}{\left(P_{CO}\right)^2} = \frac{\left(\Delta G_T^0\right)_{eq.\ 4.40} - \left(\Delta G_T^0\right)_l}{-8.3144 \cdot T} \tag{4.43}$$

which allows knowing the relation between CO and $CO_2$ to achieve a neutral atmosphere for the metal at the temperature T. Equation that, for the total pressure of 1 atmosphere, can be also expressed as follows:

$$\ln \frac{[CO]}{[CO_2]} = \frac{\left(\Delta G_T^0\right)_{eq.\ 4.40} - \left(\Delta G_T^0\right)_1}{8.3144 \cdot 2 \cdot T} \tag{4.44}$$

where [CO] and [$CO_2$] are, respectively, the volume concentrations of CO and $CO_2$ in this binary atmosphere.

See that the Eq. (4.42) is, logically, the same that would be obtained by subtracting from the reaction $2CO+O_2 \rightarrow 2CO_2$ the reaction $2x/yM+O_2 \rightarrow 2/yM_xO_y$ to obtain the reaction $2CO + 2/yM_xO_y \rightarrow 2CO_2 + 2x/yM$. The free energy of this reaction will be $(\Delta G_T)_{eq.\ 4.40} - (\Delta G_T)_l$ and its standard free energy $\left(\Delta G_T^0\right)_{eq.\ 4.40} - \left(\Delta G_T^0\right)_l$. Therefore, the following is verified in the equilibrium of this reaction:

$$(\Delta G_T)_{eq.\ 4.40} - (\Delta G_T)_l = 0 = \left(\Delta G_T^0\right)_{eq.\ 4.40} - \left(\Delta G_T^0\right)_l$$

$$+ 8.3144 \cdot T \cdot \ln \frac{(P_{CO})^2}{\left(P_{CO_2}\right)^2} \qquad (4.45)$$

Brief note: *Considerations about the PNO, Potential Neutral of Oxygen.*

*Brief note starts*:

It is necessary to remember that the thermodynamics only allows to determine the composition of the compounds in the gas that at a certain temperature and total pressure are in equilibrium either with the rest of species compounds in the gas or with the oxidized/reduced phases, if available.

Definitely, the concept that is presented here, PNO (potential that is equivalent to the Potential of Oxygen of the System, POS, developed in Verdeja et al. 2020 and 2021), is the partial pressure of oxygen in the gaseous phase that reaches the equilibrium, $P_{O_2}=$, with the other molecules of the gas: CO, $CO_2$, $H_2$, $H_2O$, $CH_4$ or more complex hydrocarbons ($C_2H_2$, ethylene; $C_2H_6$, ethane; $C_3H_8$, propane; $C_4H_{10}$, butane). If there were condensed matter in equilibrium with the gas, as it could be the case of the iron, carbon, or the oxidized species of the iron: FeO, $Fe_3O_4$, and $Fe_2O_3$, both the PNO and the POS will be identified with the partial pressure of oxygen in equilibrium with the gaseous molecules of the gas or with the matter-condensed phases if available.

*Brief note ends.*

**Exercise 4.2** The concentration of CO should be lower than 0.003 parts per million to achieve that a mixture of $CO$–$CO_2$ was oxidizing for the copper at T < 400 °C. And the CO content should be lower than 0.07% for an atmosphere of 12.5% $CO_2$, nitrogen, and CO was oxidizing for the nickel at 750 °C.

*Solution starts*

In the case of the copper, the value of the standard free energy associated to the reaction-equilibrium (see data in Table 4.1):

$$2Cu(s) + CO_2(g) \rightleftharpoons CO(g) + Cu_2O(s) \qquad (4.46)$$

is:

$$\left(\Delta G_T^0\right)_{eq.\ 4.46} = 109.0 - 0.011 \cdot T(K)kJ \cdot (mol\ Cu_2O)^{-1} \qquad (4.47)$$

The value of $\left(\Delta G_T^{\ 0}\right)_{eq.4.46}$ is the result of combining the following reactions:

$$2Cu(s) + \frac{1}{2}O_2(g) \rightarrow Cu_2O(s) \tag{4.48}$$

$$\left(\Delta G_T^0\right)_{eq.\ 4.48} = -168.0 + 0.074 \cdot T(K)kJ \cdot (mol\ Cu_2O)^{-1} \tag{4.49}$$

$$2CO(g) + O_2(g) \rightarrow 2CO_2(g) \tag{4.50}$$

$$\left(\Delta G_T^0\right)_{eq.\ 4.50} = -554.0 + 0.17 \cdot T(K)kJ \cdot (mol\ O_2)^{-1} \tag{4.51}$$

$$\left(\Delta G_T^0\right)_{eq.\ 4.46} = \left(\Delta G_T^0\right)_{eq.\ 4.48} - \frac{1}{2} \cdot \left(\Delta G_T^0\right)_{eq.\ 4.50} \tag{4.52}$$

Considering as data the value of the $\%CO_2$ volumetric in the gas: 21%, and taking into account that at 400 °C the value of the equilibrium constant of the reaction (4.46) is:

$$k_{eq} = \exp\left(-\frac{\Delta G_T^0}{8.3144 \cdot T}\right) = \exp\left(-\frac{101597}{8.3144 \cdot 673}\right) = 1.3022 \cdot 10^{-8} \tag{4.53}$$

these values of $P_{CO_2}$ and $k_{eq}$ lead to a $P_{CO}$ of:

$$k_{eq} = \frac{P_{CO}}{P_{CO_2}} \rightarrow 1.3022 \cdot 10^{-8} = \frac{P_{CO}}{21.0 \cdot 10^{-2}} \rightarrow P_{CO}$$
$$= 2.73462 \cdot 10^{-9} atm\ (0.003\ ppm\ CO) \tag{4.54}$$

The addition of partial pressures in the gas (air) is, for a total pressure of the gas (gaseous phase) of 1.0 atm (the partial pressure of oxygen tends to zero when the species $CO/CO_2$ are available in the gas):

$$P_T = P_{N_2} + P_{CO} + P_{CO_2} + P_{O_2}$$
$$= 1\ atm \rightarrow P_T \simeq P_{N_2} + P_{CO} + P_{CO_2} \tag{4.55}$$

This value indicates that $P_{CO_2} \simeq 0.21 atm$ ($P_{N_2} = 0.79 atm$ and $P_{O_2} = 0.21 atm$ are the partial pressures of the air atmosphere before reaching the equilibrium).

Consequently, the partial pressure of the carbon monoxide should be smaller than 0.003 ppm CO to have a $N_2$–CO–$CO_2$ atmosphere oxidizing for the copper at a temperature of 400 °C.

The standard free energy associated to the formation of the NiO (Table 4.1) is:

$$Ni(s) + \frac{1}{2}O_2(g) \rightarrow NiO(s) \tag{4.56}$$

$$\left(\Delta G_T^0\right)_{eq.\ 4.56} = -236.0 + 0.088 \cdot T(K)kJ \cdot (mol\ NiO)^{-1} \tag{4.57}$$

that together with the formation of $CO_2$ from the CO (reaction (4.40)) leads to the following reaction-equilibrium:

$$Ni(s) + CO_2(g) \rightarrow CO(g) + NiO(s) \qquad (4.58)$$

that has associated a standard free energy of:

$$
\begin{aligned}
\left(\Delta G_T^0\right)_{eq.\,4.58} &= \left(\Delta G_T^0\right)_{eq.\,4.56} - \frac{1}{2} \cdot \left(\Delta G_T^0\right)_{eq.\,4.40} \\
&= -236.0 + 0.088 \cdot T(K) - \frac{1}{2} \cdot (-554 + 0.17 \cdot T(K)) \\
&= 41 + 0.003 \cdot T(K) kJ \cdot (mol\ NiO)^{-1} \qquad (4.59)
\end{aligned}
$$

Considering as data the value of the $\%CO_2$ volumetric in the gas: 12.5%, and taking into account that at 750 °C the value of the equilibrium constant of the reaction (4.58) is:

$$k_{eq} = \exp\left(-\frac{\Delta G_T^0}{8.3144 \cdot T}\right) = \exp\left(-\frac{44069}{8.3144 \cdot 1023}\right) = 5.6215 \cdot 10^{-3} \qquad (4.60)$$

these values of $P_{CO_2}$ and $k_{eq}$ lead to a $P_{CO}$ of:

$$
\begin{aligned}
k_{eq} &= \frac{P_{CO}}{P_{CO_2}} \rightarrow 5.6215 \cdot 10^{-3} = \frac{P_{CO}}{12.5 \cdot 10^{-2}} \rightarrow P_{CO} \\
&= 7.268 \cdot 10^{-4} atm\ (0.07\ vol.\%\ CO) \qquad (4.61)
\end{aligned}
$$

It is necessary a $P_{CO} \simeq 7.268 \cdot 10^{-4} atm (0.07 vol.\% CO)$ for an atmosphere at 750 °C to be oxidizing in the case of the nickel.

*Solution ends.*

*Brief note: Influence of the total pressure in the solid–gas equilibriums associated to chemical reactions.*

*Brief note starts*

When there is not a volume contraction or expansion (due to an excess-defect of gases) between products-reagents of a reaction, the total pressure, $P_T$, of the system does not have influence in the value of the equilibrium constant at a certain temperature.

However, the total pressure of the system, $P_T$, can have influence in which is known as "equilibrium point". The concept of "equilibrium point" is that concentration (mole fraction or volume composition in percentage) that is reached between the molecules of gas and the condensed phases in equilibrium at a certain temperature. This concentration, "equilibrium point", can be altered/changed with the total pressure in the system that is being studied.

*Brief note ends.*

**Exercise 4.3** Chaudron's diagrams applied to understand the equilibrium in the systems Fe–O–C and Fe–O–H.

*Solution starts*

Chaudron's diagram expresses the carbon monoxide or hydrogen concentration in the gas ($CO + CO_2$ and $H_2 + H_2O$, respectively) as a function of the temperature. This allows the definition of the domains with predominance of the phases Fe–FeO–$Fe_3O_4$. We are going to calculate the Chaudron's diagrams for the Fe–O–C and Fe–O–H systems.

First, we obtain the equations for the $Fe_3O_4$–Fe, $Fe_3O_4$–FeO and FeO–Fe equilibriums in presence of carbon monoxide. The equilibrium $Fe_2O_3$–$Fe_3O_4$ is not considered because it coincides with the x-axis in the graphical representation. Data of the Appendix 1 (Appendices) is used in the calculations.

$$C(s) + \frac{1}{2}O_2(g) \rightarrow CO(g) \tag{4.62}$$

and the standard free energy of the reaction (4.62) is:

$$\left(\Delta G_T^0\right)_{eq.\ 4.62} = -118.0 - 0.084 \cdot T(K) kJ \cdot (mol\ CO)^{-1} \tag{4.63}$$

$$C(s) + O_2(g) \rightarrow C_2(g) \tag{4.64}$$

and the standard free energy of the reaction (4.64) is:

$$\left(\Delta G_T^0\right)_{eq.\ 4.64} = -395.0 + 0.001 \cdot T(K) kJ \cdot (mol\ CO_2)^{-1} \tag{4.65}$$

We obtain Eq. (4.66) from Eqs. (4.62) to (4.64):

$$2CO(s) + O_2(g) \rightarrow 2CO_2(g) \tag{4.66}$$

and the standard free energy of the reaction (4.66) is:

$$\begin{aligned}
\left(\Delta G_T^0\right)_{eq.\ 4.66} &= 2 \cdot \left(\Delta G_T^0\right)_{eq.\ 4.64} - 2 \cdot \left(\Delta G_T^0\right)_{eq..4.62} \\
&= 2 \cdot (-395.0 + 0.001 \cdot T(K)) - 2 \cdot (-118.0 - 0.084 \cdot T(K)) \\
&= -554.0 + 0.166 \cdot T(K) kJ \cdot (mol\ O_2)^{-1} \tag{4.67}
\end{aligned}$$

We start now with the equilibrium Fe–FeO-gas ($CO$–$CO_2$–$N_2$) when the total pressure of the system is 1 atmosphere and $P_{N_2} = 0$ atm.

$$2Fe(s) + O_2(g) \rightarrow 2FeO(s) \tag{4.68}$$

and the standard free energy of the reaction (4.68) is:

$$\left(\Delta G_T^0\right)_{eq.\ 4.68} = -530.0 + 0.130 \cdot T(K) kJ \cdot (mol\ O_2)^{-1} \tag{4.69}$$

We obtain Eq. (4.70) from Eqs. (4.66) to (4.68):

$$Fe(s) + CO_2(g) \rightarrow FeO(s) + CO(g) \tag{4.70}$$

and the standard free energy of the reaction (4.70) is:

$$
\begin{aligned}
\left(\Delta G_T^0\right)_{eq.4.70} &= \frac{1}{2} \cdot \left(\Delta G_T^0\right)_{eq.4.68} - \frac{1}{2} \cdot \left(\Delta G_T^0\right)_{eq.4.66} \\
&= \frac{1}{2} \cdot (-530.0 + 0.130 \cdot T(K)) - \frac{1}{2} \cdot (-554.0 + 0.166 \cdot T(K)) \\
&= 12.0 - 0.018 \cdot T(K)kJ \cdot (mol\ CO)^{-1} \tag{4.71}
\end{aligned}
$$

The equilibrium constant as a function of the temperature is:

$$
\begin{aligned}
\left(\Delta G_T^0\right)_{eq.\,4.70} &= -R \cdot T(K) \cdot \ln k_{eq} \rightarrow k_{eq} \\
&= \exp\left[-\frac{\left(\Delta G_T^0\right)_{eq.\,4.70} \cdot \frac{1000J}{1\,kJ}}{8.3144\left(\frac{J}{mol \cdot K}\right) \cdot T(K)}\right] \tag{4.72}
\end{aligned}
$$

We also know that:

$$P_T = P_{N_2} + P_{CO_2} + P_{CO} = 1\ atm \rightarrow P_{CO_2} + P_{CO} = 1\ atm \tag{4.73}$$

and ($a = 1$, for pure compounds):

$$k_{eq} = \frac{P_{CO} \cdot a_{FeO}}{P_{CO_2} \cdot a_{Fe}} = \frac{P_{CO}}{P_{CO_2}} \tag{4.74}$$

Thus, we can solve the system as we have two Eqs. (4.73) and (4.74) and two unknowns ($P_{CO}$ and $P_{CO_2}$). Finally, we obtain:

$$R = \frac{P_{CO}}{P_{CO_2}} \tag{4.75}$$

$$R_1 = \frac{P_{CO}}{P_{CO} + P_{CO_2}} \tag{4.76}$$

and the volume concentration is:

$$R_1 \cdot 100 = \frac{\%vol.CO}{\%vol.CO + \%vol.CO_2} \tag{4.77}$$

We obtain for the equilibrium Fe–FeO gas ($CO$–$CO_2$–$N_2$), the values collected in Table 4.2.

We start now with the equilibrium FeO–$Fe_3O_4$-gas ($CO$–$CO_2$–$N_2$) when the total pressure of the system is 1 atmosphere and $P_{N_2} = 0$ atm.

$$6FeO(s) + O_2(g) \rightarrow 2Fe_3O_4(s) \tag{4.78}$$

**Table 4.2** Results for the Fe–FeO-gas ($CO$–$CO_2$–$N_2$)

| Temperature (°C) | Temperature (K) | $(\Delta G_T^0)_{eq,4.70}$ | $k_{eq}$ | $P_{N_2}$ (atm) | $P_{CO}$(atm) | $P_{CO_2}$ (atm) | R | $R_1$ | %vol.CO |
|---|---|---|---|---|---|---|---|---|---|
| 387 | 660 | 0.12 | 0.978 | 0 | 0.49 | 0.51 | 0.98 | 0.49 | 49.45 |
| 527 | 800 | −2.40 | 1.43 | 0 | 0.59 | 0.41 | 1.43 | 0.59 | 58.92 |
| 627 | 900 | −4.20 | 1.75 | 0 | 0.64 | 0.36 | 1.75 | 0.64 | 63.67 |
| 727 | 1000 | −6.00 | 2.06 | 0 | 0.67 | 0.33 | 2.06 | 0.67 | 67.30 |
| 827 | 1100 | −7.80 | 2.35 | 0 | 0.70 | 030 | 2.35 | 0.70 | 70.12 |
| 927 | 1200 | −9.60 | 2.62 | 0 | 0.72 | 0.28 | 2.62 | 0.72 | 72.36 |
| 1027 | 1300 | −11.40 | 2.87 | 0 | 0.74 | 0.26 | 2.87 | 0.74 | 74.17 |
| 1127 | 1400 | −13.20 | 3.11 | 0 | 0.76 | 0.24 | 3.11 | 0.76 | 75.66 |
| 1227 | 1500 | −15.00 | 3.33 | 0 | 0.77 | 0.23 | 3.33 | 0.77 | 76.90 |

and the standard free energy of the reaction (4.78) is:

$$\left(\Delta G_T^0\right)_{eq.4.78} = -592.0 + 0.224 \cdot T(K)kJ \cdot (mol\ O)^{-1} \qquad (4.79)$$

We obtain Eq. (4.80) from Eqs. (4.66) to (4.78):

$$3FeO(s) + CO_2(g) \rightarrow Fe_3O_4(s) + CO(g) \qquad (4.80)$$

and the standard free energy of the reaction (4.80) is:

$$
\begin{aligned}
\left(\Delta G_T^0\right)_{eq.\ 4.80} &= \frac{1}{2} \cdot \left(\Delta G_T^0\right)_{eq.\ 4.78} - \frac{1}{2} \cdot \left(\Delta G_T^0\right)_{eq.\ 4.66} \\
&= \frac{1}{2} \cdot (-592.0 + 0.224 \cdot T(K)) - \frac{1}{2} \cdot (-554.0 + 0.166 \cdot T(K)) \\
&= -19.0 + 0.029 \cdot T(K)kJ \cdot (mol\ CO)^{-1} \qquad (4.81)
\end{aligned}
$$

We obtain for the equilibrium $FeO$–$Fe_3O_4$-gas ($CO$–$CO_2$–$N_2$), the values collected in Table 4.3.

We start now with the equilibrium $Fe$–$Fe_3O_4$-gas ($CO$–$CO_2$–$N_2$) when the total pressure of the system is 1 atmosphere and $P_{N_2} = 0$ atm.

$$\frac{3}{2}Fe(s) + O_2(g) \rightarrow \frac{1}{2}Fe_3O_4(s) \qquad (4.82)$$

and the standard free energy of the reaction (4.82) is:

$$\left(\Delta G_T^0\right)_{eq.\ 4.82} = -552.5 + 0.164 \cdot T(K)kJ \cdot (mol\ O)_2^{-1} \qquad (4.83)$$

We obtain Eq. (4.84) from Eqs. (4.66) to (4.82):

$$\frac{3}{4}Fe(s) + CO_2(g) \rightarrow \frac{1}{4}Fe_3O_4(s) + CO(g) \qquad (4.84)$$

and the standard free energy of the reaction (4.84) is:

$$
\begin{aligned}
\left(\Delta G_T^0\right)_{eq.\ 4.84} &= \frac{1}{2} \cdot \left(\Delta G_T^0\right)_{eq.\ 4.82} - \frac{1}{2} \cdot \left(\Delta G_T^0\right)_{eq.\ 4.66} \\
&= \frac{1}{2} \cdot (-552.5 + 0.164 \cdot T(K)) - \frac{1}{2} \cdot (-554.0 + 0.166 \cdot T(K)) \\
&= 0.8 - 0.001 \cdot T(K)kJ \cdot (mol\ CO)^{-1} \qquad (4.85)
\end{aligned}
$$

We obtain for the equilibrium $Fe$–$Fe_3O_4$-gas ($CO$–$CO_2$–$N_2$), the values collected in Table 4.4.

Finally, we represent the diagram of Chaudron for the $Fe$–$O$–$C$ system in Fig. 4.5.

**Table 4.3** Results for the FeO–Fe₃O₄-gas (CO–CO₂–N₂)

| Temperature (°C) | Temperature (K) | $(\Delta G_T^0)_{eq,4.80}$ | $k_{eq}$ | $P_{N_2}$ (atm) | $P_{CO}$ (atm) | $P_{CO_2}$ (atm) | R | $R_1$ | %vol.CO |
|---|---|---|---|---|---|---|---|---|---|
| 387 | 660 | 0.14 | 0.975 | 0 | 0.49 | 0.51 | 0.97 | 0.49 | 49.36 |
| 527 | 800 | 4.20 | 0.532 | 0 | 0.35 | 0.65 | 0.53 | 035 | 34.72 |
| 627 | 900 | 7.10 | 0.387 | 0 | 0.28 | 0.72 | 0.39 | 0.28 | 27.91 |
| 727 | 1000 | 10.00 | 0.300 | 0 | 0.23 | 0.77 | 0.30 | 0.23 | 23.10 |
| 827 | 1100 | 12.90 | 0.244 | 0 | 0.20 | 0.80 | 0.24 | 0.20 | 19.62 |
| 927 | 1200 | 15.80 | 0.205 | 0 | 0.17 | 0.83 | 0.21 | 0.17 | 17.03 |
| 1027 | 1300 | 18.70 | 0.177 | 0 | 0.15 | 0.85 | 0.18 | 0.15 | 15.06 |
| 1127 | 1400 | 21.60 | 0.156 | 0 | 0.14 | 0.86 | 0.16 | 0.14 | 13.52 |
| 1227 | 1500 | 24.50 | 0.140 | 0 | 0.12 | 0.88 | 0.14 | 0.12 | 12.30 |

**Table 4.4** Results for the Fe–Fe$_3$O$_4$-gas (CO–CO$_2$–N$_2$)

| Temperature (°C) | Temperature (K) | $(\Delta G_T^0)_{eq.4.80}$ | $k_{eq}$ | $P_{N_2}$ (atm) | $P_{CO}$ (atm) | $P_{CO_2}$ (atm) | R | $R_1$ | %vol.CO |
|---|---|---|---|---|---|---|---|---|---|
| 25 | 298 | 0.45 | 0.833 | 0 | 0.45 | 0.55 | 0.83 | 0.45 | 45.45 |
| 50 | 323 | 0.43 | 0.853 | 0 | 0.46 | 0.54 | 0.85 | 0.46 | 46.03 |
| 100 | 373 | 0.38 | 0.886 | 0 | 0.47 | 0.53 | 0.89 | 0.47 | 46.96 |
| 150 | 423 | 0.33 | 0.911 | 0 | 0.48 | 0.52 | 0.91 | 0.48 | 47.68 |
| 200 | 473 | 0.28 | 0.932 | 0 | 0.48 | 0.52 | 0.93 | 0.48 | 48.24 |
| 250 | 523 | 0.23 | 0.949 | 0 | 0.49 | 0.51 | 0.95 | 0.49 | 48.70 |
| 300 | 573 | 0.18 | 0.964 | 0 | 0.49 | 0.51 | 0.96 | 0.49 | 49.07 |
| 350 | 623 | 0.13 | 0.976 | 0 | 0.49 | 0.51 | 0.98 | 0.49 | 49.39 |
| 387 | 660 | 0.09 | 0.984 | 0 | 0.50 | 0.50 | 0.98 | 0.50 | 49.59 |

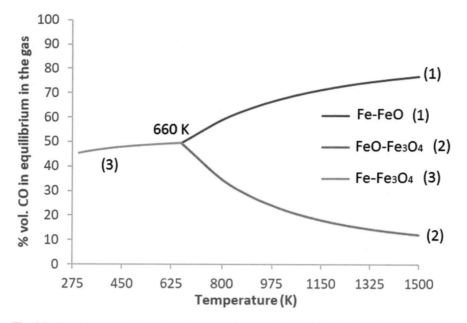

**Fig. 4.5** Fe–O–C system. Domains with predominance of Fe–FeO–$Fe_3O_4$ in equilibrium with the CO at different temperatures. It is at the triple point (660 K) where the Potential Neutral of Oxygen, the Potential of Oxygen of the System, and the oxygen equilibrium pressure are equivalent

Now, we obtain the equations for the $Fe_3O_4$–Fe, $Fe_3O_4$–FeO and FeO–Fe equilibriums in the presence of hydrogen. The equilibrium $Fe_2O_3$–$Fe_3O_4$ is not considered because in the graphical representation it coincides with the x-axis. Data of the Appendix 1 (Appendices) are used in the calculations.

We consider the reaction of the hydrogen combustion:

$$H_2(g) + \frac{1}{2}O_2(g) \rightarrow H_2O(g) \tag{4.86}$$

and the standard free energy of the reaction (4.86) is:

$$\left(\Delta G_T^0\right)_{eq.\,4.86} = -249.0 + 0.057 \cdot T(K) kJ \cdot (mol\ H_2)^{-1} \tag{4.87}$$

We start now with the equilibrium Fe–FeO-gas ($H_2$–$H_2O$–$N_2$) when the total pressure of the system is 1 atmosphere and $P_{N_2} = 0$ atm.

$$2Fe(s) + O_2(g) \rightarrow 2FeO(s) \tag{4.88}$$

and the standard free energy of the reaction (4.88) is:

$$\left(\Delta G_T^0\right)_{eq.\,4.88} = -530.0 + 0.130 \cdot T(K) kJ \cdot (mol\ O)_2^{-1} \tag{4.89}$$

We obtain Eq. (4.90) from Eqs. (4.86) to (4.88):

$$Fe(s) + H_2O(g) \rightarrow FeO(s) + H_2(g) \qquad (4.90)$$

and the standard free energy of the reaction (4.90) is:

$$
\begin{aligned}
\left(\Delta G_T^0\right)_{eq.\,4.90} &= \frac{1}{2} \cdot \left(\Delta G_T^0\right)_{eq.\,4.88} - \left(\Delta G_T^0\right)_{eq.\,4.86} \\
&= \frac{1}{2} \cdot (-530.0 + 0.130 \cdot T(K)) - (-249.0 + 0.057 \cdot T(K)) \\
&= -16.0 + 0.008 \cdot T(K) kJ \cdot (mol\ H_2)^{-1} \qquad (4.91)
\end{aligned}
$$

The equilibrium constant as a function of the temperature is:

$$
\left(\Delta G_T^0\right)_{eq.\,4.90} = -R \cdot T(K) \cdot \ln k_{eq} \rightarrow k_{eq}
$$

$$
= exp\left[ -\frac{\left(\Delta G_T^0\right)_{eq.\,4.90} \cdot \frac{1000J}{1\,kJ}}{8.3144\left(\frac{J}{mol \cdot K}\right) \cdot T(K)} \right] \qquad (4.92)
$$

We also know that:

$$P_T = P_{N_2} + P_{H_2} + P_{H_2O} = 1\ atm \rightarrow P_{H_2} + P_{H_2O} = 1\ atm \qquad (4.93)$$

and (a = 1, for pure compounds):

$$k_{eq} = \frac{P_{H_2} \cdot a_{FeO}}{P_{H_2O} \cdot a_{Fe}} = \frac{P_{H_2}}{P_{H_2O}} \qquad (4.94)$$

Thus, we can solve the system as we have two Eqs. (4.93) and (4.94) and two unknowns ($P_{H_2}$ and $P_{H_2O}$). Finally, we obtain:

$$R = \frac{P_{H_2}}{P_{H_2O}} \qquad (4.95)$$

$$R_1 = \frac{P_{H_2}}{P_{H_2} + P_{H_2O}} \qquad (4.96)$$

and the volume concentration is:

$$R_1 \cdot 100 = \frac{\%vol \cdot H_2}{\%vol.H_2 + \%vol.H_2O} \qquad (4.97)$$

We obtain for the equilibrium Fe–FeO-gas ($H_2$–$H_2O$–$N_2$) the values collected in Table 4.5.

**Table 4.5** Results for the Fe–FeO-gas ($H_2$–$H_2O$–$N_2$)

| Temperature (°C) | Temperature (K) | $(\Delta G_T^0)_{eq.4.90}$ | $k_{eq}$ | $P_{N_2}$(atm) | $P_{H_2}$(atm) | $P_{H_2O}$(atm) | R | $R_1$ | %vol.$H_2$ |
|---|---|---|---|---|---|---|---|---|---|
| 387 | 660 | −10.72 | 7.05 | 0 | 0.88 | 0.12 | 7.05 | 0.88 | 87.58 |
| 527 | 800 | −9.60 | 4.23 | 0 | 0.81 | 0.19 | 4.23 | 0.81 | 80.90 |
| 627 | 900 | −8.80 | 3.24 | 0 | 0.76 | 0.24 | 3.24 | 0.76 | 76.42 |
| 727 | 1000 | −8.00 | 2.62 | 0 | 0.72 | 0.28 | 2.62 | 0.72 | 72.36 |
| 827 | 1100 | −7.20 | 2.20 | 0 | 0.69 | 0.31 | 2.20 | 0.69 | 68.72 |
| 927 | 1200 | −6.40 | 1.90 | 0 | 0.66 | 0.34 | 1.90 | 0.66 | 65.51 |
| 1027 | 1300 | −5.60 | 1.68 | 0 | 0.63 | 0.37 | 1.68 | 0.63 | 62.67 |
| 1127 | 1400 | −4.80 | 1.51 | 0 | 0.60 | 0.40 | 1.51 | 0.60 | 60.17 |
| 1227 | 1500 | −4.00 | 1.38 | 0 | 0.58 | 0.42 | 1.38 | 0.58 | 57.95 |

We start now with the equilibrium FeO–Fe$_3$O$_4$-gas (H$_2$–H$_2$O–N$_2$) when the total pressure of the system is 1 atmosphere and $P_{N_2} = 0$ atm.

$$6FeO(s) + O_2(g) \rightarrow 2Fe_3O_4(s) \tag{4.98}$$

and the standard free energy of the reaction (4.98) is:

$$\left(\Delta G_T^0\right)_{eq.\,4.98} = -592.0 + 0.224 \cdot T(K)kJ \cdot (mol\ O_2)^{-1} \tag{4.99}$$

We obtain Eq. (4.100) from Eqs. (4.86) to (4.98):

$$3FeO(s) + H_2O(g) \rightarrow Fe_3O_4(s) + H_2(g) \tag{4.100}$$

and the standard free energy of the reaction (4.100) is:

$$
\begin{aligned}
\left(\Delta G_T^0\right)_{eq.4.100} &= \frac{1}{2} \cdot \left(\Delta G_T^0\right)_{eq.\,4.98} - \left(\Delta G_T^0\right)_{eq.\,4.86} \\
&= \frac{1}{2} \cdot (-592.0 + 0.224 \cdot T(K)) - (-249.0 + 0.057 \cdot T(K)) \\
&= -47.0 + 0.055 \cdot T(K)kJ \cdot (mol\ H_2)^{-1} \tag{4.101}
\end{aligned}
$$

We obtain for the equilibrium FeO–Fe$_3$O$_4$-gas (H$_2$–H$_2$O–N$_2$) the values collected in Table 4.6.

We start now with the equilibrium Fe–Fe$_3$O$_4$-gas (H$_2$–H$_2$O–N$_2$) when the total pressure of the system is 1 atmosphere and $P_{N_2} = 0$ atm.

$$\frac{3}{2}Fe(s) + O_2(g) \rightarrow \frac{1}{2}Fe_3O_4(s) \tag{4.102}$$

and the standard free energy of the reaction (4.102) is:

$$\left(\Delta G_T^0\right)_{eq.4.102} = -552.5 + 0.164 \cdot T(K)kJ \cdot (mol\ O_2)^{-1} \tag{4.103}$$

We obtain Eq. (4.104) from Eqs. (4.86) to (4.102):

$$\frac{3}{4}Fe(s) + CO_2(g) \rightarrow \frac{1}{4}Fe_3O_4(s) + CO(g) \tag{4.104}$$

and the standard free energy of the reaction (4.104) is:

$$
\begin{aligned}
\left(\Delta G_T^0\right)_{eq.\,4.104} &= \frac{1}{2} \cdot \left(\Delta G_T^0\right)_{eq.\,4.102} - \left(\Delta G_T^0\right)_{eq.\,4.86} \\
&= \frac{1}{2} \cdot (-552.5 + 0.164 \cdot T(K)) - (-249.0 + 0.057 \cdot T(K)) \\
&= -27.25 + 0.025 \cdot T(K)kJ \cdot (mol\ H_2)^{-1} \tag{4.105}
\end{aligned}
$$

**Table 4.6** Results for the FeO–Fe$_3$O$_4$-gas (H$_2$–H$_2$O–N$_2$)

| Temperature (°C) | Temperature (K) | $(\Delta G_T^0)_{eq.4.100}$ | $k_{eq}$ | $P_{N_2}$(atm) | $P_{H_2}$(atm) | $P_{H_2O}$(atm) | R | $R_1$ | %vol.H$_2$ |
|---|---|---|---|---|---|---|---|---|---|
| 387 | 660 | −10.70 | 7.03 | 0 | 0.88 | 0.12 | 7.03 | 0.88 | 87.54 |
| 527 | 800 | −3.00 | 1.57 | 0 | 0.61 | 0.39 | 1.57 | 0.61 | 61.09 |
| 627 | 900 | 2.50 | 0.716 | 0 | 0.42 | 058 | 072 | 0.42 | 41.72 |
| 727 | 1000 | 8.00 | 0.382 | 0 | 0.28 | 0.72 | 0.38 | 0.28 | 27.64 |
| 827 | 1100 | 13.50 | 0.229 | 0 | 0.19 | 0.81 | 0.23 | 0.19 | 18.60 |
| 927 | 1200 | 19.00 | 0.149 | 0 | 0.13 | 0.87 | 0.15 | 0.13 | 12.96 |
| 1027 | 1300 | 24.50 | 0.104 | 0 | 0.09 | 0.91 | 0.10 | 0.09 | 9.39 |
| 1127 | 1400 | 30.00 | 0.076 | 0 | 0.07 | 0.93 | 0.08 | 0.07 | 7.06 |
| 1227 | 1500 | 35.50 | 0.058 | 0 | 0.05 | 0.95 | 0.06 | 0.05 | 5.49 |

We obtain for the equilibrium Fe–Fe$_3$O$_4$-gas (H$_2$–H$_2$O–N$_2$) the values collected in Table 4.7.

Finally, we represent the diagram of Chaudron for the Fe–O–H system in Fig. 4.6.

As conclusion, it is possible to confirm, from the thermodynamics point of view, that the hydrogen is the reductant at high temperature for the iron oxides, and kinetically, it has a better performance than the carbon monoxide.

*Solution ends.*

### 4.3.2  Reduction of an Oxide M$_x$O$_y$ with H$_2$

Reasoning in a similar manner as in Sect. 4.3.1, we consider the reaction of the metal M oxidation by the oxygen, and the reaction:

$$2H_2(g) + O_2(g) \rightarrow 2H_2O(g) \tag{4.106}$$

of free energy $(\Delta G_T)_{eq.\,4.106}$ and of standard free energy $(\Delta G_T^0)_{eq.\,4.106}$. If we subtract to the second reaction the first, we will have the reaction whose standard free energy is $(\Delta G_T^0)_{eq.\,4.106} - (\Delta G_T^0)_1$. Therefore, the following relation should be verified in the equilibrium at the temperature T:

$$\left(\Delta G_T^0\right)_{eq.\,4.106} - \left(\Delta G_T^0\right)_l = -R \cdot T \cdot \ln \frac{\left(P_{H_2O}\right)^2}{\left(P_{H_2}\right)^2} \rightarrow \ln \frac{\left(P_{H_2O}\right)^2}{\left(P_{H_2}\right)^2}$$

$$= -\frac{\left(\Delta G_T^0\right)_{eq.\,4.106} - \left(\Delta G_T^0\right)_l}{R \cdot T} \tag{4.107}$$

or also:

$$\frac{\left(P_{H_2O}\right)^2}{\left(P_{H_2}\right)^2} = \exp\left[-\frac{\left(\Delta G_T^0\right)_{eq.\,4.106} - \left(\Delta G_T^0\right)_l}{R \cdot T}\right] \tag{4.108}$$

Regarding the possible oxidation of the iron to give FeO or Fe$_3$O$_4$, we indicate in the Fig. 4.7 (obtained with the data of the column R in Tables 4.2, 4.3, 4.4, 4.5, 4.6 and 4.7 and corresponding temperature) the conditions where a H$_2$–H$_2$O binary atmosphere, or a CO–CO$_2$ binary atmosphere, is reductant or oxidizing.

**Table 4.7** Results for the Fe–Fe$_3$O$_4$-gas (H$_2$–H$_2$O–N$_2$)

| Temperature (°C) | Temperature (K) | $(\Delta G_T^0)_{eq.4.105}$ | $k_{eq}$ | $P_{N_2}$ (atm) | $P_{H_2}$ (atm) | $P_{H_2O}$ (atm) | R | R$_1$ | %vol.H$_2$ |
|---|---|---|---|---|---|---|---|---|---|
| 25 | 298 | −9.80 | 2960 | 0 | 1.00 | 0.00 | 2955 | 1.00 | 99.97 |
| 50 | 323 | −19.18 | 1260 | 0 | 1.00 | 0.00 | 1262 | 1.00 | 99.92 |
| 100 | 373 | −17.93 | 324 | 0 | 1.00 | 0.00 | 324 | 1.00 | 99.69 |
| 150 | 423 | −16.68 | 115 | 0 | 0.99 | 0.01 | 115 | 0.99 | 99.13 |
| 200 | 473 | −15.43 | 50.5 | 0 | 0.98 | 0.02 | 50.51 | 0.98 | 98.06 |
| 250 | 523 | −14.18 | 26.0 | 0 | 0.96 | 0.04 | 26.04 | 0.96 | 96.30 |
| 300 | 573 | −12.93 | 15.1 | 0 | 0.94 | 0.06 | 15.07 | 0.94 | 93.78 |
| 350 | 623 | −11.68 | 9.52 | 0 | 0.90 | 0.10 | 9.52 | 0.90 | 90.50 |
| 387 | 660 | −10.75 | 7.09 | 0 | 0.88 | 0.12 | 7.09 | 0.88 | 87.64 |

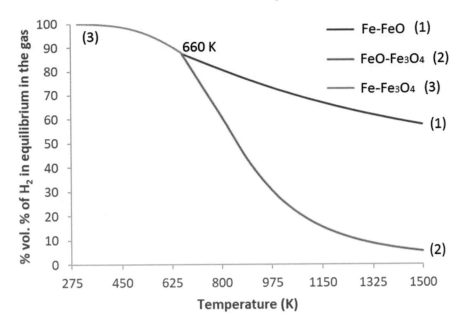

**Fig. 4.6** Fe–O–H system. Domains with predominance of Fe–FeO–Fe$_3$O$_4$ in equilibrium with the H$_2$ at different temperatures. It is at the triple point (660 K) where the Potential Neutral of Oxygen, the Potential of Oxygen of the System, and the oxygen equilibrium pressure are equivalent

### 4.3.3 Reductant or Oxidizing Aptitude of CO–CO$_2$–H$_2$–H$_2$O$_{vapor}$ Gases Mixture in Equilibrium

For the reaction:

$$2CO(g) + O_2(g) \rightarrow 2CO_2(g) \tag{4.109}$$

that has a standard free energy $\left(\Delta G_T{}^0\right)_{eq.\,4.109}$, the following relation is verified in the equilibrium:

$$\left(\Delta G_T^0\right)_{eq.4.109} = -R \cdot T \cdot \ln\left[\frac{(P_{CO_2})^2}{(P_{CO})^2 \cdot (P'_{O_2})^2}\right] \rightarrow \ln\left[\frac{(P_{CO_2})^2}{(P_{CO})^2 \cdot (P'_{O_2})^2}\right]$$

$$= \frac{\left(\Delta G_T^0\right)_{eq.\,4.109}}{-R \cdot T} \tag{4.110}$$

being $P'_{O_2}$ what determines if the metal, in this binary atmosphere, is under reductant, oxidizing or inert conditions at the temperature T. At the same time, the reaction:

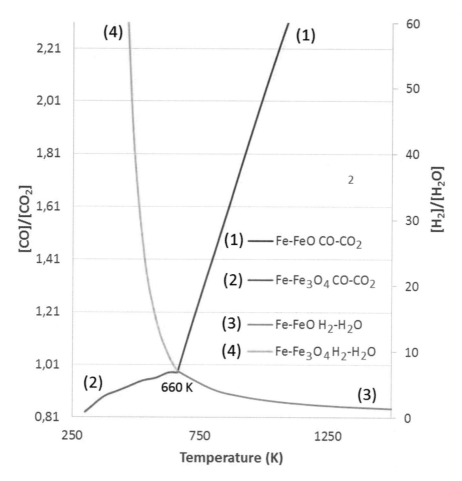

**Fig. 4.7** Oxidizing and reductant conditions of the iron in binary atmospheres $H_2$–$H_2O$ (curves in the left side of the figure) and $CO$–$CO_2$ (curves in the right side of the figure), when the total pressure is one atmosphere

$$2H_2(g) + O_2(g) \rightarrow 2H_2O(g) \tag{4.111}$$

of standard free energy $\left(\Delta G_T^0\right)_{eq.\,4.111}$, is verified in the equilibrium:

$$\left(\Delta G_T^0\right)_{eq.\,4.111} = -R \cdot T \cdot \ln\left[\frac{\left(P_{H_2O}\right)^2}{\left(P_{H_2}\right)^2 \cdot \left(P_{O_2}''\right)^2}\right] \rightarrow \ln\left[\frac{\left(P_{H_2O}\right)^2}{\left(P_{H_2}\right)^2 \cdot \left(P_{O_2}''\right)^2}\right]$$

$$= \frac{\left(\Delta G_T^0\right)_{eq.\,4.111}}{-R \cdot T} \tag{4.112}$$

being in this case $P''_{O_2}$ what determines if this binary atmosphere is reductant, oxidizing or neutral for the metal.

Imagine an atmosphere that contained these four gases; but there was no relation between the $CO$–$CO_2$ concentrations injected in the furnace and those of $H_2$–$H_2O$. We assume that they are not combined in the path along the furnace according to the following reaction:

$$2H_2(g) + CO_2(g) \rightarrow 2H_2O(g) + 2CO(g) \qquad (4.113)$$

or the reaction:

$$2H_2O(g) + 2CO(g) \rightarrow 2H_2(g) + 2CO_2(g) \qquad (4.114)$$

In this case, it would be necessary to analyze the gases at the exit of the furnace to—calculating the values of $P'_{O_2}$ and $P''_{O_2}$ separately—know the character of the atmosphere.

However, if the atmosphere contained these four gases and, additionally, at the temperature T, the concentrations (or partial pressures) of both $CO$–$CO_2$ and $H_2$–$H_2O$—that are never completely independent in a mixture of this type despite the assumptions made in the previous paragraph—were exactly those corresponding to the equilibrium of the reaction:

$$2H_2(g) + CO_2(g) \rightarrow 2H_2O(g) + 2CO(g) \qquad (4.115)$$

the values of $P'_{O_2}$ and $P''_{O_2}$ would be the same. The oxidizing, reductant or neutral character of the atmosphere could be known as a function of either $P_{CO}/P_{CO_2}$ or $P_{H_2}/P_{H_2O}$. In fact, the following equality should be verified in the equilibrium of the reaction:

$$\ln\left[\frac{\left(P_{H_2O}\right)^2 \cdot \left(P_{CO}\right)^2}{\left(P_{H_2}\right)^2 \cdot \left(P_{CO_2}\right)^2}\right] = \frac{\left(\Delta G^0_T\right)_{eq\cdot4.111} - \left(\Delta G^0_T\right)_{eq\cdot4.109}}{-R \cdot T} \qquad (4.116)$$

And, if we subtract the equations mentioned at the beginning of this paragraph, that is to say, subtracting term to term of the Eq. (4.112) the Eq. (4.110), we will obtain:

$$\ln\left[\frac{\left(P_{H_2O}\right)^2 \cdot \left(P_{CO}\right)^2 \cdot P'_{O_2}}{\left(P_{H_2}\right)^2 \cdot \left(P_{CO_2}\right)^2}\right] \cdot P''_{O_2} = \frac{\left(\Delta G^0_T\right)_{eq\cdot4.111} - \left(\Delta G^0_T\right)_{eq\cdot4.109}}{-R \cdot T} \qquad (4.117)$$

which will result in the Eq. (4.116) if the four gases are equilibrated. For this, the following equality should be verified: $P_{O_2} = P'_{O_2}$.

## 4.4  Case-Hardening and Decarburizing of Steels

Iron in contact for a long period with a binary equilibrated atmosphere of CO and $CO_2$ can remain unaltered or, on the contrary, can be superficially oxidized. However, these atmospheres can additionally produce surface decarburizing; or, on the contrary, add carbon to the surface of the steel (case-hardening) with certain concentrations of CO.

### 4.4.1  Boudouard's Equilibrium

We are going to consider an important concept in this section. This is the chemical activity of the carbon in the gas, $a_C^{gas}$ or $a_C(gas)$. This way, the capacity of reaction-activity of the carbon in a homogeneous gaseous phase is expressed by the Boudouard's equilibrium in the industrial atmospheres:

$$C(s) + CO_2(g) \rightarrow 2CO(g) \tag{4.118}$$

whose free energy is:

$$\left(\Delta G_T^0\right)_{eq.4.118} = 159.0 - 0.167 \cdot T \frac{kJ}{mol\ C} = -R \cdot T \cdot \ln\left(k_{eq}\right)_{eq.4.118} \tag{4.119}$$

that represents the gasification of the carbon by the $CO_2$, or it could be expressed by the equilibrium of formation-dissociation of the methane, $CH_4$:

$$2H_2(g) + C(s) \leftrightarrow CH_4(g) \tag{4.120}$$

$$\left(\Delta G_T^0\right)_{eq.4.120} = -95.6 + 0.113 \cdot T \frac{kJ}{mol\ C} = -R \cdot T \cdot \ln\left(k_{eq}\right)_{eq.4.120} \tag{4.121}$$

In the first case, the activity of the carbon in the gas, $a_C^{gas}$, is equal to:

$$\left(k_{eq}\right)_{eq.4.118} = \frac{P_{CO}^2}{a_C^{gas} \cdot P_{CO_2}} \rightarrow a_C^{gas} = \frac{1}{\left(k_{eq}\right)_{eq.4.118}} \cdot \frac{P_{CO^2}^2}{P_{CO_2}} \tag{4.122}$$

while in the case of the methane, $CH_4$:

$$\left(k_{eq}\right)_{eq.4.120} = \frac{P_{CH_4}^2}{a_C^{gas} \cdot P_{H_2}^2} \rightarrow a_C^{gas} = \frac{1}{\left(k_{eq}\right)_{eq.4.120}} \cdot \frac{P_{CH_4}}{P_{H_2}^2} \tag{4.123}$$

The equilibrium-reaction of Boudouard is involved in all the processes of carbothermal reduction of metals. However, it is also involved in:

1. the formation of novel carbon non-equilibrium structures, as nanotubes, fullerenes, or graphene, by nucleation and controlled growth.
2. the processes of sintering by Spark Plasma Sintering where, as an electrical conducting medium is required, the equilibriums between the carbon and the species $CO/CO_2$ are reached at temperatures within 1500 and 2000 °C.
3. It might be interesting, nowadays, considering the purification by thermal decomposition of atmospheres with high CO content with the aim of obtaining industrial atmospheres with high $H_2$ contents and low $CO/CO_2$ concentrations. The cooling of gases with CO produces the precipitation of the carbon with the support of metallic catalyzers.

On the other hand, the equilibrium of Boudouard divides the graph into two parts-zones (Fig. 4.8): one in the right side of the equilibrium curve ($a_C^{gas} = 1$) where the stability of the CO(g) is verified; the other in the left side of the equilibrium curve, which is identified with the zone of unstable gas ($CO_2 + C$).

In the formation of the methane, the equilibrium curve divides the graph into two zones (Fig. 4.9), that corresponding to the stability of the gas, $CH_4$, (in this case in the left side), and that of the carbon-$H_2$ in the right side. The higher the temperature, in the case of the methane, the larger the zone of stability of the unstable gas ($H_2$ + C).

The reaction of the Boudouard's equilibrium (gasification of the coke-carbon by an oxidizing gas, carbon dioxide) was indicated in reaction (4.118). The standard free energy of these reactions is:

$$\Delta_r G^0 = 159000 - 169 \cdot T \ \frac{J}{mol} = -R \cdot T \cdot \ln \left( \frac{P_{CO}}{P_{CO_2} \cdot a_C^{gas}} \right) \qquad (4.124)$$

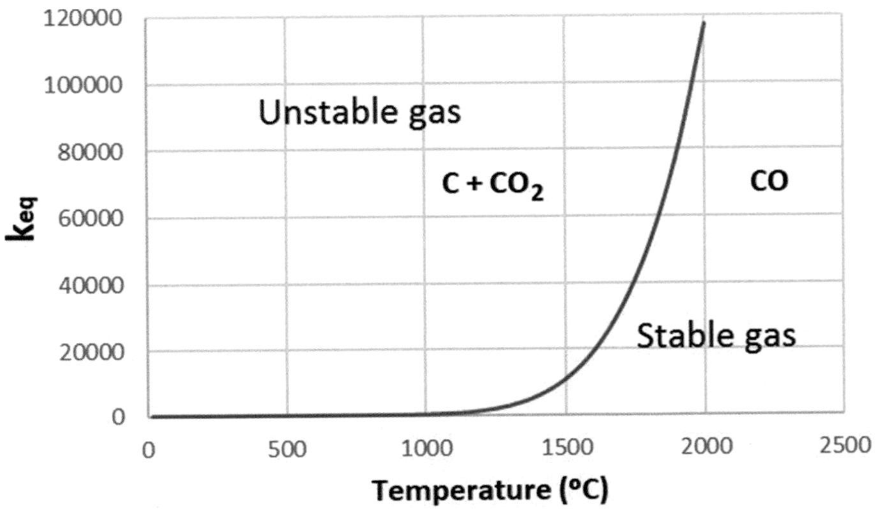

**Fig. 4.8** Boudouard equilibrium. Variation of the equilibrium constant with the temperature

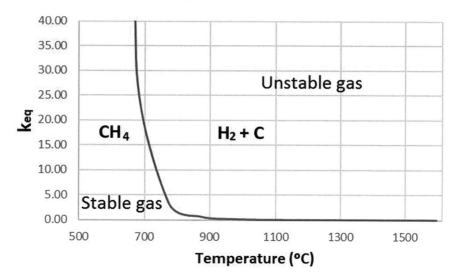

**Fig. 4.9** Methane formation. Variation of the equilibrium constant with the temperature

This is the reaction that takes place, for instance, in the zone of tuyeres of the blast furnace. The degasification of the coke is the reaction of the Boudouard's equilibrium.

There are two components in this C–O system, which are the carbon and the oxygen, two phases (one gaseous and the other solid), and, therefore, the number of degrees of freedom is also equal to two:

$$F + L = C + 2 \rightarrow 2(\text{carbon} + \text{oxygen}) + L = C(\text{gas} + \text{solid}) + 2 \rightarrow L = 2 \tag{4.125}$$

These degrees of freedom correspond to the temperature and the pressure. The total pressure of the system is fixed and equal to:

$$P_T = P_{CO} + P_{CO_2} + P_{O_2} \tag{4.126}$$

where we normally assume that $P_{O_2}$ is negligible, and:

$$P_T \simeq P_{CO} + P_{CO_2} \tag{4.127}$$

We consider a fixed value for the total pressure of the system, $P_T = 1$ atm. Thus, we obtain:

$$P_{CO} \simeq P_T - P_{CO_2} \rightarrow \text{ for } P_T = 1 \text{ atm} \rightarrow P_{CO} \rightarrow 1 - P_{CO_2} \tag{4.128}$$

On the other hand, the following equations can be obtained from the equation of the standard free energy of formation:

$$\frac{P_{CO}^2}{P_{CO_2}} = \exp\left(\frac{-159000}{R \cdot T} + \frac{169}{R}\right) \rightarrow P_{CO_2} = \frac{P_{CO}^2}{\exp\left(\frac{-159000}{R \cdot T} + \frac{169}{R}\right)} \qquad (4.129)$$

If we consider the law of Dalton, the CO partial pressure will be:

$$P_{CO} = P_T \cdot \frac{n_{CO}}{n_T} = P_T \cdot C_{CO} \qquad (4.130)$$

where $C_{CO}$ is the CO volume fraction, which will be equal to the molar fraction because all the gases occupy the same volume.

If we use all the above-indicated equations, we obtain:

$$C_{CO} = \frac{P_{CO}}{P_T} = \frac{P_T - P_{CO_2}}{P_T} = 1 - \frac{P_{CO_2}}{P_T}$$

$$= 1 - \frac{\left(\frac{P_{CO}^2}{\exp\left(\frac{-159000}{R \cdot T} + \frac{169}{R}\right)}\right)}{P_T} \qquad (4.131)$$

$$C_{CO}(\%vol) = 100 \cdot \left[1 - \left(\frac{(C_{CO}^2 \cdot P_T^2)}{P_T \cdot \exp\left(\frac{-159000}{R \cdot T} + \frac{169}{R}\right)}\right)\right]$$

$$= 100 - \left(\frac{100 \cdot C_{CO}^2 \cdot P_T}{\exp\left(\frac{-159000}{R \cdot T} + \frac{169}{R}\right)}\right) \qquad (4.132)$$

$$100 \cdot P_T \cdot C_{CO}^2 + \exp\left(\frac{-159000}{R \cdot T} + \frac{169}{R}\right) \cdot C_{CO} - 100 \cdot \exp\left(\frac{-159000}{R \cdot T} + \frac{169}{R}\right) = 0 \qquad (4.133)$$

This is a quadratic equation that is analytically solved as follows:

$$C_{CO} = \frac{-(e^A) \pm \sqrt{(e^{2A}) + (4 \cdot 100 \cdot P_T \cdot 100 \cdot e^A)}}{2 \cdot P_T \cdot 100} \qquad (4.134)$$

where:

$$e^A = \exp\left(\frac{-159000}{R \cdot T} + \frac{169}{R}\right) \qquad (4.135)$$

depends on the temperature.

In the case of a $P_T = 1$ atm, we obtain:

$$C_{CO} = \frac{-e^A \pm \sqrt{e^{2A} + 40000 \cdot e^A}}{200} \qquad (4.136)$$

We represent in Fig. 4.10 the $C_{CO}$ as a function of the temperature for different values of the total pressure of the system. As it was already mentioned, it is possible

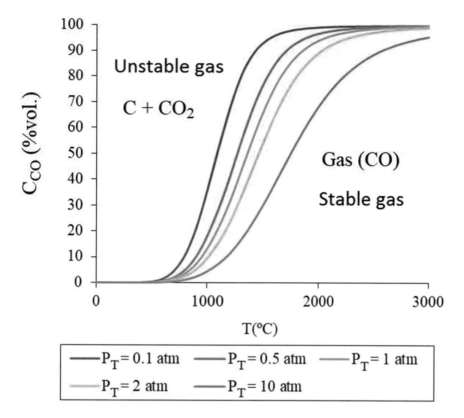

**Fig. 4.10** Volumetric relation of CO and $CO_2$ in the Boudouard equilibrium for pressures of 0.1, 0.5, 1, 2, and 10 atmospheres

to represent in Fig. 4.10 the volume fraction of CO as a function of the temperature to obtain a sigmoidal curve where it is possible to distinguish two zones:

- Zone of stability of the gas (CO).
- Zone of stability of the coke (C/$CO_2$).

If the gas of tuyeres is very rapidly cooled, it is possible to appreciate a very fine coke that precipitates as "cinder".

This reference to the total pressure of the system, $P_T$, may be surprising. The reason is that we should consider that reactions in the industrial practice are not carried out in a closed environment whose pressure could be increased by the gases. The current of gases circulates inside of the furnace and goes outside. That is to say that the pressure inside of the furnace is the atmospheric one ($P_T = 1$). However, it is necessary to take into account that, keeping the pressure equal to 1 atmosphere, if it was diluted the mixture of several active gases by means of the addition of an active gas (v. gr. nitrogen), this would be equivalent to a decrease of the total pressure of the non-diluted system. Thus, considering that $P_1 = P_T \cdot (n_1/n_T)$, if we duplicate the total volume of gases by means of nitrogen, therefore, equal to $2 \cdot n_T$, its effect would be like that obtained without doubling the volume of gases and, on the contrary using a total pressure equal of $2 \cdot P_T$. The partial pressures of each one of the active gases, $P_{CO}$ and $P_{CO_2}$, would be reduced to a half, and this would have influence in the concentrations of CO and $CO_2$ in the Boudouard's equilibrium.

Brief note: *Considerations about the symbols used to designate the variables that express the thermodynamic equilibrium.*

*Brief note starts*

We have considered within this chapter the concept of the solid–gas equilibrium or that of stable gases with different molecules that reach the state of equilibrium. The variables used to express-characterize the equilibrium, in the condensed phases (solids-melts), are the followings.

$C_i$, universal variable of concentration that means the concentration of certain component $i$ in the solution. It can be expressed in weight of molar percentage or as molar fraction.

$x_i$, molar fraction, which must be expressed as $x_i =$ (equilibrium molar fraction) when corresponds to an equilibrium value.

$\%w_i$, weight percentage, which must be expressed as $\%w_i =$ (equilibrium weight percentage) when corresponds to an equilibrium value.

We use in gases the partial pressure of the gas-gaseous molecule in equilibrium with the rest of the existing gaseous species. Nevertheless, it must be expressed as $P_{CO}^=$ or $P_{CO_2}^=$ when the equilibrium value is reached.

*Brief note ends.*

Brief note: *Some considerations that is necessary to consider when working with solid–gas heterogeneous equilibriums.*

*Brief note starts*

This brief note is presented here to clarify some thermodynamic characteristics involved in the solid–gas equilibriums for the Fe–O, Fe–O–C and Fe–O–H systems.

First, the equilibrium constant, $k_{eq}$, solute-gas, only depends on the temperature:

$$\Delta_r G^0(T) = -R \cdot T \cdot \ln k_{eq} = A + B \cdot T(K) \qquad (4.137)$$

However, it is possible to identify two situations in the solid–gas equilibriums:

1. There is not variation in volume between the products and reagents of the reaction. In this case, the relation of partial pressures of the gases (oxidized specie/reduced specie) in equilibrium with the solutes, as the equilibrium constant, depends only on the temperature independently of the total pressure of the system.

2. When there is a variation of volume between the products and reagents of the solid–gas reaction, the equilibrium constant of the system, $k_{eq}$, is only function of the temperature, but the relation of partial pressures in equilibrium with the condensed phases depends on the total pressure, $P_T$, of the system.

The concept of "equilibrium point" refers to the molar-volumetric concentration of the gases in equilibrium with the solid-condensed phases of the reaction. The composition of the "equilibrium point" expressed in unitary or percentual molar fraction, $C_{gas}^=$, is equal to:

$$C_{gas}^= = \frac{P_{gas}^=}{P_T} \qquad (4.138)$$

That is to say, the concentration of the gas in the "equilibrium point" depends on the total pressure of the system, $P_T$.

In those solid–gas equilibriums that are not affected by the volume change, the "equilibrium point" neither is affected by the value of the total pressure, $P_T$, of the system.

However, when the variation of the total pressure of the system, $P_T$, in a solid–gas equilibrium is produced by the introduction of an inert gas, $N_2$, that is not involved in the solid–gas equilibrium reactions (where there is not variation of volume), the value of the "equilibrium point" is affected by the introduction of the inert gas.

*Brief note ends.*

### 4.4.2  Decarburizing-Case Hardening in a CO–CO₂ Binary Atmosphere

We assume that a certain steel is inside of a furnace at temperature, $T_1$, greater than the temperature $A_3$ of the steel and that a binary atmosphere comprised by gases CO and $CO_2$ introduced in a certain proportion circulates inside of this furnace. During the circulation of the gases inside of the furnace, there is short time between the moment of its entrance and its exit to the exterior. It is possible, for that reason, to admit that the surface of the steel is always in contact with a gaseous mixture of composition equal to that initially considered.

On the contrary, the surface of the steel that is in contact with this gaseous current can experience modification during the time that the treatment lasts (increasing the carbon content solubilized in the austenite or, on the contrary, decarburizing itself).

In this context, the condition to reach the thermodynamic equilibrium between phases of a system, at constant pressure and temperature, will be to attain the equality between the "chemical potentials" of the components in the phases in equilibrium. In the case of the solid solutions of the diagram system of two components, Fe–C, the equality of chemical potential of the carbon between phases involves the equality of their activities. The reason is that the state of reference of the component carbon, C, in the phases in equilibrium is the same: carbon-graphite, pure from the chemical point of view. Consequently, we try to study the thermodynamic equilibrium of the carbon dissolved in a solid solution, as the austenite, and a gas, the activity of the carbon in the gas, $a_C^{gas}$, should be the same that the activity of the carbon in the steel, $a_C^{steel}$:

$$a_C^{gas} = a_C^{steel} \qquad (4.139)$$

When the carbon percentage in the periphery of the steel is not more modified, it does not matter the prolongation of its contact with the atmosphere at the temperature $T_1$, this carbon content, $C_1$, is that that obtains the equality between the values called carbon activity of the steel and the carbon activity of the atmosphere.

Carbon activity of the steel at the temperature $T_1$ is the relation that exists—the quotient—between $C_1$ and the maximum carbon content, $C_S$, that the periphery of the steel would have in solid solution if it was saturated in carbon at the temperature $T_1$ (without forming soot-cinder in its surface); that is to say, the corresponding to the maximum carbon content, $C_S$, of the austenite, at this temperature $T_1$, in the Fe–C stable system, see Fig. 4.11 (see further information about Fe–C diagram in Pero-Sanz et al. 2018).

Brief note: *Some considerations about the activity of the carbon in the austenite.*

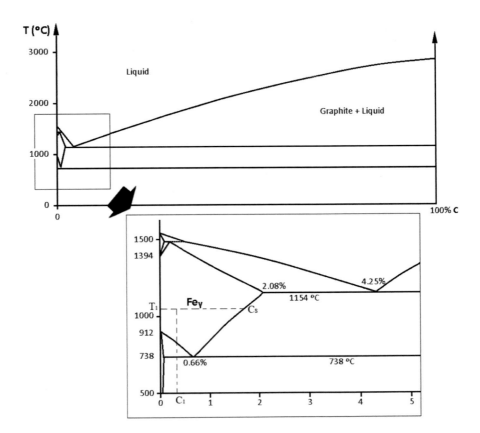

**Fig. 4.11** Fe–C stable diagram

*Brief note starts*

Some considerations arise about the activity of the carbon in the austenite (Fig. 4.11). According to the text, a first approximation to calculate the activity of the carbon in the austenite, $a_C^{Fe-\gamma}$, would be Ban-Ya et al. 1970):

$$a_C^{Fe-\gamma} = \frac{x_C}{x_C^0} = \frac{C_C}{C_C^0} \tag{4.140}$$

where $C_C$ would be identified with the concentration of the carbon in the austenite expressed as molar fraction and $C_C^0$ would be the maximum carbon concentration that can be admitted by the austenite at a certain temperature.

The problem appears because the solid solutions of the carbon in the Fe-$\gamma$ are not ideal. That is to say, that the activity of the carbon cannot be identified with the carbon concentration expressed as molar fraction.

Undoubtedly, identifying the value of the activity in the austenite with the Eq. (4.140) can be valued as a first quantitative approximation to value the activity of the carbon in the Fe-$\gamma$.

We would have in a second level the equations that are found in this chapter to estimate the activity of the carbon in the austenite as a function of the concentration and the temperature:

- Equation of Ellis et al. (1963).
- Alternative equation proposed by Verdeja et al. ( 1995).

A third level of approximation consists of incorporating to the equations used to calculate the activity of the carbon in the austenite of the Fe–C binary system (stable) the influence of the alloy elements (Mn, Si, Cr, Mo, or Ni) on the influence of the domain of prevalence of stability of the austenite.

Now we indicate the generalized equation for calculating-estimating the carbon activity in the austenite. We can use the equation proposed by Ellis, Davidson and Bosdworth (Ellis et al. 1963) to calculate the activity of the carbon in the austenite, $a_C^{Fe-\gamma}$:

$$\log a_C^{Fe-\gamma} = \log\left(\frac{x_C}{1 - x_C}\right) + \frac{2207}{T} - 0.64 \tag{4.141}$$

where $x_C$ is the molar fraction of the carbon in the Fe–C binary system and T is the temperature in Kelvin. We must take into account that the application of this equation would be only open to low alloy carbon steels (the influence of other alloy elements different from the carbon is not considered on the zone of stability of the solid solution of the carbon in the Fe-$\gamma$).

If we consider the work of Verdeja et al. (1995), it is possible to obtain other equation that allows estimating, the activity of the carbon in the austenite as a function of the temperature and composition. If we take into account that:

$$a_C = \gamma_C \cdot x_C \tag{4.142}$$

and that:

$$\ln \gamma_C = \ln \gamma_C^\infty + \varepsilon_C^C \cdot x_C \tag{4.143}$$

On the other hand, if we know that:

$$\varepsilon_C^C = 1.03 \cdot \exp\left(\frac{3436}{T}\right) \tag{4.144}$$

$$\gamma_C^\infty = 0.333 \cdot \exp\left(\frac{3581}{T}\right) \tag{4.145}$$

the activity of the carbon in the austenite will be:

$$\ln a_C^{Fe-\gamma} = \ln 0.333 + \frac{3581}{T} + 1.03 \cdot \exp\left(\frac{3436}{T}\right) \cdot x_C + \ln x_C \tag{4.146}$$

$$\ln a_C^{Fe-\gamma} = \frac{3581}{T} + 1.03 \cdot \exp\left(\frac{3436}{T}\right) \cdot x_C + \ln x_C - 1.0996 \tag{4.147}$$

It would be possible to contrast the estimations of both equations to calculate the activities of the maximum solubility of the carbon in the austenite between the eutectoid temperature of 738 °C and the eutectic temperature of 1154 °C (in this case, the activity of the carbon must be 1.00). Equally, or similarly, it is also possible to obtain the estimation for the maximum solubility of the carbon in the austenite between the eutectic temperature of 1154 °C and the peritectic temperature of 1493 °C. In this last situation, the activities estimated for the carbon in the austenite will be lower than one as the phase in equilibrium is not the carbon-graphite but the different melts with variable carbon compositions.

First, we calculate $a_C^{Fe-\gamma}$ using the equation of Ellis, Davidson and Bosdworth (Eq. 4.141) for wt. %C (0.66–2.08) and temperatures ($A_{cm}$, line 738–1154 °C) and wt. %C (0.18–2.08) and temperatures (line 1493–1154 °C). First, we calculate the molar fractions as follows:

$$x_C = \frac{\frac{\%C}{12.01}}{\frac{\%C}{12.01} + \frac{100-\%C}{55.85}} \tag{4.148}$$

and then we apply the Eq. (4.141):

$$\log a_C^{Fe-\gamma} = \log\left(\frac{x_C}{1-x_C}\right) + \frac{2207}{T} - 0.64 \rightarrow a_C^{Fe-\gamma}$$
$$= 10\left[\log\left(\frac{x_C}{1-x_C}\right) + \frac{2207}{T} - 0.64\right] \tag{4.149}$$

Results are collected in Table 4.8, using the equation proposed by Ellis, Davidson and Bosdworth.

We repeat the calculations using the Eq. (4.147) proposed by Verdeja et al. (1995). Results are collected in Table 4.9.

*Brief note ends.*

We define as carbon activity, $a_C$, of a binary atmosphere of $CO$–$CO_2$ the quotient $\left(P_{CO}^2/P_{CO_2}\right)/k_{eq}$, where $P_{CO}$ and $P_{CO_2}$ are the partial pressures of $CO$ and $CO_2$ respectively and $k_{eq}$ is the value of the constant of the Boudouard's equilibrium at this temperature $T_1$. The activity of the atmosphere will be equal to 1 if the partial pressures of $CO$ and $CO_2$ (or their volume concentrations in the mixture if the total pressure would be 1 atm) are exactly those corresponding to the Boudouard's equilibrium.

Brief note: *Different states of activity equal to one.*

*Brief note starts*

Habitually, the states of activity equal to one are identified as that of those pure compounds-elements in aggregation state more stable at a certain temperature at the total pressure of 1 atm. However, and up to these two circumstances:

1. That the thermodynamic state of the pure compounds-elements is not identified with the situation of these elements-compounds in the solid solutions or in molten state.
2. That when $x_i \to 0$ or that $\%w_i \to 0$ (field of the diluted solutions), the activity of Raoult of any element-compound is directly proportional to the concentration.

**Table 4.8** Activity of the carbon in the austenite using the equation proposed by Ellis, Davidson and Bosdworth

| %C | $x_C$ | Temperature (°C) | Temperature (K) | $\log a_C^{Fe-\gamma}$ | $a_C^{Fe-\gamma}$ |
|---|---|---|---|---|---|
| 0.66 | 0.0300 | 738 | 1011 | 0.0329 | 1.0334 |
| 0.87 | 0.0392 | 800 | 1073 | 0.0276 | 1.0657 |
| 1.18 | 0.0526 | 900 | 1173 | −0.0140 | 0.9683 |
| 1.50 | 0.0661 | 1000 | 1273 | −0.0562 | 0.9454 |
| 1.85 | 0.0806 | 1100 | 1373 | −0.0898 | 0.9141 |
| 2.08 | 0.0899 | 1154 | 1427 | −0.0987 | 0.9060 |
| 2.08 | 0.0899 | 1154 | 1427 | −0.0987 | 0.9060 |
| 1.83 | 0.0798 | 1200 | 1473 | −0.2036 | 0.8158 |
| 1.65 | 0.0724 | 1240 | 1513 | −0.2889 | 0.7491 |
| 1.29 | 0.0575 | 1300 | 1573 | −0.4516 | 0.6366 |
| 1.03 | 0.0463 | 1340 | 1613 | −0.5856 | 0.5568 |
| 0.67 | 0.0302 | 1400 | 1673 | −0.8275 | 0.4371 |
| 0.18 | 0.0083 | 1493 | 1766 | −1.4667 | 0.2307 |

**Table 4.9** Activity of the carbon in the austenite using the equation proposed by Verdeja et al. (1995)

| %C | $x_C$ | Temperature (°C) | Temperature (K) | $\ln a_C^{Fe-\gamma}$ | $a_C^{Fe-\gamma}$ |
|------|--------|------------------|-----------------|----------------------|-------------------|
| 0.66 | 0.0300 | 738  | 1011 | −0.1414 | 0.8681 |
| 0.87 | 0.0392 | 800  | 1073 | −0.0079 | 0.9921 |
| 1.18 | 0.0526 | 900  | 1173 | 0.0224  | 1.0226 |
| 1.50 | 0.0661 | 1000 | 1273 | 0.0100  | 1.0101 |
| 1.85 | 0.0806 | 1100 | 1373 | 0.0040  | 1.0040 |
| 2.08 | 0.0899 | 1154 | 1427 | 0.0296  | 1.0300 |
| 2.08 | 0.0899 | 1154 | 1427 | 0.0296  | 1.0300 |
| 1.83 | 0.0798 | 1200 | 1473 | −0.3497 | 0.7049 |
| 1.65 | 0.0724 | 1240 | 1513 | −0.6358 | 0.5295 |
| 1.29 | 0.0575 | 1300 | 1573 | −1.1528 | 0.3157 |
| 1.03 | 0.0463 | 1340 | 1613 | −1.5508 | 0.2121 |
| 0.67 | 0.0302 | 1400 | 1673 | −2.2165 | 0.1090 |
| 0.18 | 0.0083 | 1493 | 1766 | −3.8015 | 0.0223 |

We propose a second state of activity equal to one, alternative to that of the pure compound-element (activities of Raoult). This value of the activity equal to one is known as activity of Henry, $a_{H,i}$. This value of the activity of Henry, $a_{H,i}$, is identified with that of the element-compound in a solution of 1.0% in weight.

For any concentration, the activities of Raoult and Henry are:

$$a_i = x_i \cdot \gamma_i \qquad (4.150)$$

where $\gamma_i$ is the coefficient of activity of a component i of a solution (solid or liquid).

$$a_{H,i} = \%w_i \cdot f_i \qquad (4.151)$$

where $f_i$ is the coefficient of activity of Henry.

In the field of diluted solutions:

$$a_i = x_i \cdot \gamma_i^\infty \qquad (4.152)$$

where $\gamma_i^\infty$ is the coefficient of activity at infinite dilution.

$$a_{H,i} = \%w_i \qquad (4.153)$$

In the field of the binary solutions, $PM_1$ (molecular weight of solvent) and $PM_2$ (molecular weight of solute) (in the diluted solution):

$$\frac{a_{H,i}}{a_i} = \frac{\%w_i}{x_i \cdot \gamma_i^\infty} = \frac{100 \cdot PM_2}{\gamma_i^\infty \cdot PM_1} \tag{4.154}$$

From the thermodynamic point of view, the standard free energy associated to the following reaction:

$$a_i = x_i \cdot \gamma_i^\infty; \ i(s, l, g) \leftrightarrow i(dis; \ 1.00\%); \ a_{H,i} = \%w_i \tag{4.155}$$

is:

$$\Delta G^0_{reaction \ (4.155)}(T) = -R \cdot T \cdot \ln\left(\frac{a_{H,i}}{a_i}\right) = R \cdot T \cdot \ln\left(\frac{\gamma_i^\infty \cdot PM_1}{100 \cdot PM_2}\right) \tag{4.156}$$

that represents the difference of standard free energy associated to two different states of reference of activity equal to one: one, that of the pure compound-element; and the other, identified with the diluted solution of the solute-2 in a solvent-1 of molecular/atomic weights $PM_2$ and $PM_1$.

*Brief note ends.*

When it is indicated that equilibrium was reached in the reaction between the periphery of the steel and a certain atmosphere, what underlays in equalizing the carbon activities of the steel and of the atmosphere at the temperature $T_1 - C_1/C_S = (P_{CO}^2/P_{CO_2})/k_{eq}$—is the following: the carbon content assimilated by the steel, $C_1$, differs the same relatively of the saturation, $C_S$ (where the graphite would start to precipitate as powder, Fig. 4.11, as it was already indicated), as the atmosphere $P_{CO}$ of CO and $P_{CO_2}$ of $CO_2$ differs from the equilibrium conditions (where, as a result of accomplishing the Boudouard's equilibrium, graphite powder would start to appear).

From the equality $C_1/C_S = (P_{CO}^2/P_{CO_2})/k_{eq}$, we can deduce the characteristics that the binary atmosphere should have, $P_{CO}{}^2/P_{CO_2}$, to avoid at temperature $T_1$ the decarburizing of a steel whose carbon content was $C_1$. This same equation can be also used to define the atmosphere where low carbon steel could peripherally enrich in this element (case hardening).

Regarding the case hardening, it is indicated a final point to explain the possible appearance of cementite during the case hardening. We have indicated that for activities of the atmosphere lower than one, carbon powder does not precipitate in the periphery of the part; neither—with greater motivation—cementite could appear as if the austenite does not saturate in carbon in the stable Fe–C diagram, it is neither saturated in carbon in the metastable Fe–Fe$_3$C diagram (whose $A_{cm}$ curve in the Fe–Fe$_3$C diagram is located to the right of the $A_{graphite}$ curve of the Fe–C stable diagram). If peripheral cementite appears after the case hardening, it is consequence of the fact that the austenite saturated in carbon is transformed by solubility loss during the cooling: following the metastable diagram and producing this cementite that did not exist at the case hardening temperature T.

From the equality, $C_1 = C_S \cdot (P_{CO}^2/P_{CO_2})/k_{eq}$, the following can be also deduced: if a certain atmosphere is neutral at temperature $T_1$—neither decarburizing nor case hardening—for an unalloyed steel, whose carbon content was $C_1\%$, will be, on

**Table 4.10** $k_{eq}$ as a function of the temperature

| T (°C) | 750 | 800 | 850 | 900 | 950 | 1000 | 1050 | 1100 |
|--------|------|------|-------|------|------|------|------|------|
| $k_{eq}$ | 2.70 | 6.93 | 16.35 | 35.8 | 73.7 | 143 | 265 | 467 |

the contrary, decarburizing for a steel that, with this same carbon percentage $C_1\%$, has also silicon (as—as other alphagenous elements—the silicon decrease $C_S$; see Sect. 1.6).

Analogously, the equality $C_1 = C_S \cdot (P_{CO}^2/P_{CO_2})/k_{eq}$ allows advising that in a binary atmosphere of known equilibrated composition, which that could produce a carbon content $C_1\%$ in the steel, at temperature $T_1$ and the total pressure the usual in the industry (1 atmosphere), if it is maintained the same total pressure of 1 atmosphere, but the mixture CO–CO$_2$ is diluted by the addition of an inert gas as the nitrogen, reducing the partial pressures can reduce the quotient $P_{CO}^2/P_{CO_2}$ and—therefore—a steel of $C_1\%$ of carbon will decarburize in this diluted atmosphere.

It is also necessary to consider that when the total pressure is 1 atmosphere, if the temperature T is increased while maintaining invariable the CO and CO$_2$ concentrations—we should think about an atmosphere formed only by these two gases supplied in constant proportion to the furnace (always equal to $P_{CO}^2/P_{CO_2}$), and assume that these concentrations would not be modified in the trajectory of the gas until its exit from the furnace—a steel of $C_1\%$ would decarburize because— returning once again to the equality $C_1/C_S = (P_{CO}^2/P_{CO_2})/k_{eq}$—, the constant of the Boudouard's equilibrium, $k_{eq}$, increases with the temperature much more than with the increase of $C_S$.

In fact, taking as reference the stable Fe–C system, $C_S$ varies from 0.66% at 738 °C to 2.08% at 1154 °C.

Regarding the values of $k_{eq}$, they noticeably increase as a function of the temperature (see Table 4.10).

### 4.4.3  Decarburizing-Case Hardening in a Binary Atmosphere of $H_2$–$CH_4$

From the reaction:

$$C(s) + 2H_2(g) \rightarrow CH_4(g) \tag{4.157}$$

we can deduce analogous possibilities to those mentioned about the Boudouard's equilibrium regarding the duality decarburizing-case hardening. The following is verified in the equilibrium of this reaction:

$$\ln k'_{eq} = \ln\left[\frac{P_{CH_4}}{P_{H_2}2}\right] = -\frac{\Delta G^0_{eq\cdot4.157}}{R \cdot T} \tag{4.158}$$

Assuming that a steel was in prolonged contact at the temperature $T_1$ with an atmosphere where $P_{CH_4}$ was the partial pressure of $CH_4$ and $P_{H_2}$ the partial pressure of $H_2$, the carbon content, $C_2$, that the steel could achieve in the equilibrium with this atmosphere would be given by the following equality:

$$\frac{C_2}{C_S} = \frac{P_{CH_4}}{P_{H_2}^2 / k'_{eq}} \qquad (4.159)$$

where $C_S$ is the maximum carbon content that the periphery of the steel would have in solid solution if it was saturated in carbon at the temperature $T_1$ (without formation of soot-cinder in its surface).

The reasoning about the possibilities of case hardening-decarburizing is in everything analogous to that explained in the Sect. 4.4.2.

Methane is also usually employed for the vacuum case hardening. It is a non-equilibrium process: once austenitized the steel in a vacuum furnace, pure $CH_4$ is applied on the surface of the steel. The carbon transfer takes place by dissociation of the gaseous hydrocarbon in the periphery of the steel and direct absorption of the carbon by the austenite, with hydrogen release according to the following chemical reaction:

$$CH_4(g) + \text{austenite} \ \rightarrow \ C - \text{enriched austenite} + 2 \cdot H_2(g) \qquad (4.160)$$

The case hardening is produced because of this non-equilibrium reaction, whose displacement is favored by the vacuum that exists in the furnace and carbon diffusion towards inside of the part by maintenance at temperature. Apart from methane, other gaseous hydrocarbons are also employed. Habitually, an inert gas as the nitrogen is also added.

### 4.4.4 Mixture of Gases $CO$–$CO_2$–$H_2$–$CH_4$–$H_2O$ in Equilibrium

All the possible equilibrium reactions between the five components $CO$–$CO_2$–$H_2$–$CH_4$–$H_2O$ come from two fundamental equilibriums:

$$H_2(g) + CO_2(g) \rightarrow H_2O(g) + CO(g) \qquad (4.161)$$

$$CH_4(g) + CO_2(g) \rightarrow 2H_2(g) + 2CO(g) \qquad (4.162)$$

In the reaction reversible (4.161), its equilibrium constant $k_{eq}$ verifies:

$$\ln k_{eq} = \ln \left[ \frac{P_{H_2O} \cdot P_{CO}}{P_{H_2} \cdot P_{CO_2}} \right] \qquad (4.163)$$

This implicitly involves—as it was already seen in Sect. 4.3.3—that the following equilibriums are also verified:

$$2CO(g) + O_2(g) \rightarrow 2CO_2(g) \tag{4.164}$$

and:

$$2H_2(g) + O_2(g) \rightarrow 2H_2O(g) \tag{4.165}$$

and, therefore, that the oxidizing-reductant capacity of this atmosphere can be determined—with the same result—either as a function of $P_{CO}/P_{CO_2}$ or $P_{H_2}/P_{H_2O}$.

At the same time, the equilibrium constant of the reaction (4.162) is that verifies:

$$\ln k_{eq}(\text{reaction } 4.162) = \ln\left[\frac{(P_{H_2})^2 \cdot (P_{CO})^2}{P_{CH_4} \cdot P_{CO_2}}\right] \tag{4.166}$$

This equilibrated reaction indicates that the equilibriums of other two reactions (as it would be sufficient to add them term by term) also are accomplished, the Boudouard's reaction:

$$C(s) + CO_2(g) \rightarrow 2CO(g) \tag{4.167}$$

of constant:

$$k_{eq}(\text{reaction } 4.167) = \frac{(P_{CO})^2}{P_{CO_2}} \tag{4.168}$$

and the reaction:

$$CH_4(g) \rightarrow C(s) + 2H_2(g) \tag{4.169}$$

of constant:

$$k_{eq}(\text{reaction } 4.169) = \frac{P_{H_2}}{P_{CH_4}} \tag{4.170}$$

Therefore, it is deduced from reaction (4.162) that the case hardening or decarburizing capacity of this equilibrated atmosphere of five gases can be determined, with equal result, both as a function of $(P_{CO})^2/P_{CO_2}$ and of $P_{CH_4}/(P_{H_2})^2$. As if we calculate the carbon content of a steel at temperature T as a function of $(P_{CO})^2/P_{CO_2}$, it is obtained:

$$C_1 = \frac{C_S \cdot (P_{CO})^2}{P_{CO_2} \cdot k_{eq}(\text{reaction } 4.167)} \tag{4.171}$$

if we calculate as a function of the equilibrium of hydrogen and methane:

$$C_x = \frac{C_S \cdot P_{CH_4} \cdot k_{eq}(\text{reaction 4.169})}{\left(P_{H_2}\right)^2} \tag{4.172}$$

the values of $C_1$ and $C_x$ are necessarily the same. In fact, it is deduced the next from the quotient of both equalities:

$$\frac{C_1}{C_x} = \frac{\left[\frac{(P_{CO})^2 \cdot (P_{H_2})^2}{P_{CO_2} \cdot P_{CH_4}}\right]}{k_{eq}(\text{reaction 4.167}) \cdot k_{eq}(\text{reaction 4.169})}$$

$$= \frac{k_{eq}(\text{reaction 4.167}) \, k_{eq}(\text{reaction 4.169})}{k_{eq}(\text{reaction 4.169}) \, k_{eq}(\text{reaction 4.169})} = 1 \tag{4.173}$$

In the equilibrium between the five gases ($CO$–$CO_2$–$H_2$–$CH_4$–$H_2O$), apart from the reactions (4.161) and (4.162), other equilibrium reaction is verified as it is deduced from subtracting from the reaction (4.162) the reaction (4.161):

$$CH_4(g) + H_2O(g) \rightarrow 3H_2(g) + CO(g) \tag{4.174}$$

whose equilibrium constant will be:

$$k_{eq}(\text{reaction 4.174}) = \frac{k_{eq}(\text{reaction 4.162})}{k_{eq}(\text{reaction 4.161})} \tag{4.175}$$

**Exercise 4.4** Considering the phases in equilibrium in the peritectic reaction at the temperature of 1493 °C (see Verdeja et al. 2020, p. 155) of the stable Fe–C diagram, answer to the following questions:

(a)   Activity and coefficient of activity of Raoult of the carbon in the different solutions (solid or melted/liquid) that are involved in the peritectic reaction.
(b)   Activity of Raoult of the iron in the austenite (0.16% C) of the peritectic reaction.
(c)   Justify from the thermodynamics point of view, the liquidus temperature (1493 °C) reached by the iron melt in the peritectic reaction with the 0.53% C.

**Data:** We assume that the temperature and the heat of fusion of the iron are 1538 °C and 13,807 J/mol Fe, respectively. The atomic weights of the carbon and iron are 12.01 and 55.85 g/mol, respectively.

*Solution starts*

During the cooling process in the Fe–C stable system, the invariant peritectic reaction at 1493 °C can be written as follows:

$$\delta - \text{ferrite (0.09\% C)} + \text{melt (0.53\% C)} \rightarrow \gamma - \text{austenite (0.16\% C)} \tag{4.176}$$

The general criterion of the thermodynamic equilibrium between phases stablishes that, at constant temperature and pressure, the chemical potential of the components involved in the phases in equilibrium must reach the same value.

In this case, the chemical potential of the carbon (elemental component of the Fe–C binary system), available in the three phases, would be the same. As the state of reference of carbon activity equal to one is the same in the three phases in equilibrium of the peritectic reaction, the activity of Raoult of the carbon in the three phases will be the same.

We calculate the activity of Raoult of the carbon in the austenite of 0.16% C at the temperature of 1493 °C with the correlation developed in this chapter from the data showed in Verdeja (1995, p. 111). This is about the variation with the temperature of the self-interaction parameter of the carbon and the coefficient of activity at infinite dilution of the carbon in the solid solutions of austenite:

$$\ln a_C = \frac{3581}{T} + \ln x_C + \left[ 1.03 \cdot \exp\left( \frac{3436}{T} \right) \right] \cdot x_C - 1.0996 \qquad (4.177)$$

where $a_C$ is the activity of the carbon, T is the temperature and $x_C$ is the carbon molar fraction.

We obtain that, for all the phases in equilibrium that are involved in the peritectic reaction, the activity of Raoult in the γ-austenite, δ-ferrite or melt with 0.53% C is the same and takes the value of 0.0197 ($a_C = 0.0197$):

$$x_C = \frac{\frac{0.16}{12.01}}{\frac{0.16}{12.01} + \frac{100-0.16}{55.85}} = 0.0074 \qquad (4.178)$$

$$\ln a_C = \frac{3581}{(1493 + 273)} + \ln 0.0074 + \left[ 1.03 \cdot \exp\left( \frac{3436}{1493 + 273} \right) \right] \cdot 0.0074$$
$$- 1.0996 = -3.9248 \rightarrow a_C = 0.0197 \qquad (4.179)$$

However, the value of the coefficient of activity in each one of the solutions is different:

– In the γ-austenite of 0.16%:

$$x_C == \frac{\frac{0.16}{12.01}}{\frac{0.16}{12.01} + \frac{100-0.16}{55.85}} = 0.0074 \qquad (4.180)$$

$$\gamma_C = \frac{a_C}{x_C} = \frac{0.0197}{0.0074} = 2.6622 \qquad (4.181)$$

– In the δ-ferrite of 0.09% C:

$$x_C = \frac{\frac{0.09}{12.01}}{\frac{0.09}{12.01} + \frac{100-0.09}{55.85}} = 0.0042 \qquad (4.182)$$

$$\gamma_C = \frac{a_C}{x_C} = \frac{0.0197}{0.0042} = 4.6905 \qquad (4.183)$$

- In the melt of 0.53% C:

$$x_C = \frac{\frac{0.53}{12.01}}{\frac{0.53}{12.01} + \frac{100-0.53}{55.85}} = 0.0242 \qquad (4.184)$$

$$\gamma_C = \frac{a_C}{x_C} = \frac{0.0197}{0.0242} = 0.8140 \qquad (4.185)$$

Now we calculate the activity of one of the components (iron) of the binary solution, when the activity of the other (carbon) is known. We use the equation of Gibbs–Duhem, see Ballester et al. (2000, p. 149):

$$\left[\log \gamma_{Fe}\right]_{\%C=0.16} = -\int_0^{0.16} \left(\frac{x_C}{x_{Fe}}\right) \cdot d \log \gamma_C \qquad (4.186)$$

$$\left[\log \gamma_{Fe}\right]_{\%C=0.16} = -\left(\frac{x_C}{x_{Fe}}\right) \cdot \left(\log \gamma_C^* - \log \gamma_C\right) \qquad (4.187)$$

We calculate the value of the coefficient of activity of the carbon at infinite dilution, $\gamma_C^*$, of the solid solutions of austenite. We use the equation developed by Verdeja et al. (1995, p. 111):

$$\gamma_C^* = 0.333 \cdot \exp\left(\frac{3581}{T}\right) = 0.333 \cdot \exp\left(\frac{3581}{1493 + 273}\right) = 2.5298 \qquad (4.188)$$

$$\left[\log \gamma_{Fe}\right]_{\%C=0.16} = -\left(\frac{0.0074}{0.9926}\right) \cdot \left(\log 2.5298 - \log 2.6622\right) = 1.65 \cdot 10^{-4}$$
$$\rightarrow \left[\gamma_{Fe}\right]_{\%C=0.16} = 1.0004 \qquad (4.189)$$

$$a_{Fe} = \gamma_{Fe} \cdot x_{Fe} = 1.0004 \cdot 0.9926 = 0.9930 \qquad (4.190)$$

Finally, we calculate the reduction of the iron melting point, $T_M = 1811$ K, as a consequence of the dissolution of the carbon, 0.53% C. It is possible to estimate it by means of the equation of Van't Hoff (see Ballester et al. 2000, p. 121) applied to the equilibrium:

$$Fe(s) \leftrightarrow Fe(dis; Fe - C) \qquad (4.191)$$

$$\left(\frac{d \ln K}{dT}\right)_P = \frac{\Delta H}{R \cdot T^2} \qquad (4.192)$$

$$\ln a_{Fe} = -\frac{\Delta H}{R} \cdot \left(\frac{1}{T} - \frac{1}{T_M}\right) \tag{4.193}$$

$$\frac{1}{T} = -(\ln a_{Fe}) \cdot \left(\frac{R}{\Delta H}\right) + \frac{1}{T_M} \tag{4.194}$$

If we consider ideal the behavior of the solvent:

$$a_{Fe} = x_{Fe} = 1 - 0.0242 = 0.9758 \tag{4.195}$$

we have:

$$\frac{1}{T} = -(\ln 0.9758) \cdot \left(\frac{8.3144\frac{J}{mol \cdot K}}{13807\frac{J}{mol}}\right) + \frac{1}{1811K} = 5.6693 \cdot 10^{-4}K^{-1} \to T$$

$$= 1764K(1491°C) \tag{4.196}$$

As general criterion, it is possible to say that, when the solute in a solution can have positive ($\gamma_i > 1$) or negative ($\gamma_i < 1$) deviations with respect to the ideal behavior, the solvent has ideal behavior: the activity or capacity of reaction of the component in the considered phase is identified with its concentration expressed in molar fraction.

Commentaries: In the case of the peritectoid reaction, where the austenite of 0.68% is transformed at 738 °C in carbon-graphite and α-ferrite with 0.02% C, the activities of the carbon in both solid solutions reach the value one. Nevertheless, the values of the coefficients of activity of Raoult of the carbon in the austenite of the 0.68% C and in the α-ferrite with 0.02% C are, respectively:

$$x_C = \frac{\frac{0.68}{12.01}}{\frac{0.68}{12.01} + \frac{100-0.68}{55.85}} = 0.0309 \tag{4.197}$$

$$\gamma_C = \frac{a_C}{x_C} = \frac{1.0000}{0.0309} = 32.3624 \tag{4.198}$$

$$x_C = \frac{\frac{0.02}{12.01}}{\frac{0.02}{12.01} + \frac{100-0.02}{55.85}} = 0.0009 \tag{4.199}$$

$$\gamma_C = \frac{a_C}{x_C} = \frac{1.0000}{0.0009} = 1111.11 \tag{4.200}$$

We should remember that the equilibrium diagrams are the graphical representation of the thermodynamic equilibrium. The lines that delimit the field/area of stability of the austenite are calculated as a succession of points, as it was developed in the exercise in an analytical manner, with small intervals of temperature. Finally, we should consider that, two of the most used materials in engineering applications, steel, and concrete, have compounds that are not thermodynamically stable at room temperature: cementite, $Fe_3C$, and tricalcium silicate $Ca_3SiO_5$ ($3CaO \cdot SiO_2$), respectively (Pero-Sanz et al. 2019).

*Solution ends.*

**Exercise 4.5** Considering the data of the Fe–C stable diagram (see Verdeja et al. 2020, p. 155), calculate for the temperatures of 1200, 1240, 1300, 1340, and 1400 °C the thermodynamic variables that are necessarily to know the behavior of liquid-iron melt-carbon solutions in equilibrium with the solid solutions of the carbon in the $\gamma$-iron: self-interaction parameter and coefficient of activity of Raoult at infinite dilution of the carbon. Data, expressed in molar percentage of carbon, are collected in Table 4.11.

*Solution starts*

Above the eutectic temperature of 1154 °C, the equilibrium of different carbon contents melts with the austenite leads to the equality of the carbon in the two solutions. For the calculation of the carbon activity in the austenite at any temperature, it is possible to use the equation, which has been proposed in this text, as an alternative to that proposed by Ellis et al. (1963):

$$\ln a_C = \frac{3581}{T} + \ln x_C + \left[1.03 \cdot \exp\left(\frac{3436}{T}\right)\right] \cdot x_C - 1.0996 \qquad (4.201)$$

It is possible to obtain from the activity of the carbon in the austenite $a_C^{Fe-\gamma}$ the value of the coefficient of activity of the carbon in the austenite, $\gamma_C^{Fe-\gamma}$:

$$\gamma_C Fe - \gamma = \frac{a_C^{Fe-\gamma}}{x_C^{Fe-\gamma}} \qquad (4.202)$$

Finally, it is possible to calculate from the coefficient of activity in the austenite for each one of the temperatures in the enunciation the value of the self-interaction parameter of the carbon in the solid solution of austenite:

$$\varepsilon_C^C = \frac{\ln \gamma_C^{Fe-\gamma} - \ln \gamma_C^{\infty Fe-\gamma}}{x_C} \qquad (4.203)$$

We know that the value of the coefficient of activity at infinite dilution of the carbon can be calculated as follows:

$$\gamma_C^{\infty Fe-\gamma} = 0.333 \cdot \exp\left[\frac{3581}{T(K)}\right] \qquad (4.204)$$

**Table 4.11** Data necessary to solve the exercise

| $T_L$-$T_S$ (°C) | 1200 °C | 1240 °C | 1300 °C | 1340 °C | 1400 °C |
|---|---|---|---|---|---|
| %C (austenite) | 7.98 | 7.24 | 5.75 | 4.63 | 3.02 |
| %C (melt) | 15.50 | 14.13 | 11.88 | 10.20 | 7.45 |

It is possible to calculate the parameters indicated in the enunciation using the above-indicated equations. Results are collected in Table 4.12.

In the case of the solutions of the Fe–C melts, $T_L$, in equilibrium with the austenite, the following procedure of calculation is used:

1. Determination of the carbon activity in the Fe–C melts that, for a certain temperature, is the same than that of the carbon in the austenite in equilibrium with the melt.

2. Calculate the value of the coefficient of activity of the carbon in the austenite, $\gamma_C^{melt}$:

$$\gamma_C^{melt} = \frac{a_C^{Fe-\gamma}}{x_C^{melt}} \tag{4.205}$$

3. Calculate, the self-interaction parameter of the carbon in the Fe–C melts for each one of the temperatures indicated in the enunciation:

$$\varepsilon_C^C = \frac{\ln \gamma_C^{melt} - \ln \gamma_C^{\infty\,melt}}{x_C^{melt}} \tag{4.206}$$

knowing that the value of the coefficient of activity at infinite dilution of the Fe–C melts is obtained as follows (Verdeja et al. 1995, page 111):

$$\gamma_C^{\infty\,melt} = 0.904 \cdot \exp\left[\frac{1231}{T(K)}\right] \tag{4.207}$$

Proceeding as indicated above, the Table 4.13 is obtained.

Commentaries: While the coefficient of activity at infinite dilution, $\gamma_i^\infty$, of a solute in a solution (solid or liquid) depends only on the temperature, the self-interaction parameter of a solute with itself, $\varepsilon_C^C$, is function of the temperature and the solute concentration in the solution. If the interval of concentrations is not very wide, it is possible to consider as a constant value, but if the concentration varies within a wide range of concentrations, it is necessary to know/be aware of the possible mistakes

**Table 4.12** Equilibrium austenite-Fe–C melt: carbon concentration in the austenite

| $T_S$ (°C) | 1400 | 1340 | 1300 | 1240 | 1200 |
|---|---|---|---|---|---|
| $x_C^=$ | 0.0302 | 0.0463 | 0.0575 | 0.0724 | 0.0798 |
| $\%w_C^=$ | 0.67 | 1.03 | 1.29 | 1.65 | 1.83 |
| $\ln a_C$ | −2.2161 | −1.5508 | −1.1528 | −0.6358 | −0.3497 |
| $a_C$ | 0.1090 | 0.2121 | 0.3157 | 0.5295 | 0.7049 |
| $\varepsilon_C^C$ | 8.0318 | 8.6693 | 9.1517 | 9.9798 | 10.6146 |
| $\gamma_C^{Fe-\gamma}$ | 3.6089 | 4.5807 | 5.4912 | 7.3137 | 8.8333 |
| $\gamma_C^{\infty\,Fe-\gamma}$ | 2.8316 | 3.0663 | 3.2444 | 3.5509 | 3.7866 |

associated to the estimation that is carried out of the activity of a solute when it is not considered its possible variation with the concentration.

*Solution ends.*

**Exercise 4.6** According to Bodsworth and Appleton (1969, p. 139), the relation between the carbon molar fraction, $x_C$, in the austenite and its activity of Raoult, $a_C$, at the temperature of 925 °C is collected in Table 4.14.

The following results are required:

(a) Calculate the coefficient of activity of the carbon, $\gamma_C$, for each value of the concentration.

(b) Coefficient of activity at infinite dilution, $\gamma_C^\infty$, associated to the carbon solutions in the $\gamma$-Fe at 925 °C.

(c) Standard free energy associated to the reaction (4.208):

$$C(s;\ \text{graphite}) \leftrightarrow C(\text{sol.};\ \ Fe - \lambda;\ a_{H,C} = 1) \tag{4.208}$$

(d) The activity of Henry at 925 °C for a concentration of 0.55% C.

*Solution starts*

The functional relation between the carbon activity and the carbon concentration, expressed as molar fraction, is obtained by means of the coefficient of activity (thermodynamic variable that depends on the temperature and the concentration). The coefficient of activity is calculated as follows:

$$\gamma_C = \frac{a_C}{x_C} \tag{4.209}$$

Values are collected in Table 4.15.

The coefficient of activity at infinite dilution, $\gamma_C^\infty$, would be identified with the slope of the activity-concentration ($a_C - x_C$) curve when the carbon concentration in the $\gamma$-Fe tends to zero. It is possible to identify in a first approximation the value of $\gamma_C^\infty$ with that of 12.80 without making the graphical representation of the values

**Table 4.13** Equilibrium austenite-Fe–C melt: carbon concentration in the melt

| $T_S$ (°C) | 1400 | 1340 | 1300 | 1240 | 1200 |
|---|---|---|---|---|---|
| $x_C^=$ | 0.0745 | 0.1020 | 0.1188 | 0.1413 | 0.1515 |
| $\%w_C^=$ | 1.70 | 2.38 | 2.82 | 3.42 | 3.70 |
| $\ln a_C$ | −2.2161 | −1.5508 | −1.1528 | −0.6358 | −0.3497 |
| $a_C$ | 0.1090 | 0.2121 | 0.3157 | 0.5295 | 0.7049 |
| $\varepsilon_C^C$ | −3.4151 | 0.6841 | 2.4903 | 4.3056 | 5.2982 |
| $\gamma_C^{Fe-\gamma}$ | 1.4630 | 2.0793 | 2.6578 | 3.7474 | 4.6528 |
| $\gamma_C^{\infty Fe-\gamma}$ | 1.8868 | 1.9391 | 1.9771 | 2.0395 | 2.0850 |

**Table 4.14**  Values of the molar fraction and the activity

| $x_C$ | 0.005 | 0.010 | 0.015 | 0.020 | 0.025 | 0.030 |
|-------|-------|-------|-------|-------|-------|-------|
| $a_C$ | 0.064 | 0.132 | 0.205 | 0.280 | 0.361 | 0.450 |

provided in Table 4.15. This estimation of the $\gamma_C^\infty$ was made as a function of the provided experimental data: $\gamma_C^\infty \simeq (0.064/0.005) = 12.80$. In fact, the value of $\gamma_C^\infty$ will be the slope of the curve $\gamma_C/x_C$ when $x_C \to 0$. To avoid the graphical representation, we have considered as a first estimation of the value of $\gamma_C^\infty$ this result of the slope of the curve $\gamma_C/x_C$ when $x_C = 0.005$.

The variation of the standard free energy associated to the change of state of reference of the carbon in the $\gamma$-Fe (reaction (4.208)) is expressed as follows:

$$\Delta_r G^0_{\text{reaction } 4.208}(T) = R \cdot T \cdot \ln\left(\frac{a_C}{a_{H,C}}\right) = R \cdot T \cdot \ln\left(\frac{\gamma_C^\infty \cdot m_{Fe}}{m_C \cdot 100}\right) \qquad (4.210)$$

Consequently, the value associated to the standard free energy of the reaction (4.208) is:

$$\Delta_r G^0_{\text{reaction } 4.208}(T) = 8.314\frac{J}{\text{mol} \cdot K} \cdot (925 + 273)K \cdot \ln\left(\frac{12.80 \cdot 55.85}{12.01 \cdot 100}\right)$$

$$= -5167.3\frac{J}{\text{mol C}} \qquad (4.211)$$

while in the case of using the data of Sancho et al. (2000) (Table 2.15, p. 139). We consider these data of the standard free energy associated to the change of state of reference (in Joules):

$$\Delta G^0(T) = 72180 + 52.05 \cdot T \cdot \log T - 227.03 \cdot T \to \Delta G^0(1198K)$$

$$= -7842.0 J/\text{mol C}$$

The activity of Henry for the carbon in the austenite can be calculated at a certain temperature when the activity of Raoult is known. At the temperature of 1198 K for 0.55% C:

**Table 4.15**  Results of the calculation of the coefficient of activity

| $x_C$ | 0.005 | 0.010 | 0.015 | 0.020 | 0.025 | 0.030 |
|-------|-------|-------|-------|-------|-------|-------|
| %C | 0.11 | 0.22 | 0.33 | 0.44 | 0.55 | 0.66 |
| $a_C$ | 0.064 | 0.132 | 0.205 | 0.280 | 0.361 | 0.450 |
| $\gamma_C$ | 12.80 | 13.20 | 13.67 | 14.00 | 14.44 | 15.00 |

$$\frac{a_C}{a_{H,C}} = \frac{12.80 \cdot 55.85}{12.01 \cdot 100} = 0.5952 \tag{4.212}$$

$$a_{H,C} = \frac{a_C}{0.5952} = \frac{0.361}{0.5922} = 0.6065 \tag{4.213}$$

Consequently, the solutions of the carbon in the austenite also manifest a deviation with respect to the Henry's law corrected by its coefficient of activity, $f_C$:

$$a_{H,C} = f_C \cdot \%C \rightarrow f_C = 0.6065 \cdot 0.55 = 1.10 \tag{4.214}$$

Commentaries: It is possible to assign to each pair of values of the coefficient of activity and the molar fraction at 925 °C, a value of the self-interaction parameter of the carbon, $\varepsilon_C^C$, in the solid solutions of the carbon in the Fe-γ. The average of the five values that were indicated in the data of the enunciation is 4.41 at 925 °C, calculated as follows:

$$\varepsilon_C^C = \frac{\ln \gamma_C^{Fe-\gamma} - \ln \gamma_C^{*Fe-\gamma}}{x_C} \tag{4.215}$$

The activity of the carbon in the austenite at 925 °C can be calculated for a carbon concentration of 0.55%. Finally, when we consider the equation developed in the text to estimate the activity of the carbon in the austenite ($x_C = 0.025$):

$$\ln a_C = \frac{3581}{T} + \ln x_C + 1.03 \cdot x_C \cdot \exp\left(\frac{3436}{T}\right) - 1.0996 \tag{4.216}$$

the activity of the carbon in the solid solution is 0.2603.

While if the estimation is carried out by means of Ellis, Davidson, and Bosdworth ($x_C = 0.025$):

$$\ln a_C = \log\left(\frac{x_C}{1 - x_C}\right) + \frac{2207}{T} - 0.64 \tag{4.217}$$

the activity of the carbon would be 0.4085.

According to the estimation of the value of the carbon activities in the solid solutions of Fe-γ that was carried out in the text, the capacity of reaction of the carbon is sensitively lower to the estimations obtained by Ellis, Davidson, and Bosdworth, which are more aligned to the data indicated in the enunciation of the exercise.

*Solution ends.*

## 4.5   Industrial Atmospheres: C–H–O–N System

Four industrial gases are used in the iron and steelmaking industry nowadays:

- Blast furnace gas: 53% $N_2$, 25% CO, 20% $CO_2$, 2% $H_2$.
- Coke batteries gas: 57% $H_2$, 25% $CH_4$, 6% CO, 12% others.
- Linz-Donawitz gas: 75% CO and 25% $N_2$.
- Natural gas: 90% $CH_4$, 5% $H_2$, 5% others.

In all of them, the main application is to be used as energy supply in the treatment of complete austenitization that is required to carry out the hot rolling of the steel (oxidizing atmosphere). Nevertheless, the iron and steelmaking sector is considering the following alternatives in the recent times and looking towards the reduction of the emissions of $CO_2$ to the atmosphere in the steel industry:

1. Use of iron and steelmaking combustible gases, natural gas, or low-grade coals to produce industrial reductant gases able to fix the oxygen of the ore in the $CO_2$ and the $H_2O$ of the gas.
2. Optimize the heat transfer as well as the energy efficiency of the gas towards the steel, increasing the convection and radiant heat transfer mechanisms.
3. Replacement of the traditional mechanisms of heating of metals by electric or magnetic induction.

Finally, considering that even the main molecular species in the industrial gases-atmospheres are the $H_2$, CO, $N_2$, $CO_2$, and $H_2O$, there are particles in suspension and other molecules that can have a negative influence for the environment. For instance, sulphur and nitrogen oxides, $SO_x$ and $NO_x$, as well as the halogen derivates of organic molecules of hydrocarbons with benzene rings (Verdeja et al. 2020; Fernández-González et al. 2017a, b, c, d, 2018a) are compounds whose values of emission to the environment are being controlled more and more.

### 4.5.1   Introduction

From the perspective of the possibility of catching oxygen, an atmosphere can be of two different types:

(a) *Oxidizing atmospheres*, which do not exhibit capacity of catching oxygen, and even produce the oxidation of those products that are in contact with them.
(b) *Reductant atmospheres*, which have a high capacity-affinity to catch oxygen, removing the oxide layer available on the surface of the steel or reducing-producing sponge iron from an iron ore.

     On the other hand, if we pay attention to the "carbon title" (which corresponds with the activity-capacity of reaction of the carbon in the solution, $a_C^{steel}$), we can consider:

(c)   *Carbide-forming atmospheres*, which are those where the carbon activity in the gas, $a_C^{gas}$, is greater than that existing in the steel:

$$a_C^{gas} > a_C^{steel} \tag{4.218}$$

(d)   *Decarburizing atmospheres*, where the carbon dissolved in the steel is trapped by the compounds $CO_2$ and $H_2$ of the gaseous phase:

$$CO_2(g) + C(s) \rightarrow 2CO(g) \tag{4.219}$$

$$2H_2(g) + C(s) \rightarrow CH_4(g) \tag{4.220}$$

Despite the Integrated Steelworks (blast furnace + basic oxygen furnace) have available the blast furnace gas, the coke batteries gas or the Linz-Donawitz gas, the electric steelworks are one of the main natural gas consumers. However, in the recrystallization-recovery heat treatment for the crystalline structure of the steel strip (see Verdeja et al. 2021) or in the sectors dedicated to the heat treatment of the steels, synthetic atmospheres are used. They are obtained from the carbon (petroleum coke), natural gas, propane, butane, or ammonia. There are two alternatives for the obtaining of industrial synthetic atmospheres that lead to two types of atmospheres:

(e)   *Endothermic atmospheres*, where an external energy supply and a metallic catalyzer (nickel) supported by a ceramic support are necessaries for their obtaining and reaching the chemical stability of the compounds in the gas.

(f)   *Exothermic atmospheres*, where it is not necessary the supply of external energy. The partial combustion of the combustible is the energy source for the compounds in the gas to reach the thermodynamic stability.

Finally, the industrial atmospheres move around four components: carbon, hydrogen, oxygen, and nitrogen, which habitually provide stable gaseous phases where there are two possible alternatives:

– Alternative 1 (with the presence of methane in the gas, $CH_4$): CO, $H_2$, $CO_2$, $H_2O$, $N_2$, and $CH_4$.
– Alternative 2: $N_2$, CO, $H_2$, $CO_2$, and $H_2O$.

It is also necessary to remember that the costs associated to the production of endothermic atmospheres are higher than those of the exothermal atmospheres. Endothermic atmospheres are industrially used to produce sponge iron (Verdeja et al. 2020, pp. 120–155) or for the case-hardening or carburizing steels (Verdeja et al. 2020, 2021). On the other hand, the content of main molecules ($N_2$, $H_2$ and CO) in endothermic atmospheres will be: 30% $H_2$, 23% CO, 4% others ($CH_4 + CO_2 + H_2O$), and balance $N_2$. The content of main molecules ($N_2$, $H_2$, and CO) in the gas in exothermic atmospheres will be: 10% $H_2$, 10% CO, 8% $CO_2$, < 15% $H_2O$, and balance $N_2$. This type of exothermic atmospheres is usually employed to avoid the oxidation of the metals in general and of the steels in particular. Finally, it is necessary to consider that both the reductants, carburizing and deoxidizing capacities of an industrial atmosphere can be increased by elimination of both the water vapor and the $CO_2$ of the gases.

From the thermodynamic point of view, an industrial atmosphere is a homogeneous phase of four components: carbon, hydrogen, oxygen, and nitrogen. For a certain conditions of pressure and temperature, the number of degrees of freedom of the system is equal to five:

$$L = C + 2 - F = 5 \qquad (4.221)$$

It will be necessary to define five variables to have defined the relation-volumetric concentration of the different molecular species in equilibrium. Two of them, the total pressure of the system (habitually one atmosphere) and the temperature are common. The three-remaining associated to the composition of the molecules of the gas is related with the molecules where it is easier to obtain an experimental value. One of them is the water vapor in the gas through the value of the dew point. Its determination consists in measuring the temperature at which the condensation of the water vapor starts over a shiny surface (a mirror). A certain volume percentage of water vapor that contains the gas at the pressure of one atmosphere corresponds to each dew point.

The standard free energy of the reaction:

$$H_2O \text{ (liquid)} \leftrightarrow H_2O \text{ (gas)} \qquad (4.222)$$

as a function of the temperature (from 243 to 303 K) is:

$$\Delta G^0(T) = 48.038 - 0.1328 \cdot T(K) \qquad (4.223)$$

This interval of temperatures (from 243 to 303 K or $-30$ to 30 °C) is defined reasoning that it is not frequent to find an atmosphere that would contain a $P_{H_2O} >$ 0.05 atm (5 vol. %) and that was so dry that it does not reach a content smaller than 0.01% $H_2O$.

It is also convenient to develop the analytical equation that relates the dew point with the volumetric composition of the $H_2O$ in the gas (Fig. 4.12).

The other two compositions of the gaseous molecules susceptible of analysis, one of them, would be the $CO_2$: this molecule can be absorbed in a concentrated solution of caustic soda under the form of $Na_2CO_3$. The last variable to fix might be the nitrogen, $N_2$. The molecule of $N_2$ is not involved in the main reactions to reach the equilibrium between the different molecules of the gas, although it is the responsible of the appearance of the $NO_x$ in the gas. To reach the formation of an endothermic or exothermic atmosphere, it is necessary the supply of a known quantity of air and, consequently, a known proportion of nitrogen.

## 4.5.2 Endothermic and Exothermic Atmospheres

Other manner of studying the composition of the equilibrated atmospheres (those where the gas compounds reach equilibrium values) is analyzing and studying the products derived from the irreversible reaction of the hydrocarbons saturated with the air. The reaction to be considered for those hydrocarbons formed only by carbon and hydrogen would be the following:

$$C_mH_{2m+2} + N_{air} \cdot (0.21 \cdot O_2 + 0.79 \cdot N_2)(g)$$
$$\rightarrow \alpha CO_2(g) + \beta CO(g) + \gamma CH_4(g) + \delta H_2O(g)$$
$$+ \ \varepsilon H_2(g) + \eta N_2(g) \tag{4.224}$$

Knowing the characteristics of the starting hydrocarbon to obtain the endothermic atmosphere, the temperature, T, the total pressure of the system, $P_T$ (habitually of one atmosphere) and $N_{air}$, it is possible to develop a system of six equations that lead to the calculation of the six unknowns of the system: $\alpha$, $\beta$, $\gamma$, $\delta$, $\varepsilon$, and $\eta$. The equations that should be considered would be the following:

– Balance to the carbon:

$$m = \alpha + \beta + \gamma \tag{4.225}$$

– Balance to the hydrogen:

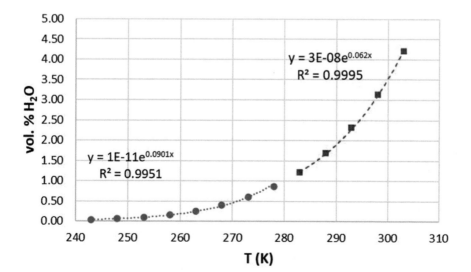

**Fig. 4.12** Representation of the dew point versus the volume composition of the $H_2O$ in the gas. The aim is to find the analytical correlations between dew point and $P_{H_2O}$ or $C_{H_2O}$

$$m + 1 = 2 \cdot \gamma + \delta + \varepsilon \tag{4.226}$$

- Balance to the oxygen:

$$0.42 \cdot N_{air} = 2 \cdot \alpha + \beta + \delta \tag{4.227}$$

- Balance to the nitrogen:

$$0.79 \cdot N_{air} = \eta \tag{4.228}$$

- Moles of gas produced, $N_{gases}$:

$$N_{gases} = \alpha + \beta + \gamma + \delta + \varepsilon + \eta \tag{4.229}$$

From the equilibrium:

$$CO_2(g) + H_2(g) \rightarrow CO(g) + H_2O(g) \tag{4.230}$$

$$k_{eq}(\text{reaction } 4.230) = \frac{\beta \cdot \varepsilon^3}{\gamma \cdot \delta \cdot N_{gases}^2} \tag{4.231}$$

From the equilibrium:

$$CH_4(g) + H_2O(g) \rightarrow CO(g) + 3H_2(g) \tag{4.232}$$

$$k_{eq}(\text{reaction } 4.232) = \frac{\beta \cdot \varepsilon^3}{\gamma \cdot \delta \cdot N_{gases}^2} \tag{4.233}$$

The values that the $CO_2$, $CO$, $CH_4$, $H_2O$ and $H_2$ molecules reach in the equilibrated gas would be:

$$\%CO_2 = \frac{\alpha}{N_{gases}} \cdot 100 \tag{4.234}$$

$$\%CO = \frac{\beta}{N_{gases}} \cdot 100 \tag{4.235}$$

$$\%CH_4 = \frac{\gamma}{N_{gases}} \cdot 100 \tag{4.236}$$

$$\%H_2O = \frac{\delta}{N_{gases}} \cdot 100 \tag{4.237}$$

$$\%H_2 = \frac{\varepsilon}{N_{gases}} \cdot 100 \tag{4.238}$$

The main difference from the physical–chemical point of view, and as it was already indicated in the Sect. 4.5.1, of the endothermic and exothermic atmosphere can be summarized in two variables:

1. $N_2$ content in the gas.
2. Activity of the carbon in the gas.

The $N_2$ content in the gas in the exothermic atmospheres is >50% and the activity of the carbon in the gas, $a_C^{gas}$, can be situated in around 0.10. Nevertheless, in the endothermic atmospheres, the $N_2$ of the gas is usually <50%, and the activity of the carbon in the gas is close to one. The endothermic atmospheres, as they operate with high carbon activities, can produce operating problems due to the carbon precipitation and reducing the porosity of the particulate matter bed in the processes of DRI obtaining.

The same manner that in the petroleum/oil refineries, heavy hydrocarbons are reformed-transformed into petrol (saturated hydrocarbons with less carbon), the reforming of industrial gases related with the iron and steelmaking process could move in the future towards the production of gases with greater hydrogen content with the corresponding elimination of the carbon (in the petrochemical industry a residue is produced, petroleum coke, whose quantity depends on the greater or smaller quantity of carbon in the processed petroleum/oil).

**Exercise 4.7** Calculate the partial pressure and volume composition of the gaseous compounds: $CH_4$, $CO$, $CO_2$, $H_2$, $H_2O$, and $O_2$ in a gas that is in equilibrium with carbon (graphite) at the temperature of 950 °C and at the total pressure of 0.80 atmospheres when the volume composition of the CO is of 40% (endothermic atmosphere).

*Solution starts*

If we know the total pressure, $P_T$, the temperature, and the volume composition of one of the compounds of the gas, in this case that of the CO, it is possible to know the volume composition of the other compounds of the gas ($CH_4$, $CO_2$, $H_2$, $H_2O$ and $O_2$). The system of three components and with two phases in equilibrium (the gas and the carbon) has zero degrees of freedom when three intensive variables of the system are defined: total pressure, temperature, and volume composition of the CO. Considering the data of the Appendix 1 (Appendices), the standard free energy associated to the formation of the CO is:

$$C(s) + \frac{1}{2}O_2(g) \leftrightarrow CO(g) \tag{4.239}$$

$$\Delta_r G^o_{reaction\ 4.239}(T) = -118.0 - 0.084 \cdot T(K)\ kJ \cdot (0.5\ mol\ O)^{-1} \tag{4.240}$$

that leads to a value of the equilibrium constant, $k_{C/CO}$, at 950 °C, of:

$$k_{C/CO}(reaction\ 4.239) = \exp\left(-\frac{\Delta_r G^o_{reaction\ 4.239}(1223\ K)}{R \cdot T}\right)$$

$$= \exp\left[-\frac{(-118 - 0.084 \cdot 1223) \text{ kJ} \cdot (0.5 \text{ mol O}_2)^{-1} \cdot 1000 \text{ J} \cdot \text{kJ}^{-1}}{8.314 \text{ J} \cdot \text{mol}^{-1} \cdot \text{K}^{-1} \cdot 1223 \text{ K}}\right]$$

$$= 2.68 \cdot 10^9 \text{(when } a_C = 1) \to k_{C/CO} = \frac{P_{CO}}{\sqrt{P_{O_2}}} \tag{4.241}$$

Consequently, solving for the partial pressure of oxygen, when the activity of the carbon is equal to one, we obtain:

$$k_{C/CO} = \frac{P_{CO}}{\sqrt{P_{O_2}}} \to \sqrt{P_{O_2}} = \frac{P_{CO}}{k_{C/CO}} \to P_{O_2} = \left(\frac{P_{CO}}{k_{C/CO}}\right)^2 = \left(\frac{0.40 \cdot 0.8}{2.68 \cdot 10^9}\right)^2$$

$$= 1.4305 \cdot 10^{-20} \text{atm} \tag{4.242}$$

Once known the oxygen partial pressure, we use the formation of carbon dioxide, Appendix 1:

$$C(s) + O_2(g) \leftrightarrow CO_2(g) \tag{4.243}$$

$$\Delta_r G^\circ_{\text{reaction 4.243}}(T) = -395.0 + 0.001 \cdot T(K) \text{ kJ} \cdot (\text{mol O})^{-1} \tag{4.244}$$

that leads to a value of the equilibrium constant, $k_{C/CO_2}$, at 950 °C, of:

$$k_{C/CO_2} \text{ (reaction 4.243)} = \exp\left(-\frac{\Delta_r G^\circ_{\text{reaction 4.243}}(1223 \text{ K})}{R \cdot T}\right)$$

$$= \exp\left[-\frac{\left(-395.0 + 0.001 \cdot 1223 \text{ kJ} \cdot (\text{mol O})^{-1}\right) \cdot 1000 \text{ J} \cdot \text{kJ}^{-1}}{8.314 \text{ J} \cdot \text{mol}^{-1} \cdot \text{K}^{-1} \cdot 1223 \text{ K}}\right]$$

$$= 6.58 \cdot 10^{16} \text{(when } a_C = 1) \to k_{C_2/CO_2} = \frac{P_{CO_2}}{P_{O_2}} \tag{4.245}$$

Consequently, solving for the partial pressure of carbon dioxide, when the activity of the carbon is equal to one, we obtain:

$$k_{C_1 CO_2} = \frac{P_{CO_2}}{P_{O_2}} \to P_{CO_2} = k_{C_2/CO_2} \cdot P_{O_2} \to P_{CO_2}$$

$$= 6.58 \cdot 10^{16} \cdot 1.4305 \cdot 10^{-20} \text{atm}$$

$$= 9.4102 \cdot 10^{-4} \text{atm} \tag{4.246}$$

We use the total pressure of the system ($P_T = 0.8$atm)

$$P_T = P_{CH_4} + P_{H_2} + P_{H_2O} + P_{CO} + P_{CO_2} + P_{O_2} \tag{4.247}$$

We divide each one of the terms of the equation by the hydrogen partial pressure, $P_{H_2}$:

$$\frac{P_T}{P_{H_2}} = \frac{P_{CH_4}}{P_{H_2}} + 1 + \frac{P_{H_2O}}{P_{H_2}} + \frac{P_{CO}}{P_{H_2}} + \frac{P_{CO_2}}{P_{H_2}} + \frac{P_{O_2}}{P_{H_2}} \tag{4.248}$$

On the other hand, we can calculate other relations of partial pressures by means of the standard free energies associated to the following reactions (Appendix 1 (Appendices)):

$$C(s) + 2 \cdot H_2(g) \leftrightarrow CH_4(g) \tag{4.249}$$

$$\Delta_r G^o_{\text{reaction } 4.249}(T) = -95.6 + 0.113 \cdot T(K) \text{ kJ} \cdot (2 \text{ mol } H_2)^{-1} \tag{4.250}$$

that leads to a value of the equilibrium constant, $k_{C/CH_4}$, at 950 °C, of:

$$k_{C/CH_4}(\text{reaction } 4.249) = \exp\left(-\frac{\Delta_r G^o_{\text{reaction } 4.249}(1223 \text{ K})}{R \cdot T}\right)$$

$$= \exp\left[-\frac{\left(-95.6 + 0.113 \cdot 1223 \text{ kJ} \cdot (2 \text{ mol } H_2)^{-1}\right) \cdot 1000 \text{ J} \cdot \text{kJ}^{-1}}{8.314 \text{ J} \cdot \text{mol}^{-1} \cdot \text{K}^{-1} \cdot 1223 \text{ K}}\right]$$

$$= 1.52 \cdot 10^{-2} \text{(when } a_C = 1) \to k_{C/CH_4}$$

$$= \frac{P_{CH_4}}{P_U} \to \frac{P_{CH_4}}{P_H} = P_{H_2} \cdot k_{C/CH_4} \tag{4.251}$$

$$H_2(s) + \frac{1}{2}O_2(g) \leftrightarrow H_2O(g) \tag{4.252}$$

$$\Delta_r G^o_{\text{reaction } 4.252}(T) = -249.0 + 0.057 \cdot T(K) \text{kJ} \cdot (\text{mol } H)^{-1} \tag{4.253}$$

that leads to a value of the equilibrium constant, $k_{H_2/H_2O}$, at 950 °C, of:

$$k_{H_2/H_2O}(\text{reaction } 4.252) = \exp\left(-\frac{\Delta_r G^o_{\text{reaction } 4.252}(1223 \text{ K})}{R \cdot T}\right)$$

$$= \exp\left[-\frac{\left(-249.0 + 0.057 \cdot 1223 \text{ kJ} \cdot (\text{mol } H)^{-1}\right) \cdot 1000 \text{ J} \cdot \text{kJ}^{-1}}{8.314 \text{ J} \cdot \text{mol}^{-1} \cdot \text{K}^{-1} \cdot 1223 \text{ K}}\right]$$

$$= 4.54 \cdot 10^7 \to k_{H_2/H_2O} = \frac{P_{H_2O}}{P_{H_2} \cdot \sqrt{P_{O_2}}} \to \frac{P_{H_2O}}{P_{H_2}} = k_{H_2/H_2O} \cdot \sqrt{P_{O_2}} \tag{4.254}$$

Consequently, we obtain a quadratic equation:

$$\frac{0.8}{P_{H_2}} = k_{C_C/CH_4} \cdot P_{H_2} + 1 + k_{H_2/H_2O} \cdot \sqrt{1.4305 \cdot 10^{-20}}$$

$$+ \frac{0.4 \cdot 0.8 + 9.4102 \cdot 10^{-4} + 1.4305 \cdot 10^{-20}}{P_{H_2}} \to 0$$

$$= k_{C_2/CH_4} \cdot P_{H_2} + 1 + k_{H_2/H_2O} \cdot \sqrt{1.4305 \cdot 10^{-20}}$$

$$+ \frac{0.4 \cdot 0.8 + 9.4102 \cdot 10^{-4} + 1.4305 \cdot 10^{-20} - 0.8}{P_{H_2}}$$

**Table 4.16** Values of the partial pressure of the components of the gas for different values of the CO partial pressure for a total pressure of 0.80 atm and 950 °C

| P(CO)-atm | P(CO$_2$)-atm | P(O$_2$)-atm | P(H$_2$)-atm | P(H$_2$O)-atm | P(CH$_4$)-atm |
|---|---|---|---|---|---|
| 0.3200 | 9.41·10$^{-4}$ | 1.43·10$^{-20}$ | 0.4731 | 2.57·10$^{-3}$ | 3.39·10$^{-3}$ |
| %CO-vol | %CO$_2$-vol | %O$_2$-vol | %H$_2$-vol | %H$_2$O-vol | %CH$_4$-vol |
| 40.00 | 0.12 | 0.00 | 59.14 | 0.32 | 0.42 |
| P(CO)-atm | P(CO$_2$)-atm | P(O$_2$)-atm | P(H$_2$)-atm | P(H$_2$O)-atm | P(CH$_4$)-atm |
| 0.4000 | 1.47·10$^{-3}$ | 2.24·10$^{-20}$ | 0.3935 | 2.67·10$^{-3}$ | 2.35·10$^{-3}$ |
| %CO-vol | %CO$_2$-vol | %O$_2$-vol | %H$_2$-vol | %H$_2$O-vol | %CH$_4$-vol |
| 50.00 | 0.18 | 0.00 | 49.19 | 0.33 | 0.29 |
| P(CO)-atm | P(CO$_2$)-atm | P(O$_2$)-atm | P(H$_2$)-atm | P(H$_2$O)-atm | P(CH$_4$)-atm |
| 0.4800 | 2.12·10$^{-3}$ | 3.22·10$^{-20}$ | 0.3138 | 2.56·10$^{-3}$ | 1.49·10$^{-3}$ |
| %CO-vol | %CO$_2$-vol | %O$_2$-vol | %H$_2$-vol | %H$_2$O-vol | %CH$_4$-vol |
| 60.00 | 0.26 | 0.00 | 39.23 | 0.32 | 0.19 |
| P(CO)-atm | P(CO$_2$)-atm | P(O$_2$)-atm | P(H$_2$)-atm | P(H$_2$O)-atm | P(CH$_4$)-atm |
| 0.5600 | 2.88·10$^{-3}$ | 4.38·10$^{-20}$ | 0.2340 | 2.23·10$^{-3}$ | 8.30·10$^{-4}$ |
| %CO-vol | %CO$_2$-vol | %O$_2$-vol | %H$_2$-vol | %H$_2$O-vol | %CH$_4$-vol |
| 70.00 | 0.36 | 0.00 | 29.25 | 0.28 | 0.11 |

$$\rightarrow 1.52 \cdot 10^{-2} \cdot P_{H_2}^2 + P_{H_2} + P_{H_2} \cdot 4.54 \cdot 10^7 \cdot \sqrt{1.4305 \cdot 10^{-20}}$$
$$+ \; 0.4 \cdot 0.8 + 9.4102 \cdot 10^{-4} + 1.4305 \cdot 10^{-20} - 0.8$$
$$= 0 \rightarrow 1.52 \cdot 10^{-2} \cdot P_{H_2}^2 + 1.0054 \cdot P_{H_2} - 0.4791 = 0 \qquad (4.255)$$

The solution of this quadratic equation leads to a value of the oxygen partial pressure, $P_{H_2}$, of 0.4731 atm. We obtain using the Eqs. (4.251) and (4.254):

$$\frac{P_{H_2O}}{P_{H_2}} = k_{H_2/H_2O} \cdot \sqrt{P_{O_2}} \rightarrow P_{H_2O} = P_{H_2} \cdot k_{H_2/H_2O} \cdot \sqrt{P_{O_2}}$$
$$= 0.4731 \cdot 4.54 \cdot 10^7 \cdot \sqrt{1.4305 \cdot 10^{-20}} = 2.5709 \cdot 10^{-3} \text{atm} \qquad (4.256)$$

$$k_{C/CH_4} = \frac{P_{CH_4}}{P_{H_2}} \rightarrow P_{CH_4} = k_{C/CH_4} \cdot P_{H_2}^2 = 1.52 \cdot 10^{-2} \cdot 0.4731^2$$
$$= 3.3925 \cdot 10^{-3} \text{atm} \qquad (4.257)$$

We collect in Table 4.16 the values of the partial pressures and volume composition of other gases in equilibrium for different values of the CO partial pressure for a total pressure of 0.8 atm and 950 °C of temperature.

Commentaries: The volume compositions of the gaseous compounds in equilibrium with the carbon at the total pressure of 0.80 atm and 950 °C have the maximum activity of the carbon in the gas:

$$a_C^{gas} = a_C^{steel} = 1 \qquad (4.258)$$

Nevertheless, the relation of partial pressures of CO and $CO_2$ to reach an activity equal to one also depends on the volume composition of the hydrogen and methane in the gas through the $C–CH_4–H_2$ equilibrium (Table 4.17).

The presence of nitrogen is habitually detected in the carbide-forming and reductant atmospheres (endothermic atmospheres), reformed from natural gas or saturated hydrocarbons. As the molecule of $N_2$ is not involved in any of the above-indicated equilibriums, the result-equilibrium partial pressures are correct. In this case, the system has four components (C–H–O–N), the number of degrees of freedom is four, that is to say, it is necessary to define four intensive variables to reach a complete knowledge of the system (total pressure, temperature, composition of CO and $N_2$ in the gas). Considering an atmosphere with $P_T$ (1 atmosphere) and with a nitrogen partial pressure $P_{N_2}$ of 0.20 atm, which has influence in the known "equilibrium point" (volume composition of the gases equilibrated or in equilibrium). For example, for the first of the four studied cases at 950 °C and with a partial pressure of CO in the gas of 0.3200 atm, a nitrogen partial pressure of 0.2000 atm and a total pressure of 1.0000 atm lead to Table 4.18.

As it was expressed in the table of results, the partial pressure of oxygen and the value of its composition in the gas are considered equal to zero. To the same practical effects, we also consider equal to zero the corresponding moles of oxygen involved in the balances of matter that are considered in the generation of the endothermic or exothermic atmospheres (Sect. 4.5.2). The value of the oxygen partial pressure that is indicated in the tables of the equilibrated gas is identified with the value that was known in the text as: Potential Neutral of Oxygen (PNO), Potential of Oxygen of the System (POS), or Partial Pressure of Oxygen in Equilibrium (POE), $P_{O_2}^=$.

In the equilibrated atmosphere that is considered in the exercise: 950 °C, 1 atm of pressure and with CO and $N_2$ contents of 32.00 and 20.00 vol. %, respectively, it is possible to estimate, for a certain quantity of air (number of air moles), the quantity of equilibrated gas that could be obtained (number of moles) and the corresponding quantity of saturated hydrocarbon that was used considering the number of carbon moles in the equilibrated gas. In this case, if we start from 10 mol of air:

$$n_{N_2} = n_{air} \cdot 0.79 = 7.90 \text{ moles} \tag{4.259}$$

The total quantity of gas moles is:

$$n_T = n_{N_2} \cdot \frac{100}{\text{vol.\% } N_2} = 7.90 \cdot \frac{100}{20} = 39.50 \text{ moles} \tag{4.260}$$

**Table 4.17** Relation of partial pressures and partial pressures of hydrogen and methane

| $P(CO)/P(CO_2)$ | 340.06 | 272.04 | 226.70 | 194.32 |
|---|---|---|---|---|
| $P(H_2)$-atm | 0.4731 | 0.3935 | 0.3138 | 0.2340 |
| $P(CH_4)$-atm | $3.39 \times 0^{-3}$ | $2.35 \times 10^{-3}$ | $1.49 \times 10^{-3}$ | $8.30 \times 10^{-4}$ |

**Table 4.18** Situation when the total pressure is 1 atm and the nitrogen partial pressure is 0.20 atm

| P(CO)-atm | P(CO$_2$)-atm | P(O$_2$)-atm | P(H$_2$)-atm | P(H$_2$O)-atm | P(N$_2$)-atm | P(CH$_4$)-atm |
|-----------|---------------|--------------|--------------|---------------|--------------|---------------|
| 0.3200 | 9.41·10$^{-4}$ | 1.43·10$^{-20}$ | 0.4731 | 2.57·10$^{-3}$ | 0.2000 | 3.39·10$^{-3}$ |
| %CO-vol | %CO$_2$-vol | %O$_2$-vol | %H$_2$-vol | %H$_2$O-vol | %N2-vol | %CH$_4$-vol |
| 32.00 | 0.0941 | 0.00 | 47.31 | 0.2570 | 20.00 | 0.3390 |

The number of carbon moles is:

$$
\begin{aligned}
n_C = n_T \cdot \frac{\text{vol.\% CO}}{100} + n_T \cdot \frac{\text{vol.\% CO}_2}{100} + n_T \cdot \frac{\text{vol.\% CH}_4}{100} \\
= 39.50 \cdot \frac{32.00}{100} + 39.50 \cdot \frac{0.0941}{100} + 39.50 \cdot \frac{0.3390}{100} \\
= 12.81 \text{ moles}
\end{aligned}
\tag{4.261}
$$

*Solution ends.*

**Exercise 4.8** The volume composition of a gas at the temperature of 950 °C and at the total pressure of one atmosphere is: 50% N$_2$, 20% CO, 20% CO$_2$, and 10% H$_2$. Calculate what would be the composition in equilibrium under these same conditions of pressure and temperature (exothermic atmosphere).

*Solution starts*

We start in this exercise from a dry gas, exempt of moisture, produced by partial combustion, in defect of oxygen to not transform all the carbon into CO$_2$ (exothermal atmospheres). The obtained volume composition, for the active compounds of the gas: 20% CO, 20% CO$_2$, and 10% H$_2$ can be more or less far from the equilibrium point. In the case of the CO/CO$_2$ system, the quantity of CO could be increased at the expense of that of CO$_2$, while in the H$_2$/H$_2$O system, the proportion of hydrogen would decrease with the appearance of moisture.

It is necessary to find four independent equations to reach the composition of the four active species in equilibrium: CO, CO$_2$, H$_2$, and H$_2$O. The first one can be obtained from the equilibrium between the four compounds:

$$
CO(g) + H_2O(g) \leftrightarrow H_2(g) + CO_2(g)
\tag{4.262}
$$

which is obtained by combination of:

$$
H_2(g) + \frac{1}{2}O_2(g) \leftrightarrow H_2O(g)
\tag{4.263}
$$

$$
\Delta_r G^\circ_{\text{reaction 4.263}}(T) = -249.0 + 0.057 \cdot T(K) kJ \cdot (\text{mol } H_2)^{-1}
\tag{4.264}
$$

$$
C(s) + \frac{1}{2}O_2(g) \leftrightarrow CO(g)
\tag{4.265}
$$

$$\Delta_r G^o_{\text{reaction 4.265}}(T) = -118.0 - 0.084 \cdot T(K)kJ \cdot (0.5 \text{mol } O_2)^{-1} \qquad (4.266)$$

$$C(s) + O_2(g) \leftrightarrow CO_2(g) \qquad (4.267)$$

$$\Delta_r G^o_{\text{reaction 4.267}}(T) = -395.0 + 0.001 \cdot T(K)kJ \cdot (\text{mol } O_2)^{-1} \qquad (4.268)$$

It is possible to obtain the standard free energy associated to the reaction (4.262) from the data of the Appendix 1 (Appendices) and the corresponding value of the equilibrium constant at 950 °C:

$$\Delta_r G^o_{\text{reaction 4.262}}(T) = - [-249.0 + 0.057 \cdot T(K)] + [-395.0 + 0.001 \cdot T(K)]$$
$$- [-118.0 - 0.084 \cdot T(K)]$$
$$= -28.0 + 0.028 \cdot T(K)kJ \cdot (\text{mol } H_2)^{-1} \qquad (4.269)$$

$$\Delta_r G^o_{\text{reaction 4.262}}(1223 \text{ K}) = 6244 \text{ kJ} \cdot (\text{mol } H_2)^{-1} \qquad (4.270)$$

$$k_{eq}(\text{reaction 4.262}) = \exp\left(-\frac{\Delta_r G^o_{\text{reaction 4.262}}(1223 \text{ K})}{R \cdot T}\right)$$
$$\rightarrow k_{eq}(\text{reaction 4.262}) = 0.5412 = \frac{P_{CO_2} \cdot P_{H_2}}{P_{CO} \cdot P_{H_2O}} \qquad (4.271)$$

The second equation is the result of the addition of the partial pressures of the active compounds in the gas ($P_{N_2}$ accounts for the remaining 0.5 atm):

$$P_{CO_2} + P_{CO} + P_{H_2} + P_{H_2O} = 0.5 \text{ atm} \qquad (4.272)$$

We obtain the other two equations that would complete the system using the atomic mass balance (atoms-gram) of carbon, $n_C$, oxygen, $n_O$, and hydrogen, $n_H$, in the gas. The number of carbon, oxygen, and hydrogen atoms, whether there is not any entrance or exit of matter from the system, is constant. The number of moles of the active species of the system experiences a variation in the number of moles: $n_{CO_2}$, $n_{CO}$, $n_{H_2}$ and $n_{H_2O}$. That is to say:

$$n_C = n_{CO_2} + n_{CO} \qquad (4.273)$$

$$n_H = 2 \cdot n_{H_2} + 2 \cdot n_{H_2O} \qquad (4.274)$$

$$n_O = n_{CO} + 2 \cdot n_{CO_2} + n_{H_2O} \qquad (4.275)$$

On the other hand, if we know the data of the initial composition of the gas indicated in the enunciation (20% CO, 20% $CO_2$, and 10% $H_2$), it is possible to stablish the following relations:

$$n_C = n_T \cdot \frac{\text{vol.}\% \ CO}{100} + n_T \cdot \frac{\text{vol.} \ CO_2}{100} = n_T \cdot (0.20 + 0.20)$$

$$= 0.4 \cdot n_T \xrightarrow{n_{H_2}=0.1\cdot n_T} n_C = 0.4 \cdot \frac{n_{H_2}}{0.1} = 4 \cdot n_{H_2} \qquad (4.276)$$

$$n_O = n_T \cdot \frac{\text{vol.}\% \ CO}{100} + 2 \cdot n_T \cdot \frac{\text{vol.}\% \ CO_2}{100} = n_T \cdot (0.20 + 0.40)$$

$$= 0.6 \cdot n_T \xrightarrow{n_{H_2}^{\rightarrow}=0.1\cdot n_T} n_O = 0.6 \cdot \frac{n_{H_2}}{0.1} = 6 \cdot n_{H_2} \qquad (4.277)$$

Consequently, the two independent equations that should be provided would be (we should consider that as the process progresses towards the equilibrium $n_{H_2}$ transforms into $n_{H_2} + n_{H_2O}$):

$$n_C = n_{CO_2} + n_{CO} \xrightarrow{n_C=4\cdot n_{H_2}} 4 \cdot n_{H_2} = n_{CO_2} + n_{CO} \to n_{CO} + n_{CO_2}$$

$$= 4 \cdot \left(n_{H_2} + n_{H_2O}\right) \qquad (4.278)$$

$$n_O = n_{CO} + 2 \cdot n_{CO_2} + n_{H_2O} \xrightarrow{n_O=6\cdot n_{H_2}} 6 \cdot n_{H_2} = n_{CO} + 2 \cdot n_{CO_2} + n_{H_2O}$$

$$\to n_{CO} + 2 \cdot n_{CO_2} + n_{H_2O} = 6 \cdot \left(n_{H_2} + n_{H_2O}\right) \qquad (4.279)$$

If we know, due to the general theory of the perfect gases, that the partial pressure of a gas is proportional to the number of moles, the two previous equations are transformed into:

$$P_{CO} + P_{CO_2} = 4 \cdot \left(P_{H_2} + P_{H_2O}\right) \qquad (4.280)$$

$$P_{CO} + 2 \cdot P_{CO_2} + P_{H_2O} = 6 \cdot \left(P_{H_2} + P_{H_2O}\right) \qquad (4.281)$$

Considering Eq. (4.272) we transform Eq. (4.280) into:

$$P_{CO} + P_{CO_2} = 4 \cdot \left(P_{H_2} + P_{H_2O}\right) \xrightarrow{P_{CO_2}+P_{CO}=0.5-\left(P_{H_2}+P_{H_2O}\right)} 0.5 - \left(P_{H_2} + P_{H_2O}\right)$$

$$= 4 \cdot \left(P_{H_2} + P_{H_2O}\right) \to 0.5 = 5 \cdot P_{H_2} + 5 \cdot P_{H_2O} \to P_{H_2}$$

$$= 0.10 - P_{H_2O} \qquad (4.282)$$

Using Eqs. (4.272), (4.281) and (4.282), we obtain:

$$P_{CO} + 2 \cdot P_{CO_2} + P_{H_2O} = 6 \cdot \left(P_{H_2} + P_{H_2O}\right) \to P_{CO} + 2 \cdot P_{CO_2} - 6 \cdot P_{H_2} - 5 \cdot P_{H_2O}$$

$$= 0 \xrightarrow{P_{CO}=0.5-\left(P_{H_2}+P_{H_2O}+P_{CO_2}\right)} 0.5 - \left(P_{H_2} + P_{H_2O} + P_{CO_2}\right)$$

$$+ 2 \cdot P_{CO_2} - 6 \cdot P_{H_2} - 5 \cdot P_{H_2O}$$

$$= 0 \rightarrow P_{CO_2} - 6 \cdot P_{H_2O} - 7 \cdot P_{H_2}$$

$$= -0.5 \xrightarrow{P_2=0.10-P_{H_2O}} P_{CO_2} - 6 \cdot P_{H_2O} - 7 \cdot \left(0.10 - P_{H_2O}\right)$$

$$= -0.5 \rightarrow P_{CO_2} = 0.2 - P_{H_2O} \tag{4.283}$$

Finally, we obtain Eq. (4.282) with the Eqs. (4.283) and (4.272):

$$P_{CO_2} + P_{CO} + P_{H_2} + P_{H_2O} = 0.5 \xrightarrow{P_{H_2}=0.10-P_{H_2O}} P_{CO_2} + P_{CO} + 0.10 - P_{H_2O} + P_{H_2O}$$

$$= 0.5 \xrightarrow{P_{CO_2}=0.2-P_{H_2O}} 0.2 - P_{H_2O} + P_{CO} = 0.4 \rightarrow P_{CO}$$

$$= 0.2 + P_{H_2O} \tag{4.284}$$

We obtain the partial pressures of the active compounds of the gas as a function of the water partial pressure with the Eqs. (4.282)–(4.284). We replace their value in the Eq. (4.271):

$$k_{eq}(\text{reaction } 4.262) = 0.5412 = \frac{P_{CO_2} \cdot P_{H_2}}{P_{CO} \cdot P_{H_2O}} \xrightarrow[\substack{P_{CO_2}=0.2-P_{H_2O} \\ P_{CO}=0.2+P_{H_2O}}]{P_{H_2}=0.10-P_{H_2O}} 0.5412$$

$$= \frac{\left(0.2 - P_{H_2O}\right) \cdot \left(0.10 - P_{H_2O}\right)}{\left(0.2 + P_{H_2O}\right) \cdot P_{H_2O}} \tag{4.285}$$

We obtain a quadratic equation if we develop the previous Eq. (4.285):

$$0.5412 \cdot (0.2 + P_{H_2O}) \cdot P_{H_2O} = \left(0.2 - P_{H_2O}\right) \cdot \left(0.10 - P_{H_2O}\right)$$

$$\rightarrow 0.1082 \cdot P_{H_2O} + 0.5412 \cdot P_{H_2O}^2$$

$$= 0.02 - 0.2 \cdot P_{H_2O} - 0.1 \cdot P_{H_2O} + P_{H_2O}^2 \rightarrow 0$$

$$= -0.1082 \cdot P_{H_2O} - 0.5412 \cdot P_{H_2O}^2 + 0.02 - 0.2 \cdot P_{H_2O}$$

$$- 0.1 \cdot P_{H_2O} + P_{H_2O}^2 \rightarrow 0$$

$$= 0.4588 \cdot P_{H_2O}^2 - 0.4082 \cdot P_{H_2O} + 0.02 \tag{4.286}$$

If we solve the Eq. (4.286), we obtain the partial pressure $P_{H_2O}$ and volume composition and using the other equations we obtain the values for the other compounds of the gas. Table 4.19 collects the results.

Commentaries: We show a new method of calculation in this exercise by means of which we indicate how a system that is far from the equilibrium can move towards the thermodynamic equilibrium.

On the other hand, as in the exercise of endothermic atmospheres, it is possible to estimate for a certain quantity of air (number of moles of air), the quantity of equilibrated gases obtained. In this case, if we consider 20 mol of air:

$$n_{N_2} = n_{air} \cdot 0.79 = 15.80 \text{ moles} \tag{4.287}$$

The total quantity of gas moles is:

$$n_T = n_{N_2} \cdot \frac{100}{\text{vol.}\% \, N_2} = 15.80 \cdot \frac{100}{50} = 31.60 \text{ moles} \tag{4.288}$$

The number of carbon moles is:

$$n_C = n_T \cdot \frac{\text{vol.}\% \, CO}{100} + n_T \cdot \frac{\text{vol.}\% \, CO_2}{100} + n_T \cdot \frac{\text{vol.}\% \, CH_4}{100}$$
$$= 31.60 \cdot \frac{25.20}{100} + 31.60 \cdot \frac{14.80}{100} + 31.60 \cdot \frac{0.00}{100}$$
$$= 12.64 \text{ moles} \tag{4.289}$$

In this case, to reach a number of carbon moles of 12.640 in the equilibrated atmosphere similar to that reached by the endothermic atmosphere (12.811 mol C), we would need 20 mol of air and the quantity of gases produced, 31.60 mol, does not reach the 39.50 mol of the endothermic atmosphere.

On the other hand, the carbon activity in the gas, $a_C^{gas}$, can be calculated by means of the Boudouard equilibrium (gasification of the carbon by the $CO_2$):

$$CO_2(g) + C(s) \leftrightarrow 2CO(g) \tag{4.290}$$

$$\Delta_r G^o_{\text{reaction } 4.290}(T) = 2 \cdot [-118.0 - 0.084 \cdot T(K)] - [-395.0 + 0.001 \cdot T(K)] kJ$$
$$\cdot \left( \text{mol C}^{-1} = 159 - 0.169 \cdot T(K) kJ \cdot (molC)^{-1} \right.$$
$$\rightarrow k_{eq}(\text{reaction } 4.290) = \exp\left(-\frac{\Delta_r G^o_{\text{reaction } 4.290}(1223K)}{R \cdot T}\right)$$
$$= 109 \rightarrow k_{eq}(\text{reaction } 4.290) = \frac{P^2_{CO}}{a_C^{gas} \cdot P_{CO_2}} \tag{4.291}$$

obtained from Eq. (4.291):

$$a_C^{gas} = \frac{P^2_{CO}}{P_{CO_2}} \cdot \frac{1}{k_{eq}} = \frac{0.252^2}{0.148} \cdot \frac{1}{109} = 3.94 \cdot 10^{-3} \tag{4.292}$$

**Table 4.19** Volume composition of the gas

| $P_{H_2O}$(atm) | $P_{CO_2}$ (atm) | $P_{H_2}$ (atm) | $P_{CO}$ (atm) | $P_{O_2}$ (atm) | $P_{N_2}$ (atm) |
|---|---|---|---|---|---|
| 0.0520 | 0.1480 | 0.0480 | 0.2520 | $5.69 \cdot 10^{-16}$ | 0.5000 |
| vol. % $H_2O$ | vol. % $CO_2$ | vol. % $H_2$ | vol. % CO | vol. % $O_2$ | vol. % $N_2$ |
| 5.20 | 14.80 | 4.80 | 25.20 | 0.00 | 50.00 |

Using the correlation developed in this chapter to estimate the activity of the carbon in the Fe-$\gamma$ as a function of the carbon molar fraction in the austenite and the temperature, the dissolution of the carbon corresponding to an activity, $a_C^{gas}$, of $3.94 \cdot 10^{-3}$ is (950 °C):

$$\ln a_C = \frac{3581}{T} + \ln x_C + 1.03 \cdot x_C \cdot \exp\left(\frac{3436}{T}\right) - 1.0996$$

$$\rightarrow \ln x_C + 1.03 \cdot x_C \cdot \exp\left(\frac{3436}{T}\right) = 1.0996 + \ln a_C - \frac{3581}{T}$$

$$\rightarrow x_C = 0.00060425 \tag{4.293}$$

$$x_C = \frac{\frac{\% C}{12.01}}{\frac{\% C}{12.01} + \frac{100 - \% C}{55.85}} = 0.00060425 \rightarrow 0C = 0.013\% \, C \tag{4.294}$$

That is to say, the exothermic atmosphere of the exercise, weakly deoxidizing, is decarburizing even for certain qualities of ferritic steels.

*Solution ends.*


## 4.5.3  Atmospheres Obtained from NH₃

Endothermic atmospheres are obtained by dissociation of the $NH_3$ with heat supply and in presence of a catalyzer. Some have the objective of achieving the nitriding of steels. Others have simply the aim of obtaining pure atmospheres of hydrogen and nitrogen for applications where it is not convenient the presence of carbon in the gases.

Obviously, the atmospheres considered in the Sect. 4.5.1 are not adequate to nitride because nitriding requires the supply of nitrogen in nascent state to the surface of the steel. It is not sufficient with the contact of the steel with a current of gaseous nitrogen.

Nascent nitrogen is obtained by the dissociation of a current of anhydrous $NH_3$ that is passed over the load of steel parts at the temperature of treatment, close to 500 °C (and always <590 °C to avoid the formation of nitroaustenite, which is the eutectoid of the Fe–N system). The iron ($Fe_4N$) at this temperature is a catalyzer for the dissociation of the ammonia:

$$NH_3(g) \rightarrow \frac{3}{2}H_2(g) + \frac{1}{2}N_2(g) \tag{4.295}$$

$$\Delta_r G^0(T) = 49.3 - 0.109 \cdot T(K) kJ \cdot (mol\ NH)^{-1} \tag{4.296}$$

It is necessary to control the quantity, m, of ammonia supplied with the purpose of avoiding the excessive quantity of $H_2$ obtained in the dissociation because the steel could be decarburized according to the following reaction:

$$2H_2(g) + C(s) \rightarrow CH_4(g) \uparrow \qquad (4.297)$$

It is convenient in the practice that the value $\alpha$ of the ammonia dissociation was between 15 and 30%.

The value of $\alpha$ is calculated from the ammonia flow (number of moles, m, of $NH_3$ that circulates through the furnace at this temperature $T_1$ of the heat treatment) and the equilibrium constant, $k_{eq}$. It is calculated as follows:

$$\frac{\left(\frac{3}{2} \cdot m \cdot \alpha\right)^3 \cdot \left(\frac{m}{2 \cdot \alpha}\right)}{[m \cdot (1 - \alpha)]^3} = \frac{1.5 \cdot m^2 \cdot \alpha}{(1 - \alpha)^2} = k_{eq}(\text{reaction 4.298, for 500 °C}) \qquad (4.298)$$

which also allows, once determined the value of $\alpha$, to know the percentage of residual $NH_3$, as:

$$2NH_3(g) \rightarrow 3H_2(g) + N_2(g) \qquad (4.299)$$

$$m \cdot (1 - \alpha) = \frac{3}{2} \cdot m \cdot \alpha + \frac{m}{2} \cdot \alpha \qquad (4.300)$$

It is also possible to obtain by thermal dissociation atmospheres whose purpose is not to nitride but simply obtain nitrogen and hydrogen. In this way, liquid $NH_3$, vaporized by heating, is passed through a catalyzer. However, iron is not used as catalyzer (because it swells when nitrides and this would little by little seal the catalysis chamber). The catalysis, in retort exothermically heated, is not carried at 500 °C but at 900 °C. Dissociated gas is next cooled in a hydro coolant to remove the $NH_3$. In fact, the $NH_3$ content is small: the residual content in $NH_3$ in the dissociation at 900 °C does not exceed the 0.02 vol. % (while in the catalysis at 500 °C, the volume of residual $NH_3$ is around 0.2%).

The catalytic decomposition of the $NH_3$ at 900 °C provides a gaseous mixture whose composition is almost 25% $N_2$ and 75% $H_2$. The consumption of $NH_3$ is around 0.355 kg per $m^3$ of produced gases. The purity of the gaseous mixture depends on the quality of the $NH_3$. It is also necessary for certain applications to depurate the gas by absorption in molecular sieves before putting them inside of the furnace.

This type of atmosphere is used, due to its reductant behavior, in extractive metallurgy. It is also used in heat treatments, both of very oxidable metals as in stainless steels with the aim of avoiding the formation of oxides.

In the case of endothermic atmospheres of $NH_3$ indicated in the Sect. 4.5.2, the atmosphere was obtained by dissociation of ammonia by means of heat supply. In the case of the exothermic atmospheres, it is simply burning $NH_3$ with air in presence of a catalyzer to speed up the combustion. The installation has a dryer of the combustion products to remove all the vapor of $H_2O$ and a purifier of the obtained gases (to remove all the possible traces of oxygen, habitually <2%).

The atmosphere obtained in this manner comes from the reaction:

$$NH_3 + N_{air} \cdot (0.21 \cdot O_2 + 0.79 \cdot N_2) \rightarrow \delta \cdot H_2O + \varepsilon \cdot H_2 + \eta \cdot N_2 \qquad (4.301)$$

if we consider as reference the dissociation of 1 mol of $NH_3$, reaction (4.295), and we obtain from the balance to the hydrogen, oxygen, and nitrogen the next:

$$\varepsilon = 1.5 - 0.42 \cdot N_{air} \qquad (4.302)$$

and

$$\eta = 0.5 + 0.79 \cdot N_{air} \qquad (4.303)$$

Therefore, the percentage of dry hydrogen—which is $100 \cdot \varepsilon/(\varepsilon + \eta)$—is equal to:

$$\%\text{dry hydrogen} = 100 \cdot \frac{1.5 - 0.42 \cdot N_{air}}{2 + 0.37 \cdot N_{air}} \qquad (4.304)$$

which is independent of the temperature whether the combustion is complete. For the common values of $N_{air}$, within 2 and 3, the dry gas would have a percentage of hydrogen of around 25 to 8% in the mixture comprised by this gas and the nitrogen.

**Exercise 4.9** Calculate the partial pressure and the volume composition of the $NH_3$ in equilibrium with the compounds of a gas formed by $N_2$, $H_2$, $H_2O$, $O_2$, and $NH_3$ at the temperatures of 500 and 1000 °C when the total pressure and the volume composition of the $H_2$ and $H_2O$ were 1 atm, 20.00% and 0.20%, respectively.

*Solution starts*

The dissociation of the ammonia, $NH_3$, data in Appendix 1 (Appendices):

$$NH_3(g) \leftrightarrow \frac{3}{2}H_2(g) + \frac{1}{2}N_2(g) \qquad (4.305)$$

$$\Delta_r G^o_{reaction\ 4.305}(T) = 49.3 - 0.109 \cdot T(K)kJ \cdot (mol\ NH_3)^{-1}$$

$$\rightarrow k_{NH_3/N_2}(reaction\ 4.305) = \exp\left(-\frac{\Delta_r G^o_{reaction\ 4.305}(T)}{R \cdot T}\right)$$

$$\rightarrow k_{NH_3/N_2}(reaction\ 4.305) = \frac{P_{H_2}^{3/2} \cdot P_{N_2}^{1/2}}{P_{NH_3}} \qquad (4.306)$$

This reaction is thermodynamically favorable:

– at 500 °C:

$$\Delta_r G^o_{\text{reaction 4.305}}(773K) = 49.3 - 0.109 \cdot 773$$
$$= -34.957 kJ \cdot (\text{mol } NH_3)^{-1} \qquad (4.307)$$

– at 1000 °C:

$$\Delta_r G^o_{\text{reaction 4.305}}(1273K) = 49.3 - 0.109 \cdot 1273$$
$$= -89.457 kJ \cdot (\text{mol } NH)^{-1} \qquad (4.308)$$

while the equilibrium constants at the above-indicated temperatures are:
– at 500 °C:

$$k_{NH_3/N_2}(\text{reaction 4.305}) = \exp\left[-\frac{(-34.957 \text{ kJ} \cdot (\text{mol } NH_3)^{-1}) \cdot 1000 \text{ J} \cdot kJ^{-1}}{8.314 \text{ J} \cdot mol^{-1} \cdot K^{-1} \cdot 773 \text{ K}}\right]$$
$$= 230.2 \qquad (4.309)$$

– at 1000 °C:

$$k_{NH_3/N_2}(\text{reaction 4.305}) = \exp\left[-\frac{(-89.457 \text{ kJ} \cdot (\text{mol } NH_3)^{-1}) \cdot 1000 \text{ J} \cdot kJ^{-1}}{8.314 \text{ J} \cdot mol^{-1} \cdot K^{-1} \cdot 1273 \text{ K}}\right]$$
$$= 4684.0 \qquad (4.310)$$

On the other hand, the total pressure of the system, $P_T$, is equal to:

$$P_T = P_{N_2} + P_{H_2} + P_{H_2O} + P_{O_2} + P_{NH_3} \qquad (4.311)$$

The hydrogen and water vapor partial pressures can be determined from the following equilibrium with the data of the Appendix 1:

$$H_2(g) + \frac{1}{2}O_2(g) \leftrightarrow H_2O(g) \qquad (4.312)$$

$$\Delta_r G^o_{\text{reaction 4.312}}(T) = -249.0 + 0.057 \cdot T(K) kJ \cdot (\text{mol } H_2)^{-1}$$
$$\rightarrow k_{H_2/H_2O}(\text{reaction 4.312}) = \exp\left(-\frac{\Delta_r G^o_{\text{reaction 4.312}}(T)}{R \cdot T}\right)$$
$$\rightarrow k_{H_2/H_2O}(\text{reaction 4.312}) = \frac{P_{H_2O}}{P_{H_2} \cdot \sqrt{P_{O_2}}} \rightarrow \sqrt{P_{O_2}}$$
$$= \frac{P_{H_2O}}{P_{H_2} \cdot k_{H_2/H_2O}(\text{reaction 4.312})} \rightarrow P_{O_2}$$
$$= \left[\frac{P_{H_2O}}{P_{H_2} \cdot k_{H_2/H_2O}(\text{reaction 4.312})}\right]^2 \qquad (4.313)$$

At 500 °C

$k_{H_2/H_2O}$(reaction 4.312)

$$= \exp\left[-\frac{\left(-249.0 + 0.057 \cdot 773 \text{ kJ} \cdot (\text{mol H}_2)^{-1}\right) \cdot 1000 \text{ J} \cdot \text{kJ}^{-1}}{8314 \text{ J} \cdot \text{mol} - 1 \cdot \text{K}^{-1} \cdot 773 \text{ K}}\right]$$

$$= 7.06 \cdot 10^{13} \tag{4.314}$$

$$P_{O_2} = \left[\frac{P_{H_2O}}{P_{H_2} \cdot k_{H_2/H_2O}(\text{reaction 4.312})}\right] = \left[\frac{0.002}{0.2 \cdot 7.06 \cdot 10^{13}}\right]^2$$

$$= 2.01 \cdot 10^{-32} \text{atm} \tag{4.315}$$

At 1000 °C

$k_{H_2/H_2O}$(reaction 4.312)

$$= \exp\left[-\frac{\left(-249.0 + 0.057 \cdot 1273 \text{ kJ} \cdot (\text{mol H}_2)^{-1}\right) \cdot 1000 \text{ J} \cdot \text{kJ}^{-1}}{8.314 \text{ J} \cdot \text{mol}^{-1} \cdot \text{K}^{-1} \cdot 1273 \text{ K}}\right]$$

$$= 1.74 \cdot 10^7 \tag{4.316}$$

$$P_{O_2} = \left[\frac{P_{H_2O}}{P_{H_2} \cdot k_{H_2/H_2O}(\text{reaction 4.312})}\right]^2$$

$$= \left[\frac{0.002}{0.2 \cdot 1.74 \cdot 10^7}\right]^2 = 3.31 \cdot 10^{-19} \text{atm} \tag{4.317}$$

We calculate the $NH_3$ partial pressure in the gas considering the equilibrium of the ammonia dissociation:

$$k_{NH_3/N_2}(\text{reaction 4.312}) = \frac{P_{H_2}^{3/2} \cdot P_{N_2}^{1/2}}{P_{NH_3}} \rightarrow P_{N_2}^{1/2}$$

$$= \frac{k_{NH_3/N_2}(\text{reaction 4.312}) \cdot P_{NH_3}}{P_{H_2}^{3/2}} \rightarrow P_{N_2}$$

$$= \left[\frac{k_{NI_3/N_2}(\text{reaction 4.312}) \cdot P_{NII_3}}{P_{H_2}^{3/2}}\right]^2 \tag{4.318}$$

where the nitrogen partial pressure is:

$$P_{N_2} = P_T - P_{H_2} + P_{H_2O} + P_{O_2} + P_{NH_3} \tag{4.319}$$

Replacing Eq. (4.318) in Eq. (4.319), we have:

$$\left[\frac{k_{NH_3/N_2}(\text{reaction } 4.312) \cdot P_{NH_3}}{P_{H_2}^{3/2}}\right]^2 = P_T - \left(P_{H_2} + P_{H_2O} + P_{O_2} + P_{NH_3}\right)$$

$$\rightarrow \ k_{NH_3/N_2}(\text{reaction } 4.312)^2 \cdot P_{NH_3}^2$$

$$= P_T \cdot P_{H_2}^3 - \left(P_{H_2} + P_{H_2O} + P_{O_2}\right) \cdot P_{H_2}^3 - P_{NH_3} \cdot P_{H_2}^3 \rightarrow 0$$

$$= k_{NH_3/N_2}(\text{reaction } 4.312)^2 \cdot P_{NH_3}^2 + P_{NH_3} \cdot P_{H_2}^3$$

$$+ \left(P_{H_2} + P_{H_2O} + P_{O_2} - P_T\right) \cdot P_{H_2}^3 \qquad (4.320)$$

We solve for $P_{NH_3}$. The values of the other variables are known and, therefore, we collect the values of the partial pressures in Table 4.20 for the temperature of 500 °C and in Table 4.21 for the temperature of 1000 °C (for a total pressure of one atmosphere).

Commentaries: Atmospheres of ammonia-dry gas are usually employed in the batch and continuous annealing heat treatments with the aim of reaching a shiny finishing of the sheet free of superficial carbon.

On the other hand, nitriding atmospheres to obtain superficial hard layers usually contain ammonia-gas at temperatures of around 500 °C (see Pero-Sanz 2004, pp. 398–401).

*Solution ends.*


### 4.5.3.1  Nickel as Catalyzer of Endothermic Atmosphere

The obtaining of equilibrated atmospheres from industrial gases or saturated hydro-carbons requires to reach high concentrations of carbon monoxide in the gas, CO, which produces that, when cooling the gases of the flame at temperatures of around

**Table 4.20**  Composition of the gas for the temperature of 500 °C

| P(NH$_3$)-atm | P(H$_2$)-atm | P(N$_2$)-atm | P(O$_2$)-atm | P(H$_2$O)-atm |
|---|---|---|---|---|
| $3.47 \cdot 10^{-4}$ | 0.2000 | 0.7977 | 0.0000 | 0.0020 |
| %NH$_3$-vol | %H$_2$-vol | %N$_2$-vol | %O$_2$-vol | %H$_2$O-vol |
| 0.03 | 20.00 | 79.77 | 0.00 | 0.20 |

**Table 4.21**  Composition of the gas for the temperature of 1000 °C

| P(NH$_3$)-atm | P(H$_2$)-atm | P(N$_2$)-atm | P(O$_2$)-atm | P(H$_2$O)-atm |
|---|---|---|---|---|
| $1.71 \cdot 10^{-5}$ | 0.2000 | 0.7980 | 0.0000 | 0.0020 |
| %NH$_3$-vol | %H$_2$-vol | %N$_2$-vol | %O$_2$-vol | %H$_2$O-vol |
| 0.00 | 20.00 | 79.80 | 0.00 | 0.20 |

1000 °C, carbon could precipitate (Boudouard equilibrium). The rate of the gasification processes of the carbon precipitated either by the carbon dioxide or by the water vapor can be accelerated/promoted by the presence of metal, concretely the nickel. The Mond process exists in the nickel metallurgy, which is used as a method for the purification of the metal through the formation-decomposition of the nickel carbonyl, $Ni(CO)_4$ (Sancho et al. 2000, p. 307).

It also seems that in all the processes of obtaining graphene, nanotubes, or graphene oxide, metals are used to achieve the nucleation-development of these non-equilibrium carbon structures that, recently, are gaining a special attention by the international scientific community.

### 4.5.4 The Future of the Industrial Atmospheres

As it was already mentioned in previous parts of the Sect. 4.5, the application of the industrial atmospheres, consequence of their properties, goes beyond the iron and steelmaking sector and have their application in the metallurgical, chemical, food, and pharmaceutical industries. However, the word that appears in specialized conferences or in the Research and Development programs of the public or private agencies of research promotion is the decarbonization. In the eyes of the authors, the interpretation of the word can give origin to the following alternatives:

(A) Promoting and funding all the research that could lead to zero emissions of $CO_2$ to the atmosphere.

(B) Increasing the efficiency of the current production processes and researching some other new process that were able to control—or not increase—the emissions of $CO_2$.

We should think that a drastic reduction of the levels of $CO_2$ in the atmosphere could have influence in the "carbon cycle" or "$CO_2$ cycle" and, consequently, in the production of food. We think that, even when it is important to control the $CO_2$ emissions, it could be more important to control-eliminate other molecules that, as a consequence of the generation-production of industrial gases must be subjected to a greater and more strict control: suspended particle matter, $SO_x$, $NO_x$, or dioxins (Fernández-González et al. 2017a, b, c, d, 2018a).

So far, the practice followed in the utilization of the industrial gases is that, once they have accomplished their purpose, they are incorporated-driven to the traditional systems of production by combustion of electric power. Obviously, this happens without the previous removal of those elements that result potentially dangerous for the environment. The two alternatives that could be considered in the future are:

A. Incorporating the industrial gases to the conventional processes of production of electric power by combustion, always and when the quantity of molecules-compounds that are deleterious for the environment was reduced.

B. Recycling the industrial gases to the processes reducing before the quantities of $CO_2$ and $H_2O$ necessaries to maintain the suitable properties of the gas.

In both alternatives, the weight of the thermal energy-energy costs of the process is considered essential. Under these circumstances, going deeper in the application of the concentrated solar energy to heat-cold of the industrial gases could be important (Fernández-González et al. 2018b, c, d, 2019a, b, c, 2021; Fernández-González 2019).

It is possible that the future tendency of the industrial atmosphere could lead to an increase of the $H_2$ content and a decrease of the CO content. The decarbonization of the gases will not be only associated to the removal of $CO_2$ but also of CO. The elimination of the CO in the gases can be carried out in different manners:

(A)    By its conversion into $CO_2$ and $H_2$ by oxidation with water vapor:

$$CO(g) + H_2O(g) \leftrightarrow H_2(g) + CO_2(g) \qquad (4.321)$$

(B)    By cooling-thermal destabilization of the CO-rich gas (Boudouard equilibrium):

$$2CO(g) \rightarrow C(s) + CO_2(g) \qquad (4.322)$$

The depuration-scrubbing of the $CO_2$ in gases can be carried out by wet route or dry route. In the wet route, the gas is bubbled up in an aqueous solution of monoethanolamide, which retains the carbon dioxide as bicarbonate. Then, the solution of bicarbonate amine is regenerated by boiling and, the released $CO_2$ can be collected-stored-captured. The energy required to treat the solution of amine bicarbonate might be captured from the latent heats of the gases in the combustion processes or in the processes of heating inert gaseous fluids with concentrated solar energy.

The dry depuration scrubbing of gases is carried out by means of the equipment known as molecular sieves, which are formed by synthetic zeolites that are able to simultaneously absorb the carbon dioxide and the water vapor (zeolites are aluminum silicates that form "reticular structures" suitable for retaining molecules of $CO_2$ and $H_2O$). Consequently, either from the coke batteries gas or from reformed natural gas, it is possible to obtain a phase with high $H_2$ and CO contents. The CO removal (by transformation into $CO_2$ or into carbon), the capture of water vapor and the capture of carbon dioxide lead to a gas with high hydrogen content. This is the case of the Exercise 4.10 (Verdeja et al. 2020, pp. 125–127), where a gas with high hydrogen content (99% $H_2$ and 1% $CO_2$) is used to produce direct reduction iron (DRI).

**Exercise 4.10** Hydrogen with 1% $CO_2$ is used, 40,000 kg·h⁻¹, to reduce the hematite according to the following reaction:

$$Fe_2O_3(s) + 3H_2(g) \rightarrow 3H_2O(g) + 2Fe(s) \qquad (4.323)$$

In the circuit, the hydrogen is recycled and the release, P, of carbon dioxide, $CO_2$, is performed to avoid that its concentration was higher than 2.5 wt. % in the gas. If the relation of recirculated gas flow to the reduction reactor, R, with respect to the feeding of reductant gas, F, is 4/1, calculate the quantity and composition of the gases that is necessary to purge or remove from the circuit.

**Data**: Although it is assumed that the treated mineral is 100% hematite, the metallization yield by the gas is of 95%.

*Solution starts*

The current entrance of 100% hematite iron ore and the stoichiometry of the reduction reaction provide the composition of the solid charge that leaves the reduction with gas-DRI reactor, iron sponge and non-transformed hematite.

– Obtained sponge iron:

$$\frac{40000 \text{ kg FeO}_3}{h} \cdot \frac{2 \cdot 55.85 \text{ kg Fe}}{(2 \cdot 55.85 + 3 \cdot 16) \text{ kg Fe}_2O_3} \cdot 0.95 = 26579 \text{ kg Fe} \cdot h^{-1}$$

$$(4.324)$$

– Non-transformed hematite ore:

$$\frac{40000 \text{ kg Fe}_2O_3}{h} \cdot 0.05 = 2000 \text{ kg Fe}_2O_3 \cdot h^{-1} \qquad (4.325)$$

The requirements of hydrogen for the hematite reduction are calculated as follows:

$$\frac{40000 \text{ kg FeO}_3}{h} \cdot \frac{3 \cdot 2 \text{ kg H}_2}{(2 \cdot 55.85 + 3 \cdot 16) \text{ kg Fe}_2O_3} \cdot 0.95 = 1428 \text{ kg H}_2 \cdot h^{-1}$$

$$(4.326)$$

The feeding of gas supplied to the system, F, has a composition in mass of 99% $H_2$ and 1% $CO_2$. For that reason, the quantity of gas, F, is:

$$1428 \text{ kg H}_2 \cdot h^{-1} \cdot \frac{100 \text{ kg of reductant gas F}}{99 \text{ kg H}_2}$$
$$= 1442 \text{ kg of reductant gas F} \cdot h^{-1} \qquad (4.327)$$

The ratio indicated in the enunciation is used to calculate the current of gas recirculated, R, to the reduction reactor, where R/F is equal to four: the mass flow of R is of 5768 kg·h$^{-1}$ of gas.

With the above-calculated values, the calculation of the gas current that leaves the reduction reactor can be performed because the values of the three currents that enter-leave the DRI reactor are known. The mass flow of the current of gas, G-DRI, enriched with water vapor that leaves the furnace is:

$$G\text{-DRI} = \frac{40000 \text{ kg Fe}_2O_3}{h} + \frac{5768 \text{ kg ggas}}{h} - \frac{2000 \text{ kg Fe}_2O_3}{h} - \frac{26579 \text{ kg Fe}}{h}$$
$$= \frac{17189 \text{ kg}}{h} \qquad (4.328)$$

It is incorporated the quantity of water vapor consequence of the reduction process to this gas current of exit from the reactor, DRI:

$$\frac{26579 \text{ kg Fe}}{h} \cdot \frac{3 \cdot 18 \text{ kg } H_2O}{2 \cdot 55.85 \text{ kgFe}} = 12849 \text{ kg } H_2O \cdot h^{-1} \qquad (4.329)$$

Equally, it is necessary to subtract to this current of exit from the reactor the quantity of hydrogen consumed by the reduction of the total quantity of hydrogen of the current, R; thus, the mass flow of hydrogen in G-DRI is:

$$\frac{5768 \text{ kg gas R}}{h} \cdot 0.975 - \frac{1428 \text{ kgH}_2}{h} = 4196 \text{ kgH}_2 \cdot h^{-1} \qquad (4.330)$$

Consequently, the mass percentage composition of the G-DRI gas contains 74.75% $H_2O$, 24.41% $H_2$ and 0.84% $CO_2$. In a first stage, all the water vapor from the G-DRI condense: 12,849 kg $H_2O \cdot h^{-1}$ to later subject the 4340 kg of dry gas $\cdot h^{-1}$ to a partial discharge of $CO_2$ (milk of lime): 14 kg $CO_2 \cdot h^{-1}$. Finally, the 4326 kg of dry gas $\cdot h^{-1}$ and purge of $CO_2$ are joined to the 1442 kg $\cdot h^{-1}$ of the feeding of reductant gas, F, to provide the 5768 kg $\cdot h^{-1}$ of recycled gas, R, to the reduction furnace.

Commentaries: The following iron-making operations might be pointed out in a facility able to produce 233,000 tons of iron sponge per year (26,579 kg Fe $\cdot h^{-1}$):

–   Recycling of all the gas that leaves the reduction unity.
–   Requirement of purging of all the oxidizing gases (water vapor and carbon dioxide), as an increase in their concentration will cause a reduction of the driving force of the reaction and a diminishing of the reduction rate.

Finally, it is necessary to point out that, habitually, all the poor gas, which is obtained in a DRI furnace, is used in a thermal power station located close to the iron and steelmaking installations. Consequently, the DRI plants are usually connected to a thermal power station.

*Solution ends.*

Cavaliere (2019, pp. 13–22) has estimated that the iron and steelmaking industry generates 2.0 Gt/year of $CO_2$. This quantity is similar to that estimated for human breathing contribution 2.5 Gt/year of $CO_2$ considering a current world population of approximately 7500 million people. This quantity can be calculated considering that humans typically breathe 12–20 times/min, and the air that humans exhale is roughly 78 vol. % $N_2$, 18 vol. % $O_2$, 0.96 vol. % Ar and 4 vol. % $CO_2$ (depending on the fitness of every particular person) (Jana and Majumder 2010). In normal quiet breathing human exchange about 0.5 litres of air per breathing (Jana and Majumder 2010), assuming 18 breaths per minute, we have:

$$\frac{0.04l\ CO_2}{air \cdot human} \cdot \frac{0.5l\ air}{per\ breath} \cdot \frac{18\ breaths}{min} \cdot \frac{44\ gCO_2}{22.4l} \cdot \frac{60\ min}{1\ h} \cdot \frac{24\ h}{1\ day}$$
$$\cdot \frac{365\ days}{1\ year} \cdot 7500 \cdot 10^6\ humans \cdot \frac{1\ Gt}{10^{15}\ g} = 2.78 \frac{GtCO_2}{year} \qquad (4.331)$$

This represents 6.4–8.0% of all the $CO_2$ emissions every year (Sondergard 2009). If the pets are included, this percentage increases up to 10%, if the animals raised for food and clothing are included, the $CO_2$ emissions increase out to 15–20% (Sondergard 2009). The global yearly $CO_2$ emissions are approximately 35–45 Mt. This gives us an idea about the human contribution with respect to the iron and steelmaking carbon dioxide emissions. Within this context, it would be interesting to have a more accurate measurement or estimations of the emissions associated to breathing of other organisms (apart from humans) for a better knowledge of the $CO_2$ emissions associated to all the organisms living in the Earth. The level of accurateness in the values of $CO_2$ emissions in the iron and steelmaking sector (there is even data about the contribution of every stage of the process), which could be extended to other industrial sectors as that of concrete, non-ferrous metallurgy, ceramics or transports, among others. This would be useful to know the contribution of each sector (and stage of the process) and where the economic and technological efforts would be focused with the aim of reducing carbon dioxide emissions.

On another note, Cavaliere (2019) also estimates that 95% of the emissions are associated to the production of pig iron in the blast furnace. Industrial gases together with the coke batteries gases could contribute to reduce the $CO_2$ emissions. With the utilization of the iron and steelmaking gases in the reduction-production of DRI with the support of the $H_2$, the transformation of the CO into $H_2$ and the capture of the $CO_2$ could noticeably reduce the carbon dioxide emissions in the blast furnace (blast furnace + basic oxygen furnace route).

Finally, it is convenient to study and acquire a deeper knowledge, especially in the atmospheres that contain $H_2$, about those mixtures of gases with hydrogen that are not explosives. There is an interval of hydrogen volume concentration in the atmosphere for a certain CO, $CO_2$, $H_2O$, and $N_2$ composition that makes it dangerous due to explosion risks, while hydrogen concentrations above or below the above-indicated interval lead to the formation of stable atmospheres.

Consequently, all the processes of transformation of the CO into $H_2$ or the removal of $CO_2$ and $H_2O$ from the gases involve operations/processes of heating–cooling, and definitely, energy costs. Concentrated solar energy could be, if it is technologically developed, an instrument for the recycling and scrubbing-purification of the gases in a profitable option from the economics point of view in front of the processes of synthesis of hydrogen by water splitting-electrolysis.

## 4.6  Complementary Considerations

It seems convenient to point out that the thermodynamic equilibriums that were presented along this chapter do not mention anything about the time required to reach such equilibriums. The thermodynamic equilibrium and the kinetics of the reactions—as it is well known—are different concepts. Therefore, the equilibrium constant, $k_{eq}$, of one reaction not only depends on the temperature. However, although the $k_{eq}$ would not vary and the concentrations neither too, it is possible to have variation—due to the incidence of other factors—the rate of the reactions and, consequently, the time required to reach the equilibrium.

For that reason, for instance, it is necessary to use the suitable catalyzers and, therefore, controlled atmospheres should not be generated in the same furnace that they will be used but in a place with the catalyzer. It is also convenient to mention that, at the temperature T, the rate when certain chemical reaction tends to the equilibrium usually considerably decreases when it approaches it.

On another note, it seems convenient to advise that the controlled atmospheres involve toxicity risks (v. gr. CO, $NH_3$, etc.). Thus, the user must know the biological risks (whose explanation exceeds the scope of this chapter), the aid that should be rendered in the event of intoxication and the measurements that should be adopted to protect the environment. Moreover, it is necessary to remember that many mixtures of combustible gases can—when mixed with the oxygen—generate explosive mixtures if the content in the flammable gas is comprised within the lower and upper limits of flammability. This should be considered especially in the case of the hydrogen.

Finally, we are going to conclude with some complementary considerations about the concept of decarbonization. If the concept of decarbonization in the iron and steelmaking industry is interpreted as something empty of meaning instead as technologic-scientific objective, it will not be never reached. The complete elimination of the carbon dioxide emissions, apart from contradicting the carbon cycle in the nature, is neither a problem that could be solved unilaterally by certain countries, regions, or continents. If the concept of decarbonization is considered as a scientific-technologic objective, it would be possible to reach, in the short and medium term, great importance achievements for the sector as:

1. Improvement of the energy efficiency of all the iron-making and steel-making operations and processes.
2. The introduction of the renewable energy in the electric steelworks.
3. The increase of the participation of the concentrated solar energy and biomass in the iron and steelmaking operations and processes.
4. The utilization of the iron and steelmaking gases in the production of direct reduction iron by the solid–gas mechanisms.

All the previously-indicated advances-progresses would lead to a stabilization and reduction of the net carbon dioxide emissions to the environment. The treatment of the iron and steelmaking gases associated to the production of BOF-BF steel could progress towards the generation of highly reductant atmospheres (with high hydrogen content) and low carburizing potential (low carbon monoxide contents). Definitely, with the treatment of the iron and steelmaking gases, apart from fixing the carbon (by removal of the carbon dioxide from the gases as calcium carbonate) and removing-condensing the water vapor, would also lead to a noticeable reduction of the emissions of molecules of organic or inorganic characteristics that are detrimental for the environment.

# References

Ballester A, Verdeja LF, Sancho JP (2000) Metalurgia Extractiva. Volumen 1. Fundamentos. Ed. Síntesis, Madrid, Spain

Ban-Ya S, Elliot JF, Chipman J (1970) Thermodynamics of austenitic Fe-C alloys. Metallurg Trans 1:1313–1320

Bodsworth C, Appleton AS (1969) Problèmes de Thermodynamique Chimique. Ed. Dunod, Paris, France

Cavaliere P (2019) Clean ironmaking and steelmaking process. Springer, Cham, Switzerland

Ellis T, Davidson IM, Bosdworth C (1963) Some thermodynamic properties of carbon in austenite. J Iron Steel Inst 201:582–587

Fernández-González D, Ruiz-Bustinza I, Mochón J, González-Gasca C, Verdeja LF (2017a) Iron ore sintering: raw materials and granulation. Miner Process Extr Metall Rev 38(1):36–46

Fernández-González D, Ruiz-Bustinza I, Mochón J, González-Gasca C, Verdeja LF (2017b) Iron ore sintering: process. Miner Process Extr Metall Rev 38(4):215–227

Fernández-González D, Ruiz-Bustinza I, Mochón J, González-Gasca C, Verdeja LF (2017c) Iron ore sintering: quality indices. Miner Process Extr Metall Rev 38(4):254–264

Fernández-González D, Ruiz-Bustinza I, Mochón J, González-Gasca C, Verdeja LF (2017d) Iron ore sintering: environment, automatic and control techniques. Miner Process Extr Metall Rev 38(4):238–249

Fernández-González D, Piñuela-Noval J, Verdeja LF (2018a) Iron ore agglomeration technologies. In: Shatokha V(ed) Iron ores and iron oxide materials. Intechopen, Londres, pp 61–80 (Chapter 4).

Fernández-González D, Prazuch J, Ruiz-Bustinza I, González-Gasca C, Piñuela-Noval J, Verdeja LF (2018b) Iron metallurgy via concentrated solar energy. Metals 8(11): art. 873

Fernández-González D, Ruiz-Bustinza I, González-Gasca C, Piñuela-Noval J, Mochón-Castaños J, Sancho-Gorostiaga J, Verdeja LF (2018c) Concentrated solar energy applications in materials science and metallurgy. Sol Energy 170(8):520–540

Fernández-González D, Prazuch J, Ruiz-Bustinza I, González-Gasca C, Piñuela-Noval J, Verdeja LF (2018d) Solar synthesis of calcium aluminates. Sol Energy 171(9):658–666

Fernández-Gonzalez D (2019) Aplicaciones de la energía solar concentrada en metalurgia y ciencia de los materiales, PhD thesis, Universidad de Oviedo

Fernández-González D, Prazuch J, Ruiz-Bustinza I, González-Gasca C, Piñuela-Noval J, Verdeja LF (2019a) Transformations in the Si–O–Ca system: silicon–calcium via solar energy. Sol Energy 181(3):414–423

Fernández-González D, Prazuch J, Ruiz-Bustinza I, González-Gasca C, Piñuela-Noval J, Verdeja LF (2019b) The treatment of basic oxygen furnace (BOF) slag with concentrated solar energy. Sol Energy 180(3):372–380

Fernández-González D, Prazuch J, Ruiz-Bustinza I, González-Gasca C, Piñuela-Noval J, Verdeja LF (2019c) Transformations in the Mn–O–Si system using concentrated solar energy. Sol Energy 181(5):148–152

Fernández-González D, Prazuch J, Ruiz-Bustinza I, González-Gasca C, Gómez-Rodríguez C, Verdeja LF (2021) Recovery of copper and magnetite from copper slag using concentrated solar Power (CSP). Metals 11(7):art. 1032

Jana BK, Majunder M (2010) Impact of climate change on natural resource management. Springer, Heidelberg, Germany

Massalski TB, Murray JL, Bennett LH, Bakker H (1986) Binary alloy phase diagrams. American Society for Metals, Metals Park, Ohio, USA

Pero-Sanz JA, Quintana MJ, Verdeja LF (2017) Solidification and solid-state transformations of metals and alloys, 1st edn. Elsevier, Boston, USA

Pero-Sanz JA, Fernández-González D, Verdeja LF (2018) Physical metallurgy of cast irons. Springer International Publishing, Cham, Switzerland

Pero-Sanz JA, Fernández-González D, Verdeja LF (2019) Structural materials: properties and selection. Springer International Publishing, Cham, Switzerland

Pero-Sanz JA (2004) Aceros. Metalurgia Física, Selección y Diseño, Ed. CIE Dossat 2000, Madrid, Spain

Sancho JP, Verdeja, LF, Ballester A (2000) Metalurgia Extractiva. Volumen II. Procesos de Obtención. Ed. Síntesis, Madrid, Spain

Sondergard SE (2009) Climate balance: a balanced and realistic view of the climate, change. Tate Publishing & Enterprises, LLC, Ocklahoma, USA

Verdeja JI, Fernández-González D, Verdeja LF (2020) Operations and basic processes in ironmaking. Springer International Publishing, Cham, Switzerland

Verdeja LF, Fernández-González D, Verdeja JI (2021) Operations and basic processes in steelmaking. Springer International Publishing, Cham, Switzerland

Verdeja LF, Alfonso A, Huerta MA (1995) Aplicación del diagrama Fe–C estable al cálculo de parámetros termodinámicos de aceros y fundiciones. Revista de Minas, 11 and 12:109–114

# Appendices

## Appendix 1

Standard free energies for some iron and steelmaking reactions, $\Delta_r G^0$.

| $\Delta_r G^0 (kJ \cdot mol^{-1}) = A + B \cdot T(K)$ | Linear correlation coefficient | | | | Ref. (*) |
|---|---|---|---|---|---|
| | $-A$ | B | $(\pm)$ kJ | °C | |
| *Aluminum* | | | | | |
| $2Al(s) + 3/2O_2(g) = Al_2O_3(s)$ | 1672.0 | 0.313 | 6.9 | 25–659 | 1,2,3 |
| $2Al(l) + 3/2O_2(g) = Al_2O_3(s)$ | 1680.0 | 0.324 | 4.5 | 659–1700 | 1,2,3,4 |
| $Al(s) + 1/2N_2(g) = AlN(s)$ | 319.0 | 0.101 | 6.6 | 25–659 | 1,2,3 |
| $Al(l) + 1/2N_2(g) = AlN(s)$ | 328.0 | 0.112 | 19.8 | 659–1700 | 1,2,3,4 |
| $4Al(s) + 3C(s) = Al_4C_3(s)$ | 213.0 | 0.042 | 4.1 | 25–659 | 1,2,3 |
| $4Al(l) + 3C(s) = Al_4C_3(s)$ | 283.0 | 0.107 | 4.5 | 659–1700 | 1,2,3 |
| $2Al(l) + 1/2O_2(g) = Al_2O(g)$ | 193.0 | −0.046 | 0.2 | 1227–1727 | 2,3,4 |
| $Al(l) + 1/2O_2(g) = AlO(g)$ | −42.5 | −0.059 | 0.1 | 1227–1727 | 2,3 |
| *Barium* | | | | | |
| $Ba(s) + 1/2O_2(g) = BaO(s)$ | 554.0 | 0.094 | 11.1 | 25–704 | 1,2,3 |
| $Ba(l) + 1/2O_2(g) = BaO(s)$ | 561.0 | 0.102 | 8.1 | 704–1638 | 1,2,3,4 |
| $3Ba(s) + N_2(g) = Ba_3N_2(s)$ | 352.0 | 0.234 | 8.9 | 25–704 | 1,3 |
| $BaO(s) + SiO_2(s) = BaSiO_3(s)$ | 135.0 | 0.004 | 22.4 | 25–1300 | 1,3 |
| *Beryllium* | | | | | |
| $Be(s) + 1/2O_2(g) = BeO(s)$ | 604.0 | 0.098 | 8.3 | 25–1283 | 1,2,3 |
| $Be(l) + 1/2O_2(g) = BeO(s)$ | 605.0 | 0.100 | 8.4 | 1283–1700 | 1,2,3,4 |
| $3Be(s) + N_2(g) = Be_3N_2(s)$ | 578.0 | 0.175 | 14.9 | 25–700 | 1,2,3 |

(continued)

J. I. Verdeja González et al., *Physical Metallurgy and Heat Treatment of Steel*,
Topics in Mining, Metallurgy and Materials Engineering,
https://doi.org/10.1007/978-3-031-05702-1

(continued)

| $\Delta_r G^0 (kJ \cdot mol^{-1}) = A + B \cdot T(K)$ | Linear correlation coefficient | | | | Ref. (*) |
|---|---|---|---|---|---|
| | $-A$ | B | $(\pm)$ kJ | °C | |
| *Boron* | | | | | |
| $2B(s) + 3/2O_2(g) = B_2O_3(s)$ | 1272.0 | 0.264 | 0.8 | 25–450 | 1,2,3 |
| $2B(s) + 3/2O_2(g) = B_2O_3(l)$ | 1225.0 | 0.208 | 3.5 | 450–1700 | 1,2,3 |
| $B(s) + 1/2N_2(g) = BN(s)$ | 252.0 | 0.088 | 1.0 | 25–900 | 1,2,3 |
| $4B(s) + C(s) = B_4C(s)$ | 60.6 | 0.003 | 6.2 | 25–900 | 1,2,3 |
| $B(s) + 1/2N_2(g) = BN(g)$ | −648.0 | −0.112 | | 1227–1727 | 4 |
| $BaO(s) + SiO_2(s) = BaSiO_3(s)$ | 135.0 | 0.004 | 22.4 | 25–1300 | 1,3 |
| *Calcium* | | | | | |
| $Ca(s) + 1/2O_2(g) = CaO(s)$ | 633.0 | 0.103 | 1.5 | 25–850 | 1,2,3 |
| $Ca(l) + 1/2O_2(g) = CaO(s)$ | 640.0 | 0.109 | 2.9 | 850–1487 | 1,2,3,4 |
| $Ca(g) + 1/2O_2(g) = CaO(s)$ | 788.0 | 0.193 | 1.6 | 1487–1700 | 1,2,3,4 |
| $Ca(s) + 1/2S_2(g) = CaS(s)$ | 517.0 | 0.072 | 27.7 | 25–850 | 1,2,3 |
| $Ca(l) + 1/2S_2(g) = CaS(s)$ | 549.0 | 0.106 | 3.5 | 850–1487 | 1,2,3,4 |
| $Ca(g) + 1/2S_2(g) = CaS(s)$ | 696.0 | 0.189 | 3.1 | 1487–1700 | 2,4 |
| $3Ca(s) + N_2(g) = Ca_3N_2(s)$ | 437.0 | 0.155 | 31.6 | 25–850 | 1,3 |
| $Ca(s)a + 2C(s) = CaC_2(s)$ | 57.0 | −0.024 | 0.3 | 25–400 | 1,3 |
| $Ca(s)b + 2C(s) = CaC_2(s)$ | 49.4 | −0.035 | 0.5 | 400–850 | 1,3 |
| $Ca(l) + 2C(s) = CaC_2(s)$ | 60.6 | −0.025 | 1.2 | 850–1487 | 1,3,4 |
| $Ca(g) + 2C(s) = CaC_2(s)$ | 212.0 | 0.058 | 13.3 | 1487–1900 | 1,3,4 |
| $Ca(s) + Si(s) = CaSi(s)$ | 150.0 | 0.002 | – | 25–850 | 1 |
| $Ca(l) + Si(s) = CaSi(s)$ | 107.0 | −0.028 | – | 850–1444 | 1 |
| $2Ca(l) + Si(s) = Ca_2Si(s)$ | 178.0 | −0.019 | – | 850–1444 | 1 |
| $3CaO(s) + Al_2O_3(s) = Ca_3Al_2O_6(s)$ | 18.9 | −0.028 | 11.1 | 25–1550 | 1,3,5 |
| $12CaO(s) + 7Al_2O_3(s) = Ca_{12}Al_{14}O_{33}(s)$ | 72.8 | −0.210 | – | 25–1500 | 1 |
| $CaO(s) + Al_2O_3(s) = CaAl_2O_4(s)$ | 20.4 | 0.017 | 4.2 | 25–1600 | 1,3,5 |
| $CaO(s) + CO_2(g) = CaCO_3(s)$ | 168.0 | 0.143 | – | 25–880 | 1 |
| $2CaO(s) + Fe_2O_3(s) = Ca_2Fe_2O_5(s)$ | 38.3 | −0.010 | – | 600–1435 | 1 |
| $2CaO(s) + Fe_2O_3(s) = Ca_2Fe_2O_5(l)$ | −31.5 | −0.051 | – | 1435–1600 | 1 |
| $4CaO(s) + P_2(g) + 5/2O_2(g) = Ca_4P_2O_9(s)$ | 2348.0 | 0.600 | – | 1300–1600 | 1 |
| $3CaO(s) + P_2(g) + 5/2O_2(g) = Ca_3P_2O_8(s)$ | 2306.0 | 0.600 | – | 1300–1600 | 1 |
| $2CaO(s) + SiO_2(s) = Ca_2SiO_4(s)$ | 103.0 | −0.020 | 18.7 | 25–400 | 1,5 |
| $CaO(s) + SiO_2(s) = CaSiO_3(s)\alpha$ | 88.7 | 0.001 | – | 25–1210 | 1 |
| $CaO(s) + SiO_2(s) = CaSiO_3(s)\beta$ | 82.9 | −0.003 | – | 1210–1543 | 1 |
| $CaO(s) + SiO_2(s) = CaSiO_3(s)$ | 76.7 | −0.003 | – | 25–727 | 5 |

(continued)

(continued)

| $\Delta_r G^0 (kJ \cdot mol^{-1}) = A + B \cdot T(K)$ | Linear correlation coefficient | | | | Ref. (*) |
|---|---|---|---|---|---|
| | $-A$ | B | $(\pm)$ kJ | °C | |
| $3CaO(s) + 2SiO_2(s) = Ca_3Si_2O_7(s)$ | 202.0 | −0.013 | – | 25–727 | 5 |
| $3CaO(s) + SiO_2(s) = Ca_3SiO_5(s)$ | 115.0 | −0.023 | – | 25–727 | 5 |
| $CaO(s) + 2Al_2O_3(s) = CaAl_4O_7(s)$ | 16.8 | −0.034 | 3.8 | 25–1765 | 3,5 |
| $CaO(s) + 6Al_2O_3(s) = CaAl_{12}O_{19}(s)$ | 22.5 | −0.032 | – | 25–727 | 5 |
| $2CaO(s) + SiO_2(s) + Al_2O_3(s) = Ca_2SiAl_2O_7(s)$ | 96.0 | −0.050 | 31.5 | 25–1590 | 3,5 |
| $CaO(s) + 2SiO_2(s) + Al_2O_3(s) = CaSi_2Al_2O_8(s)$ | 55.1 | −0.035 | 35.4 | 25–1553 | 3,5 |
| $CaO(s) + MgO(s) + 2SiO_2(s) = CaMgSi_2O_6(s)$ | 37.6 | −0.015 | 29.6 | 25–1392 | 3 |
| $CaO(s) + MgO(s) + SiO_2(s) = CaMgSiO_4(s)$ | 99.2 | −0.009 | 14.3 | 25–1227 | 3,5 |
| $2CaO(s) + MgO(s) + 2SiO_2(s) = Ca_2MgSi_2O_7(s)$ | 157.0 | −0.030 | 24.8 | 25–1454 | 3,5 |
| $3CaO(s) + MgO(s) + 2SiO_2(s) = Ca_3MgSi_2O_8(s)$ | 204.0 | −0.032 | – | | 5 |
| *Carbon* | | | | | |
| $C(s) + 2H_2(g) = CH_4(g)$ | 95.6 | 0.113 | 1.6 | 25–2000 | 1,3,4 |
| $C(s) + 1/2O_2(g) = CO(g)$ | 118.0 | −0.084 | 3.7 | 25–2000 | 1,2,3,4 |
| $C(s) + O_2(g) = CO_2(g)$ | 395.0 | 0.001 | 5.1 | 25–2000 | 1,2,3,4 |
| $C(s) + 1/2S_2(g) = CS(g)$ | −176.0 | −0.067 | 10.7 | 1600–1800 | 1,2,3,4 |
| $CO(g) + 1/2O_2(g) = CO_2(g)$ | 277.0 | 0.085 | | | |
| $C(s) + S_2(g) = CS_2(g)$ | −26.5 | −0.050 | 43.4 | 25–1300 | 1,2,3 |
| $CO(g) + 1/2S_2(g) = COS(g)$ | 132.0 | 0.065 | 98.2 | 25–1200 | 1,2,3 |
| $2C(s) + H_2(g) = C_2H_2(g)$ | −217.0 | −0.049 | – | 1227–1727 | 4 |
| *Cerium* | | | | | |
| $2Ce(s) + 3/2O_2(g) = Ce_2O_3(s)$ | 1805.0 | 0.309 | 2.0 | 25–804 | 1,3 |
| $2Ce(l) + 3/2O_2(g) = Ce_2O_3(s)$ | 1810.0 | 0.325 | 68.4 | 804–1700 | 1,3,4 |
| $Ce(s) + O_2(g) = CeO_2(s)$ | 1053.0 | 0.207 | 35.5 | 25–804 | 1,3 |
| $Ce(l) + O_2(g) = CeO_2(s)$ | 1046.0 | 0.211 | 52.8 | 804–1700 | 1,3,4 |
| *Chromium* | | | | | |
| $2Cr(s) + 3/2O_2(g) = Cr_2O_3(s)$ | 1129.0 | 0.264 | 11.6 | 25–1898 | 1,2,3 |
| $2Cr(s) + 1/2N_2(g) = Cr_2N(s)$ | 117.0 | 0.067 | 11.4 | 25–1898 | 1,2,3 |
| $Cr(s) + 1/2N_2(g) = CrN(s)$ | 113.0 | 0.075 | 4.8 | 25–1898 | 1,2,3 |
| $23Cr(s) + 6C(s) = Cr_{23}C_6(s)$ | 352.0 | −0.045 | 55.1 | 25–1400 | 1,2,3 |
| $7Cr(s) + 3C(s) = Cr_7C_3(s)$ | 164.0 | −0.026 | 10.5 | 25–1200 | 1,2,3 |
| $3Cr(s) + 2C(s) = Cr_3C_2(s)$ | 84.7 | −0.011 | 5.4 | 25–1700 | 1,2,3 |

(continued)

(continued)

| $\Delta_r G^0 (kJ \cdot mol^{-1}) = A + B \cdot T(K)$ | Linear correlation coefficient | | | | Ref. (*) |
|---|---|---|---|---|---|
| | $-A$ | B | $(\pm)$ kJ | °C | |
| *Cobalt* | | | | | |
| $Co(s) + 1/2O_2(g) = CoO(s)$ | 235.0 | 0.074 | 1.4 | 25–1495 | 1,2,3 |
| $3CoO(s) + 1/2O_2(g) = Co_3O_4(s)$ | 198.0 | 0.158 | 10.3 | 25–1000 | 1,2,3 |
| $9Co(s) + 4S_2(g) = Co_9S_8(s)$ | 1316.0 | 0.634 | 18.7 | 25–778 | 1,3 |
| $2Co(s) + C(s) = Co_2C(s)$ | −16.5 | −0.009 | | 25–900 | 1 |
| *Copper* | | | | | |
| $2Cu(s) + 1/2O_2(g) = Cu_2O(s)$ | 168.0 | 0.074 | 2.9 | 25–1084 | 1,2,3 |
| $2Cu(l) + 1/2O_2(g) = Cu_2O(s)$ | 182.0 | 0.082 | 1.9 | 1084–1230 | 1,2,3 |
| $1/2Cu_2O(s) + 1/4O_2(g) = CuO(s)$ | 72.1 | 0.054 | 3.5 | 25–1000 | 1,2,3 |
| $2Cu(l) + 1/2O_2(g) = Cu_2O(l)$ | 129.0 | 0.049 | 6.0 | 1229–1727 | 2,3,4 |
| $Cu(l) + 1/2O_2(g) = CuO(l)$ | 154.0 | 0.088 | 1.6 | 1447–1727 | 2,4 |
| *Hydrogen* | | | | | |
| $H_2(g) + 1/2O_2(g) = H_2O(g)$ | 249.0 | 0.057 | 1.0 | 25–1700 | 1,2,3,4 |
| $H_2(g) + 1/2S_2(g) = H_2S(g)$ | 87.3 | 0.045 | 1.4 | 25–1500 | 1,2,3 |
| $3/2H_2(g) + 1/2N_2(g) = NH_3(g)$ | 49.3 | 0.109 | 0.5 | 25–700 | 1,3 |
| *Iron* | | | | | |
| $Fe(s) + 1/2O_2(g) = FeO(s)-\alpha$ | 265.0 | 0.065 | 6.0 | 25–1371 | 1,2,3 |
| $0.95Fe(s) + 1/2O_2(g) = Fe_{0.95}O(s)$ | 264 | 0.066 | – | 25–1371 | – |
| $Fe(s) + 1/2O_2(g) = Wustita(s)-\beta$ | 31.2 | 0.019 | – | 25–1371 | 5 |
| $Fe(l) + 1/2O_2(g) = FeO(l)$ | 244.0 | 0.050 | 4.4 | 1537–1700 | 1,2 |
| $3Fe(s) + 2O_2(g) = Fe_3O_4(s)$ | 1105.0 | 0.328 | 14.0 | 25–560 | 1,2,3 |
| $3FeO(s) + 1/2O_2(g) = Fe_3O_4(s)$ | 296.0 | 0.112 | 14.6 | 560–1371 | 1,2,3 |
| $2/3Fe_3O_4(s) + 1/6O_2(g) = Fe_2O_3(s)$ | 80.2 | 0.045 | 2.0 | 25–1400 | 1,2,3 |
| $Fe(s) + 1/2S_2(g) = FeS(s)\alpha$ | 164.0 | 0.080 | 6.0 | 25–140 | 1,2,3 |
| $Fe(s) + 1/2S_2(g) = FeS(s)\beta$ | 138.0 | 0.038 | 12.3 | 140–906 | 1,2,3 |
| $FeS(s)b + 1/2S_2(g) = FeS_2(s)$ | 135.0 | 0.134 | 23.9 | 300–800 | 1,2,3 |
| $4Fe(s) + 1/2N_2(g) = Fe_4N(s)$ | 7.7 | 0.044 | 1.9 | 25–600 | 1,3 |
| $3Fe(s) + 1/2P_2(g) = Fe_3P(s)$ | 224.0 | 0.066 | 5.6 | 25–1170 | 1,3 |
| $3Fe(s) + C(s) = Fe_3C(s)$ | −26.5 | −0.023 | 0.7 | 25–190 | 1,3 |
| $3Fe(s) + C(s) = Fe_3C(s)$ | −28.1 | −0.027 | 0.5 | 190–840 | 1,3 |
| $3Fe(s) + C(s) = Fe_3C(s)$ | −9.8 | −0.010 | – | 840–1537 | 1,3 |
| $FeO(s) + Al_2O_3(s) = FeAl_2O_4(s)$ | 47.1 | 0.017 | 15.3 | 25–1371 | 1,3 |
| $FeO(s) + Cr_2O_3(s) = FeCr_2O_4(s)$ | 33.2 | −0.006 | 18.9 | 35–1371 | 1,3 |
| $2FeO(s) + SiO_2(s) = Fe_2SiO_4(s)$ | 55.0 | 0.028 | 0.1 | 25–1217 | 1,3,5 |
| $2FeO(s) + SiO_2(s) = Fe_2SiO_4(l)$ | −62.0 | −0.048 | – | 1217–1371 | 1 |
| $2FeO(l) + SiO_2(s) = Fe_2SiO_4(l)$ | −14.4 | −0.019 | – | 1371–1700 | 1 |

(continued)

(continued)

| $\Delta_r G^0 (kJ \cdot mol^{-1}) = A + B \cdot T(K)$ | Linear correlation coefficient | | | | Ref. (*) |
|---|---|---|---|---|---|
| | $-A$ | B | $(\pm) kJ$ | °C | |
| $FeO(s) + TiO_2(s) = FeTiO_3(s)$ | 11.2 | −0.003 | 7.4 | 900–1371 | 1,3 |
| *Lanthanum* | | | | | |
| $2La(s) + 3/2O_2(g) = La_2O_3(s)$ | 1823.0 | 0.283 | 33.1 | 25–700 | 1,3 |
| $La(s) + 1/2N_2(g) = LaN(s)$ | 299.0 | 0.097 | 3.3 | 25–700 | 1,3 |
| *Magnesium* | | | | | |
| $Mg(s) + 1/2O_2(g) = MgO(s)$ | 600.0 | 0.105 | 1.6 | 25–650 | 1,2,3 |
| $Mg(l) + 1/2O_2(g) = MgO(s)$ | 609.0 | 0.115 | 2.5 | 650–1120 | 1,2,3 |
| $Mg(g) + 1/2O_2(g) = MgO(s)$ | 732.0 | 0.205 | 5.6 | 1120–1700 | 1,2,3,4 |
| $Mg(s) + 1/2S_2(g) = MgS(s)$ | 391.0 | 0.070 | 28.5 | 25–650 | 1,2,3 |
| $Mg(l) + 1/2S_2(g) = MgS(s)$ | 419.0 | 0.103 | 1.6 | 650–1120 | 1,2,3 |
| $Mg(g) + 1/2S_2(g) = MgS(s)$ | 548.0 | 0.196 | 3.2 | 1120–1700 | 1,2,3,4 |
| $3 Mg(s) + N_2(g) = Mg_3N_2(s)$ | 460.0 | 0.199 | 3.0 | 25–650 | 1,2,3 |
| $3 Mg(l) + N_2(g) = Mg_3N_2(s)$ | 487.0 | 0.228 | 3.9 | 650–1120 | 1,2,3 |
| $MgO(s) + CO_2(g) = MgCO_3(s)$ | 106.0 | 0.158 | 14.1 | 25–700 | 1,2,3 |
| $3MgO(s) + P_2(g) + 5/2O_2(g) = Mg_3P_2O_8(s)$ | 2047.0 | 0.576 | 12.2 | 1000–1250 | 1,2,3 |
| $2MgO(s) + SiO_2(s) = Mg_2SiO_4(s)$ | 60.1 | −0.002 | 7.1 | 650–1120 | 1,2,3,5 |
| $MgO(s) + SiO_2(s) = MgSiO_3(s)$ | 37.8 | 0.003 | 2.8 | 25–1300 | 1,2,3,5 |
| $Mg(g) + 2C(s) = MgC_2(s)$ | 30.9 | 0.072 | 37.9 | 122–1727 | 2,4 |
| $3 Mg(g) + N_2(g) = Mg_3N_2(l)$ | 865.0 | 0.507 | 2.1 | 1227–1727 | 2,4 |
| $MgO(s) + Al_2O_3(s) = MgAl_2O_4(s)$ | 17.0 | −0.015 | 7.7 | 25–1727 | 3,5 |
| *Manganese* | | | | | |
| $Mn(s) + 1/2O_2(g) = MnO(s)$ | 383.0 | 0.073 | 1.0 | 25–1244 | 1,3 |
| $Mn(l) + 1/2O_2(g) = MnO(s)$ | 402.0 | 0.086 | 3.2 | 1244–1700 | 1,3,4 |
| $Mn(s) + 1/2S_2(g) = MnS(s)$ | 274.0 | 0.066 | 4.3 | 25–1244 | 1,3 |
| $Mn(l) + 1/2S_2(g) = MnS(s)$ | 291.0 | 0.078 | 9.0 | 1244–1530 | 1,3,4 |
| $Mn(l) + 1/2S_2(g) = MnS(l)$ | 265.0 | 0.064 | 9.3 | 1530–1700 | 1,3,4 |
| $3Mn(s) + C(s) = Mn_3C(s)$ | 14.0 | −0.001 | 0.2 | 25–740 | 1,3 |
| $MnO(s) + SiO_2(s) = MnSiO_3(s)$ | 25.4 | 0.004 | 5.7 | 25–1300 | 1,3,5 |
| $3Mn(l) + 2O_2(g) = Mn_3O_4(\beta)$ | 1411.0 | 0.366 | 7.0 | 1243–1560 | 3,4 |
| $7Mn(l) + 3C(s) = Mn_7C_3(s)$ | 65.8 | −0.058 | – | 1243–1727 | 4 |
| $2MnO(s) + SiO_2(s) = Mn_2SiO_4(s)$ | 55.6 | 0.016 | 9.0 | 25–1345 | 3,5 |
| $MnO(s) + Al_2O_3(s) = MnAl_2O_4(s)$ | 45.2 | 0.010 | 2.0 | 25–1727 | 3,5 |
| *Molybdenum* | | | | | |
| $Mo(s) + O_2(g) = MoO_2(s)$ | 584.0 | 0.177 | 1.4 | 25–1000 | 1,3 |
| $MoO_2(s) + 1/2O_2(g) = MoO_3(s)$ | 158.0 | 0.074 | 1.6 | 25–1000 | 1,2,3 |

(continued)

(continued)

| $\Delta_r G^0 (kJ \cdot mol^{-1}) = A + B \cdot T(K)$ | Linear correlation coefficient | | | | Ref. (*) |
|---|---|---|---|---|---|
| | $-A$ | B | $(\pm)$ kJ | °C | |
| $2Mo(s) + 3/2S_2(g) = Mo_2S_3(s)$ | 560.0 | 0.242 | 6.7 | 850–1200 | 1,2,3 |
| $2Mo(s) + 1/2N_2(g) = Mo_2N(s)$ | 82.2 | 0.090 | 0.1 | 25–1000 | 1,3 |
| $2Mo(s) + C(s) = Mo_2C(s)$ | 36.8 | 0.004 | 17.3 | 25–1000 | 1,3 |
| $Mo(s) + C(s) = MoC(s)$ | −40.4 | −0.058 | – | 1227–1727 | 4 |
| $Mo(s) + 3/2O_2(g) = MoO_3(g)$ | 392.0 | 0.057 | 87.9 | 1280–1727 | 2,3,4 |
| $Mo(s) + S_2(g) = MoS_2(s)$ | 379.0 | 0.181 | 41.4 | 1227–1727 | 2,3,4 |
| *Nickel* | | | | | |
| $Ni(s) + 1/2O_2(g) = NiO(s)$ | 236.0 | 0.088 | 1.7 | 25–1452 | 1,3 |
| $Ni(l) + 1/2O_2(g) = NiO(s)$ | 249.0 | 0.093 | 0.5 | 1452–1900 | 1,3,4 |
| $Ni(s) + 1/2S_2(g) = NiS(s)$ | 135.0 | 0.057 | 24.2 | 300–580 | 1,2,3 |
| $3Ni(s) + C(s) = Ni_3C(s)$ | −63.5 | −0.055 | – | 25,700 | 1 |
| *Niobium* | | | | | |
| $2Nb(s) + 2O_2(g) = Nb_2O_4(s)$ | 1560.0 | 0.336 | – | 1227–1727 | 4 |
| $2Nb(s) + 5/2O_2(g) = Nb_2O_5(s)$ | 1860.0 | 0.404 | 5.8 | 1227–1453 | 2,3,4 |
| $2Nb(s) + 5/2O_2(g) = Nb_2O_5(l)$ | 1741.0 | 0.337 | 6.4 | 1453–1727 | 2,3,4 |
| *Phosphorus* | | | | | |
| $1/2P_2(g) + 1/2O_2(g) = PO(g)$ | 104.0 | −0.001 | 11.3 | 1227–1727 | 2,4 |
| $2P_2(g) + 5O_2(g) = P_4O_{10}(g)$ | 3125.0 | 0.980 | 38.9 | 1227–1727 | 2,3,4 |
| *Platinum* | | | | | |
| $Pt(s) + 1/2S_2(g) = PtS(s)$ | 16.0 | −0.005 | 0.3 | 1227–1727 | 3 |
| *Silicon* | | | | | |
| $Si(s) + 1/2O_2(g) = SiO(g)$ | 102.0 | −0.072 | 6.4 | 25–1410 | 1,2,3 |
| $Si(l) + 1/2O_2(g) = SiO(g)$ | 157.0 | −0.047 | 5.5 | 1410–1700 | 1,2,3,4 |
| $Si(s) + O_2(g) = SiO_2(s)cristob$ | 903.0 | 0.175 | 3.6 | 400–1410 | 1,2,3 |
| $Si(l) + O_2(g) = SiO_2(s)cristob$ | 945.0 | 0.198 | 5.7 | 1410–1700 | 1,2,3,4 |
| $3Si(s) + 2N_2(g) = Si_3N_4(s)$ | 738.0 | 0.326 | 14.2 | 25–1410 | 1,2,3 |
| $3Si(l) + 2N_2(g) = Si_3N_4(s)$ | 874.0 | 0.406 | 3.0 | 1410–1700 | 1,2,3,4 |
| $Si(s) + C(s) = SiC(s)\beta$ | 66.3 | 0.006 | 11.2 | 25–1410 | 1,2,3 |
| $Si(l) + C(s) = SiC(s)\beta$ | 113.0 | 0.036 | 11.1 | 1410–1700 | 1,2,3,4 |
| $2SiO_2(s) + 3Al_2O_3(s) = Si_2Al_6O_{13}(s)$ | −22.0 | −0.032 | 5.3 | 25–1850 | 3,5 |
| *Sulphur* | | | | | |
| $1/2S_2(g) = S(g)$ | −213.0 | −0.057 | 1.3 | 25–700 | 1,2,3 |
| $1/2S_2(g) + 1/2O_2(g) = SO(g)$ | 61.7 | −0.006 | 3.7 | 25–1700 | 1,2,3,4 |
| $1/2S_2(g) + O_2(g) = SO_2(g)$ | 361.0 | 0.072 | 1.3 | 25–1700 | 1,2,3,4 |
| $1/2S_2(g) + 3/2O_2(g) = SO_3(g)$ | 455.0 | 0.151 | 10.7 | 25–1500 | 1,2,3 |

(continued)

(continued)

| $\Delta_r G^0 (kJ \cdot mol^{-1}) = A + B \cdot T(K)$ | Linear correlation coefficient | | | | Ref. (*) |
|---|---|---|---|---|---|
| | $-A$ | B | $(\pm)$ kJ | °C | |
| *Tantalum* | | | | | |
| $2Ta(s) + 5/2O_2(g) = Ta_2O_5(s)$ | 1991.0 | 0.395 | 5.0 | 1227–1727 | 2,3 |
| $2Ta(s) + C(s) = Ta_2C(s)$ | 179.0 | 0.007 | 35.9 | 1227–1727 | 3,4 |
| $Ta(s) + C(s) = TaC(s)$ | 157.0 | 0.006 | 6.5 | 1227–1727 | 3,4 |
| $Ta(s) + 1/2N_2(g) = TaN(s)$ | 235.0 | 0.078 | 2.5 | 1227–1727 | 3,4 |
| *Thorium* | | | | | |
| $Th(s) + O_2(g) = ThO_2(s)$ | 1226.0 | 0.185 | 3.5 | 25–1500 | 1,3 |
| $3Th(s) + 2N_2(g) = Th_3N_4(s)$ | 1298.0 | 0.364 | 11.0 | 25–1700 | 1,3 |
| $Th(s) + 2C(s) = ThC_2(s)$ | 156.0 | 0.001 | 29.0 | 25–2000 | 1,3 |
| *Tin* | | | | | |
| $Sn(l) + O_2(g) = SnO_2(s)$ | 584.0 | 0.213 | 1.7 | 500–700 | 1,3 |
| $Sn(l) + 1/2O_2(g) = SnO(s)$ | 268.0 | 0.089 | 13.1 | 1227–1727 | 3,4 |
| $Sn(l) + 1/2O_2(g) = SnO(g)$ | 14.0 | −0.045 | 11.5 | 1227–1727 | 3,4 |
| $Sn(l) + O_2(g) = SnO_2(l)$ | 511.0 | 0.168 | – | 1227–1727 | 4 |
| *Titanium* | | | | | |
| $Ti(s) + 1/2O_2(g) = TiO(s)$ | 530.0 | 0.092 | 18.3 | 300–1700 | 1,2,3 |
| $2TiO(s) + 1/2O_2(g) = Ti_2O_3(s)$ | 448.0 | 0.087 | 32.0 | 25–1700 | 1,2,3 |
| $3/2Ti_2O_3(s) + 1/4O_2(g) = Ti_3O_5(s)$ | 176.0 | 0.032 | 7.3 | 400–1700 | 1,2,3 |
| $1/3Ti_3O_5(s) + 1/6O_2(g) = TiO_2(s)$ | 116.0 | 0.031 | 15.5 | 25–1850 | 1,2,3 |
| $Ti(s)\alpha + 1/2N_2(g) = TiN(s)$ | 337.0 | 0.095 | 1.6 | 25–882 | 1,2,3 |
| $Ti(s)\beta + 1/2N_2(g) = TiN(s)$ | 336.0 | 0.094 | 1.3 | 882–1200 | 1,2,3 |
| $Ti(s)\alpha + C(s) = TiC(s)$ | 183.0 | 0.011 | 0.9 | 25–882 | 1,2,3 |
| $Ti(s)\beta + C(s) = TiC(s)$ | 187.0 | 0.014 | 1.3 | 882–1700 | 1,2,3 |
| $Ti(s) + 1/2O_2(g) = TiO(g)$ | −32.4 | −0.079 | 1.3 | 1227–1667 | 3,4 |
| *Vanadium* | | | | | |
| $V(s) + 1/2O_2(g) = VO(s)$ | 425.0 | 0.081 | 11.0 | 600–1500 | 1,2,3 |
| $2VO(s) + 1/2O_2(g) = V_2O_3(s)$ | 393.0 | 0.065 | 58.2 | 550–1112 | 1,2,3 |
| $V(s) + C(s) = VC(s)$ | 102.0 | 0.009 | 1.7 | 900–1100 | 1,3 |
| $2 V(s) + 2O_2(g) = V_2O_4(s)$ | 1380.0 | 0.294 | 2.7 | 1227–1545 | 2,4 |
| $2 V(s) + 2O_2(g) = V_2O_4(l)$ | 1256.0 | 0.225 | 1.9 | 1545–1727 | 2,4 |
| $2 V(s) + 5/2O_2(g) = V_2O_5(l)$ | 1442.0 | 0.320 | 9.1 | 1227–1727 | 2,3,4 |
| $V(s) + 1/2N_2(g) = VN(s)$ | 195.0 | 0.079 | 28.8 | 1227–1727 | 2,3,4 |
| *Wolfram* | | | | | |
| $W(s) + C(s) = WC(s)$ | 38.8 | 0.002 | 1.3 | 25–1700 | 1,3 |
| $W(s) + O_2(g) = WO_2(s)$ | 567.0 | 0.164 | 2.3 | 1227–1727 | 2,3,4 |

(continued)

(continued)

| $\Delta_r G^0 (kJ \cdot mol^{-1}) = A + B \cdot T(K)$ | Linear correlation coefficient | | | | Ref. (*) |
|---|---|---|---|---|---|
| | $-A$ | B | $(\pm)$ kJ | °C | |
| $W(s) + 3/2O_2(g) = WO_3(s)$ | 815.0 | 0.232 | 3.8 | 1227–1470 | 2,3,4 |
| $W(s) + 3/2O_2(g) = WO_3(l)$ | 736.0 | 0.186 | 4.0 | 1470–1727 | 2,3,4 |
| *Zirconium* | | | | | |
| $Zr(s)\alpha + O_2(g) = ZrO_2(s)$ | 1092.0 | 0.191 | 11.3 | 25–870 | 1,2,3 |
| $Zr(s)\beta + O_2(g) = ZrO_2(s)$ | 1076.0 | 0.177 | 11.9 | 1205–1865 | 1,2,3,4 |
| $Zr(s)\alpha + 1/2N_2(g) = ZrN(s)$ | 365.0 | 0.094 | 2.6 | 25–862 | 1,2,3 |
| $Zr(s)\beta + 1/2N_2(g) = ZrN(s)$ | 365.0 | 0.093 | 3.9 | 862–1200 | 1,2,3 |
| $Zr(s) + C(s) = ZrC(s)$ | 196.0 | 0.010 | 11.6 | 25–1900 | 1,2,3 |

(*) In the elaboration of this table, we took as reference the work of the Association of Iron and Steel Engineers (AISE) previously included in the database of Janaf, Knacke, Pehlke, and Gaye (according to Verdeja, L.F., Alfonso, A., and Suárez, M.: Energías libres de formación de compuestos siderúrgicos. Revista de Minas, No. 15–16, 109–112,1997). The numbers of the last column are contributions to the values of A and B:
1. Lankford WT (1985) The making, shaping and treating of steel. Association of Iron and Steel Engineers, 10ª edición. Pittsburgh, EEUU.
2. Chase MW Jr, Davies CA, Downey JR Jr, Frurip DJ, McDonald RA, Syverud AN (1986) JANAF thermochemical tables. American chemical Society y American Institute of Physics for the National Bureau Standards, 3ª edition, New York, USA.
3. Knacke O, Kubaschewski O, Hesselmann K (1991) Thermochemical properties of inorganic substances. Springer. 2ª edition, Berlin, Germany.
4. Pehlke RD (1975) Unit processes of extractive metallurgy. Elsevier, 2ª edición. New York, USA.
5. Gaye HY, Welfringer J (1984) Modelling of the thermodynamic properties of complex metallurgical slags. Proceedings second international symposium on metallurgical slags and fluxes. Editors Fine, U.A. and Gaskell, D.R., Metallurgical Society of AIME, Warrendale, USA.

# Appendix 2

Standard enthalpies of several iron and steelmaking reactions, $\Delta_r H^0$.

| $\Delta_r H^0 (kJ \cdot mol^{-1}) = A + B \cdot T(K)$ | Linear correlation coefficient | | | | Ref. (*) |
|---|---|---|---|---|---|
| | $-A$ | B | $(\pm)$ kJ | °C | |
| *Aluminum* | | | | | |
| $2Al(s) + 3/2O_2(g) = Al_2O_3(s)$ | 1676.0 | 0.0016 | 0.6 | 25–659 | 1,2 |
| $2Al(l) + 3/2O_2(g) = Al_2O_3(s)$ | 1710.0 | 0.0149 | 0.2 | 659–1700 | 1,2 |
| $Al(s) + 1/2N_2(g) = AlN(s)$ | 318.0 | −0.0026 | 0.4 | 25–659 | 1,2 |
| $Al(l) + 1/2N_2(g) = AlN(s)$ | 331.0 | 0.0014 | 03 | 659–1700 | 1,2 |
| $4Al(s) + 3C(s) = Al_4C_3(s)$ | 212.0 | −0.0005 | 3.5 | 25–659 | 1,2 |
| $4Al(l) + 3C(s) = Al_4C_3(s)$ | 257.0 | −0.0058 | 7.7 | 659–1700 | 1,2 |

(continued)

(continued)

| $\Delta_r H^0 (kJ \cdot mol^{-1}) = A + B \cdot T(K)$ | Linear correlation coefficient | | | | Ref. (*) |
|---|---|---|---|---|---|
| | $-A$ | B | $(\pm)$ kJ | °C | |
| $2Al(l) + 1/2O_2(g) = Al_2O(g)$ | 157.0 | −0.0203 | 0.1 | 1227–1727 | 1,2 |
| $Al(l) + 1/2O_2(g) = AlO(g)$ | −55.0 | −0.0074 | 0.1 | 1227–1727 | 1,2 |
| *Barium* | | | | | |
| $Ba(s) + 1/2O_2(g) = BaO(s)$ | 547.0 | −0.0016 | 0.5 | 25–704 | 1,2 |
| $Ba(l) + 1/2O_2(g) = BaO(s)$ | 564.0 | 0.0040 | 0.6 | 704–1638 | 1,2 |
| *Beryllium* | | | | | |
| $Be(s) + 1/2O_2(g) = BeO(s)$ | 608.0 | −0.0003 | 0.2 | 25–1283 | 1,2 |
| $Be(l) + 1/2O_2(g) = BeO(s)$ | 627.0 | 0.0069 | 3.5 | 1283–1700 | 1,2 |
| $3Be(s) + N_2(g) = Be_3N_2(s)$ | 599.0 | 0.0311 | 10.5 | 25–700 | 1,2 |
| *Boron* | | | | | |
| $2B(s) + 3/2O_2(g) = B_2O_3(s)$ | 1273.0 | 0.0020 | 0.4 | 25–450 | 1,2 |
| $2B(s) + 3/2O_2(g) = B_2O_3(l)$ | 1251.0 | 0.0172 | 0.2 | 450–1700 | 1,2 |
| $B(s) + 1/2N_2(g) = BN(s)$ | 251.0 | −0.0012 | 0.8 | 25–900 | 1,2 |
| $4B(s) + C(s) = B_4C(s)$ | 62.7 | 0.0003 | 0.2 | 25–900 | 1,2 |
| $B(s) + 1/2O_2(g) = BO(g)$ | −8.0 | −0.0108 | 0.1 | 1227–1727 | 1,2 |
| *Calcium* | | | | | |
| $Ca(s) + 1/2O_2(g) = CaO(s)$ | 636.0 | 0.0043 | 0.1 | 25–850 | 1,2 |
| $Ca(l) + 1/2O_2(g) = CaO(s)$ | 647.0 | 0.0030 | 0.1 | 850–1487 | 1,2 |
| $Ca(g) + 1/2O_2(g) = CaO(s)$ | 823.0 | 0.0184 | 0.1 | 1487–1700 | 1,2 |
| $Ca(s) + 1/2S_2(g) = CaS(s)$ | 503.0 | −0.0068 | 32.2 | 25–850 | 1,2 |
| $Ca(l) + 1/2S_2(g) = CaS(s)$ | 548.0 | 0.0016 | 0.1 | 850–1487 | 1,2 |
| $Ca(g) + 1/2S_2(g) = CaS(s)$ | 724.0 | 0.0168 | 0.1 | 1487–1700 | 1,2 |
| *Carbon* | | | | | |
| $C(s) + 1/2O_2(g) = CO(g)$ | 104.0 | −0.0075 | 0.2 | 25–2000 | 1,2 |
| $C(s) + O_2(g) = CO_2(g)$ | 393.0 | −0.0017 | 1.0 | 25–2000 | 1,2 |
| $C(s) + 1/2S_2(g) = CS(g)$ | −224.0 | −0.0074 | 0.1 | 1600–1800 | 1,2 |
| $CO(g) + 1/2S_2(g) = COS(g)$ | 142.0 | −0.0196 | 9.2 | 25–1200 | 1,2 |
| *Chromium* | | | | | |
| $2Cr(s) + 3/2O_2(g) = Cr_2O_3(s)\beta$ | 1143.0 | 0.0180 | 3 | 25–1898 | 1,2 |
| $2Cr(s) + 1/2N_2(g) = Cr_2N(s)$ | 128.0 | 0.0071 | 0.6 | 25–1898 | 1,2 |
| $Cr(s) + 1/2N_2(g) = CrN(s)$ | 120.0 | 0.0088 | 0.2 | 25–1898 | 1,2 |
| $23Cr(s) + 6C(s) = Cr_{23}C_6(s)$ | 342.0 | 0.0458 | 2 | 25–1400 | 1,2 |
| $7Cr(s) + 3C(s) = Cr_7C_3(s)$ | 167.0 | 0.0209 | 0.4 | 25–1200 | 1,2 |
| $3Cr(s) + 2C(s) = Cr_3C_2(s)$ | 89.9 | 0.0151 | 0.3 | 25–1700 | 1,2 |

(continued)

(continued)

| $\Delta_r H^0 (kJ \cdot mol^{-1}) = A + B \cdot T(K)$ | Linear correlation coefficient | | | | Ref. (*) |
|---|---|---|---|---|---|
| | $-A$ | B | $(\pm)$ kJ | °C | |
| *Cobalt* | | | | | |
| $Co(s) + 1/2O_2(g) = CoO(s)$ | 241.0 | 0.0110 | 0.3 | 25–1495 | 1,2 |
| $3CoO(s) + 1/2O_2(g) = Co_3O_4(s)$ | 192.0 | −0.0298 | 5.0 | 25–1000 | 1,2 |
| *Copper* | | | | | |
| $2Cu(l) + 1/2O_2(g) = Cu_2O(s)$ | 10.5 | −0.1230 | 9 | 1084–1230 | 1,2 |
| $1/2Cu_2O(s) + 1/4O_2(g) = CuO(s)$ | 72.4 | 0.0057 | 0.1 | 25–1000 | 1,2 |
| $2Cu(l) + 1/2O_2(g) = Cu_2O(l)$ | 149.0 | 0.0158 | 0.2 | 1229–1727 | 1,2 |
| *Hydrogen* | | | | | |
| $H_2(g) + 1/2O_2(g) = H_2O(g)$ | 245.0 | −0.0034 | 0.2 | 25–1700 | 1,2 |
| $H_2(g) + 1/2S_2(g) = H_2S(g)$ | 82.0 | −0.0097 | 0.2 | 25–1500 | 1,2 |
| $3/2H_2(g) + 1/2N_2(g) = NH_3(g)$ | 40.8 | −0.0176 | 0.3 | 25–700 | 1,2 |
| *Iron* | | | | | |
| $Fe(s) + 1/2O_2(g) = FeO(s)-\alpha$ | 271.0 | 0.0080 | 3 | 25–1371 | 1,2 |
| $3Fe(s) + 2O_2(g) = Fe_3O_4(s)$ | 1129.0 | 0.0338 | 3.6 | 25–560 | 1,2 |
| $3FeO(s) + 1/2O_2(g) = Fe_3O_4(s)$ | 292.0 | 0.0035 | 4.4 | 560–1371 | 1,2 |
| $2/3Fe_3O_4(s) + 1/6O_2(g) = Fe_2O_3(s)$ | 78.6 | −0.0011 | 1 | 25–1400 | 1,2 |
| $Fe(s) + 1/2S_2(g) = FeS(s)\alpha$ | 175.0 | 0.0188 | 4 | 25–140 | 1,2 |
| $Fe(s) + 1/2S_2(g) = FeS(s)\beta$ | 170.0 | 0.0149 | 4.6 | 140–906 | 1,2 |
| $FeS(s)b + 1/2S_2(g) = FeS_2(s)$ | 135.0 | −0.0066 | 3 | 300–800 | 1,2 |
| $4Fe(s) + 1/2N_2(g) = Fe_4N(s)$ | 10.7 | −0.0006 | 0.4 | 25–600 | 1,2 |
| *Magnesium* | | | | | |
| $Mg(s) + 1/2O_2(g) = MgO(s)$ | 600.0 | 0.0024 | 2 | 25–650 | 1,2 |
| $Mg(l) + 1/2O_2(g) = MgO(s)$ | 607.0 | 0.0007 | 2.4 | 650–1120 | 1,2 |
| $Mg(g) + 1/2O_2(g) = MgO(s)$ | 755.0 | 0.0157 | 3 | 1120–1700 | 1,2 |
| $Mg(s) + 1/2S_2(g) = MgS(s)$ | 376.0 | −0.0074 | 32.2 | 25–650 | 1,2 |
| $Mg(l) + 1/2S_2(g) = MgS(s)$ | 607.0 | 0.0007 | 2.4 | 650–1120 | 1,2 |
| $Mg(g) + 1/2S_2(g) = MgS(s)$ | 563.0 | 0.0140 | 05 | 1120–1700 | 1,2 |
| $3\,Mg(s) + N_2(g) = Mg_3N_2(s)$ | 461.0 | −0.0012 | 0.5 | 25–650 | 1,2 |
| $3\,Mg(l) + N_2(g) = Mg_3N_2(s)$ | 477.0 | −0.0108 | 1.3 | 650–1120 | 1,2 |
| $MgO(s) + CO_2(g) = MgCO_3(s)$ | 112.0 | 0.0057 | 6.4 | 25–700 | 1,2 |
| $3MgO(s) + P_2(g) + 5/2O_2(g) = Mg_3P_2O_8(s)$ | 2155.0 | 0.0902 | 0.3 | 1000–1250 | 1,2 |
| $2MgO(s) + SiO_2(s) = Mg_2SiO_4(s)$ | 66.0 | −0.0027 | 4 | 650–1120 | 1,2 |
| $MgO(s) + SiO_2(s) = MgSiO_3(s)$ | 37.2 | −0.0032 | 1.9 | 25–1300 | 1,2 |
| $MgO(s) + Al_2O_3(s) = MgAl_2O_4(s)$ | 31.5 | 0.0068 | 20.7 | 25–1727 | 1,2 |

(continued)

(continued)

| $\Delta_r H^0 (kJ \cdot mol^{-1}) = A + B \cdot T(K)$ | Linear correlation coefficient | | | | Ref. (*) |
|---|---|---|---|---|---|
| | $-A$ | B | $(\pm)$ kJ | °C | |
| *Molybdenum* | | | | | |
| $MoO_2(s) + 1/2O_2(g) = MoO_3(s)$ | 159.0 | 0.0043 | 0.1 | 25–1000 | 1,2 |
| $2Mo(s) + 3/2S_2(g) = Mo_2S_3(s)$ | 597.0 | 0.1430 | 195 | 850–1200 | 1,2 |
| $Mo(s) + 3/2O_2(g) = MoO_3(g)$ | 337.0 | −0.0080 | 0.3 | 1280–1727 | 1,2 |
| $Mo(s) + S_2(g) = MoS_2(s)$ | 407.0 | 0.0112 | 0.1 | 1227–1727 | 1,2 |
| *Nickel* | | | | | |
| $Ni(s) + 1/2S_2(g) = NiS(s)$ | 98.7 | −0.0356 | 38.4 | 300–580 | 1,2 |
| *Niobium* | | | | | |
| $2Nb(s) + 5/2O_2(g) = Nb_2O_5(s)$ | 1909.0 | 0.0300 | 1.2 | 1227–1453 | 1,2 |
| $2Nb(s) + 5/2O_2(g) = Nb_2O_5(l)$ | 1898.0 | 0.0821 | 1.3 | 1453–1727 | 1,2 |
| *Phosphorus* | | | | | |
| $2P_2(g) + 5O_2(g) = P_4O_{10}(g)$ | 3228.0 | 0.0618 | 0.8 | 1227–1727 | 1,2 |
| *Silicon* | | | | | |
| $Si(s) + 1/2O_2(g) = SiO(g)$ | 97.6 | −0.0060 | 1.1 | 25–1410 | 1,2 |
| $Si(l) + 1/2O_2(g) = SiO(g)$ | 147.0 | −0.0084 | 1.1 | 1410–1700 | 1,2 |
| $Si(s) + O_2(g) = SiO_2(s)cristob$ | 912.0 | 0.0082 | 0.7 | 400–1410 | 1,2 |
| $Si(l) + O_2(g) = SiO_2(s)cristob$ | 962.0 | 0.0093 | 1 | 1410–1700 | 1,2 |
| $3Si(s) + 2N_2(g) = Si_3N_4(s)$ | 741.0 | −0.0121 | 0.4 | 25–1410 | 1,2 |
| $3Si(l) + 2N_2(g) = Si_3N_4(s)$ | 934.0 | 0.0296 | 3 | 1410–1700 | 1,2 |
| $Si(s) + C(s) = SiC(s)\beta$ | 72.6 | 0.0006 | 0.9 | 25–1410 | 1,2 |
| $Si(l) + C(s) = SiC(s)\beta$ | 125.0 | 0.0017 | 0.8 | 1410–1700 | 1,2 |
| *Sulphur* | | | | | |
| $1/2S_2(g) = S(g)$ | −211.0 | 0.0058 | 0.1 | 25–700 | 1,2 |
| $1/2S_2(g) + 1/2O_2(g) = SO(g)$ | 57.8 | −0.0008 | 0.4 | 25–1700 | 1,2 |
| $1/2S_2(g) + O_2(g) = SO_2(g)$ | 363.0 | 0.0009 | 0.1 | 25–1700 | 1,2 |
| $1/2S_2(g) + 3/2O_2(g) = SO_3(g)$ | 459.0 | −0.0025 | 0.5 | 25–1500 | 1,2 |
| *Tantalum* | | | | | |
| $2Ta(s) + 5/2O_2(g) = Ta_2O_5(s)$ | 2068.0 | 0.0424 | 0.2 | 1227–1727 | 1,2 |
| $Ta(s) + C(s) = TaC(s)$ | 146.0 | 0.0027 | 0.2 | 1227–1727 | 1,2 |
| *Titanium* | | | | | |
| $Ti(s) + 1/2O_2(g) = TiO(s)$ | 546.0 | 0.0082 | 0.2 | 300–1700 | 1,2 |
| $2TiO(s) + 1/2O_2(g) = Ti_2O_3(s)$ | 441.0 | 0.0180 | 1 | 25–1700 | 1,2 |
| $3/2Ti_2O_3(s) + 1/4O_2(g) = Ti_3O_5(s)$ | 165.0 | −0.0176 | 5.9 | 400–1700 | 1,2 |
| $1/3Ti_3O_5(s) + 1/6O_2(g) = TiO_2(s)$ | 120.0 | −0.0094 | 5.4 | 25–1850 | 1,2 |
| $Ti(s)\alpha + 1/2N_2(g) = TiN(s)$ | 339.0 | 0.0027 | 0.4 | 25–882 | 1,2 |

(continued)

(continued)

| $\Delta_r H^0 \left( kJ \cdot mol^{-1} \right) = A + B \cdot T(K)$ | Linear correlation coefficient | | | | Ref. (*) |
|---|---|---|---|---|---|
| | $-A$ | B | $(\pm)$ kJ | °C | |
| $Ti(s)\beta + 1/2N_2(g) = TiN(s)$ | 345.0 | 0.0068 | 1.4 | 882–1200 | 1,2 |
| $Ti(s)\alpha + C(s) = TiC(s)$ | 185.0 | 0.0022 | 0.2 | 25–882 | 1,2 |
| $Ti(s)\beta + C(s) = TiC(s)$ | 186.0 | 0.0004 | 1.4 | 882–1700 | 1,2 |
| *Vanadium* | | | | | |
| $V(s) + 1/2O_2(g) = VO(s)$ | 438.0 | 0.0136 | 0.4 | 600–1500 | 1,2 |
| $2VO(s) + 1/2O_2(g) = V_2O_3(s)$ | 350.0 | −0.0048 | 3 | 550–1112 | 1,2 |
| $2\,V(s) + 5/2O_2(g) = V_2O_5(l)$ | 1482.0 | 0.0237 | 1.1 | 1227–1727 | 1,2 |
| $V(s) + 1/2N_2(g) = VN(s)$ | 219.0 | 0.0062 | 0.2 | 1227–1727 | 1,2 |
| *Wolfram* | | | | | |
| $W(s) + O_2(g) = WO_2(s)$ | 606.0 | 0.0227 | 0.7 | 1227–1727 | 1,2 |
| $W(s) + 3/2O_2(g) = WO_3(s)$ | 852.0 | 0.0221 | 0.4 | 1227–1470 | 1,2 |
| $W(s) + 3/2O_2(g) = WO_3(l)$ | 817.0 | 0.0443 | 0.4 | 1470–1727 | 1,2 |
| *Zirconium* | | | | | |
| $Zr(s)\alpha + O_2(g) = ZrO_2(s)$ | 1101.0 | 0.0080 | 1.7 | 25–870 | 1,2 |
| $Zr(s)\beta + O_2(g) = ZrO_2(s)$ | 1095.0 | 0.0068 | 1.8 | 1205–1865 | 1,2 |
| $Zr(s)\alpha + 1/2N_2(g) = ZrN(s)$ | 368.0 | 0.0042 | 1.6 | 25–862 | 1,2 |
| $Zr(s)\beta + 1/2N_2(g) = ZrN(s)$ | 375.0 | 0.0078 | 2.4 | 862–1200 | 1,2 |
| $Zr(s) + C(s) = ZrC(s)$ | 203.0 | 0.0035 | 5.9 | 25–1900 | 1,2 |

(*) The numbers indicated in the last column of the table are the contributions that were considered to calculate the constants A and B of the linear regression line to obtain the standard enthalpies of the above-mentioned reactions:
1. Chase MW Jr, Davies CA, Downey JR Jr, Frurip DJ, McDonald RA, Syverud AN (1986) JANAF thermochemical tables. American Chemical Society y American Institute of Physics for the National Bureau Standards, 3ª edition, New York, USA.
2. Knacke O, Kubaschewski O, Hesselmann K (1991) Thermochemical properties of inorganic substances. Springer. 2ª edition, Berlin, Germany.

Printed in the United States
by Baker & Taylor Publisher Services